THIRD EDITION

FIRE PROTECTION HYDRAULICS AND WATER SUPPLY

WILLIAM F. CRAPO

JONES & BARTLETT
LEARNING

World Headquarters

Jones & Bartlett Learning
5 Wall Street
Burlington, MA 01803
978-443-5000
info@jblearning.com
www.jblearning.com

Jones & Bartlett Learning books and products are available through most bookstores and online booksellers. To contact Jones & Bartlett Learning directly, call 800-832-0034, fax 978-443-8000, or visit our website, www.jblearning.com.

Substantial discounts on bulk quantities of Jones & Bartlett Learning publications are available to corporations, professional associations, and other qualified organizations. For details and specific discount information, contact the special sales department at Jones & Bartlett Learning via the above contact information or send an email to specialsales@jblearning.com.

09583-8

Production Credits

Chief Executive Officer: Ty Field
Chief Product Officer: Eduardo Moura
Vice President, Publisher: Kimberly Brophy
Vice President of Sales, Public Safety Group: Matthew Maniscalco
Director of Sales, Public Safety Group: Patricia Einstein
Executive Editor: William Larkin
Senior Acquisitions Editor: Janet Maker
Associate Editor: Michelle Hochberg
Production Editor: Lori Mortimer
Senior Marketing Manager: Brian Rooney

Production Services Manager: Colleen Lamy
Senior Production Specialist: Carolyn Downer
VP, Manufacturing and Inventory Control: Therese Connell
Composition: S4Carlisle Publishing Services
Cover Design: Kristin E. Parker
Rights & Media Research Assistant: Robert Boder
Media Development Editor: Shannon Sheehan
Cover Image: Courtesy of City of Harrisonburg Virginia Fire Department
Printing and Binding: LSC Communications
Cover Printing: LSC Communications

Library of Congress Cataloging-in-Publication Data
Names: Crapo, William F., author.
Title: Fire protection hydraulics and water supply / William F. Crapo.
Other titles: Hydraulics for firefighting
Description: Third edition. | Burlington, MA : Jones & Bartlett Learning,
  [2015] | ?2017 | Previous editions entitled: Hydraulics for firefighting.
  | Includes bibliographical references and index.
Identifiers: LCCN 2015034853| ISBN 9781284058529 | ISBN 1284058522
Subjects: LCSH: Fire extinction—Water-supply. | Fire streams. | Hydraulics.
Classification: LCC TH9311 .C73 2015 | DDC 628.9/252—dc23

6048

Printed in the United States of America
21 20 19   10 9 8 7 6 5 4 3 2

BRIEF CONTENTS

TABLE OF CONTENTS

Chapter 7

Pump Theory and Operation 97

Chapter 8

Theory of Drafting and Pump Testing 125

Chapter 9

Fire Streams 143

Chapter 10

Calculating Pump Discharge Pressure 165

William F. Crapo began his career in the fire service as a volunteer fire fighter with the Brentwood Volunteer Fire Department, in Brentwood, Maryland, in 1965. As a volunteer he filled several line and staff positions, including Fire Chief.

In 1973, he was appointed to the District of Columbia Fire Department (DCFD), beginning a career that spanned 20 years. Most of those 20 years were in the firefighting division, serving as a Fire Fighter, Sergeant, Lieutenant (Shift Commander), Captain (Company Commander), and Acting Battalion Fire Chief. During his career in the DCFD, Mr. Crapo was privileged to serve in some of the District's most prestigious engine companies, including two years as Lieutenant at Engine 4, and later as Captain for two years at Engine 11. Additionally, he served in the Office of the Fire Marshal for a total of three years.

After retiring from DCFD, Mr. Crapo worked for several years in the Fire Marshal's Office of the Fairfax County, Virginia, Department of Fire and Rescue. He retired, after 10 years of service, from the Harrisonburg, Virginia, Fire Department at the rank of Deputy Fire Chief.

Mr. Crapo graduated from the University of Maryland, College Park (UMD) with a BA in Fire Science and an AA in Management. He holds an additional AA in Fire Science from Prince George's Community College, Largo, Maryland.

He also served in the United States Air Force, including a tour of duty in Vietnam.

ACKNOWLEDGMENTS

Reviewers

Derrick S. Clouston, MPA, EFO, CFO, CTO
Kaplan University

Gary M. Courtney, AAS, AD
Fire Technologies Program
Lakes Region Community College
Laconia, New Hampshire

Danial Cremeans, EFO
Durham Technical Community College
Durham, North Carolina

Glenn Davis, MS, CFO
Assistant Chief of Training (ret.), Helena Fire
    Department
Helena, Montana

Tom Gaddie, Battalion Chief (ret.), AS, BS
Program Chair, Fire Science Technology
Lanier Technical College
Oakwood, Georgia

Anthony Gianantonio, FO, BS
Battalion Chief, Palm Bay Fire-Rescue Department
Palm Bay, Florida

David R. Hauger, BS
Westmoreland County Community College, Public
    Safety Training Center (ret.)
Smithton, Pennsylvania

Terry L. Heyns, PhD
Professor of Fire Science and Education (ret.)
Lake Superior State University
Marie, Michigan

A. Maurice Jones, Jr., CFO
Supervisor, Fire Protection Systems
Alexandria Fire Department
Alexandria, Virginia

Frederick J. Knipper
Director, Fire & Life Safety Division
Duke University–Duke Health System
Durham, North Carolina

Byron Mathews
Chief of Fire Prevention
Cheyenne Fire and Rescue
Cheyenne, Wyoming

Keith Padgett, MS, EFO, CFO
Fire and EMS Academic Program Director
Columbia Southern University
Orange Beach, Alabama

James M. Stedman, MPA
Battalion Chief (ret.), Chicago Fire Department
Program Director, Fire Science Administration,
    Penn Foster College

Brian A. Wade, BS
Training Specialist, NC Office of State Fire Marshal
Kinston, North Carolina

Intent of This Text

It has been said that the person who knows *how* will always have a job, but the person who knows *why* will be his or her boss. This text is written for the person who wants to know why. Intended for associate-level courses in hydraulics within the fire science curriculum, this text begins with the principles of hydraulics, followed by the introduction of formulas essential to understanding pump operations, and then finally delves into the more complex calculations required for pump operations on the scene. It is intended to be a logical progression of knowledge that can be easily understood by readers with all levels of experience, enabling even the newest fire fighter to understand the laws, principles, and formulas involved, but without sacrificing the technical accuracy or detail demanded by the needs of today's fire service.

The contents of this text meet and exceed the objectives and outcomes of the National Fire Academy's Fire and Emergency Services Higher Education (FESHE) Associate's Non-Core course Fire Protection Hydraulics and Water Supply. To underscore the current job requirements for a pump operator, requisite knowledge elements of NFPA 1002, *Standard for Fire Apparatus Driver/Operator Professional Qualifications*, are also addressed.

In this text, the *science* of hydraulics has been deliberately taken on in an effort to explain the *why* behind its various principles and formulas, which will help students gain a more thorough understanding of the discipline. Much effort has also been put into showing how various formulas and principles found in any physics book are adapted to our study of hydraulics.

A goal of this text is to provide an in-depth knowledge of hydraulics. Hydraulics is more than simply knowing how to find a pump pressure that works or what an impeller is. To have a true understanding of hydraulics, a person must understand the appropriate laws and fundamentals of physics and chemistry. It also requires the ability to apply the correct formula to find an answer, regardless of which variables have been given. Only through an in-depth knowledge of the subject can one find the answers to both common and unusual, or more complex, problems.

Most hydraulics texts today have oversimplified the topic of hydraulics. In large part this oversimplification can be traced back to the notion that, because water supply and pump operations on the fireground have a number of variables that cannot be well controlled, any attempt to be accurate is useless. In reality, the exact opposite should be our guiding principle. Because those variables do exist, it is even more important to control them and be as exact as possible where we can.

One consideration in writing this text was an update of many of the common hydraulics formulas. For example, in the past many texts have been written with the assumption that 1 cubic foot (ft^3) of water weighs 62.5 pounds (lb). While using 62.5 lb as the weight of water, a more accurate weight, 62.4 lb, was known. It was just more convenient to round up to 62.5 lb. This issue may seem insignificant, but it has a ripple effect through many of the formulas necessary to hydraulics. Although realistically any differences may be minor, it is important to be as accurate as possible. The formulas in this text reflect the more accurate weight of 1 ft^3 of water as 62.4 lb. Another reason to use the more accurate weight in this text is that today we have the benefit of inexpensive scientific calculators that can easily handle the precise numbers. In short, there is just no logical reason not to be accurate.

Organization of Contents

The chapters build on cumulative knowledge to assist the learning process. Throughout the text, various applicable laws of physics are mentioned to illustrate that hydraulics is part of that larger discipline.

- **Chapter 1** lays a foundation for studying the chapters that follow. It outlines facts and details about water, which are essential to understanding the remainder of the text.
- **Chapters 2 through 6** focus on the principles necessary to the study of hydraulics. They also contain most of its related formulas, which provide the basis for calculating pump discharge pressure in Chapters 10 and 11. In presenting the formulas, Chapters 2 through 6 put them in perspective: where appropriate, the origin of the formula is illustrated and its relationship to fire service hydraulics is demonstrated.
- **Chapters 7 and 8** are dedicated to understanding the pump. They are intended to demystify the operating principles of the pump at a level that few people, other than engineers, commonly study, but in terms any fire fighter should easily grasp.
- **Chapter 9** discusses several principles and formulas not accounted for in the preceding chapters. In a sense, this chapter ties up some loose ends before moving into pump discharge pressure calculations.
- **Chapters 10 and 11** apply the knowledge gained in the prior nine chapters to solve pump problems. New principles needed to solve problems are also introduced, such as calculating pump discharge pressure for parallel lines, wyed lines, and Siamesed lines. Fireground pump discharge pressure calculations are also included.
- **Chapter 12** introduces basic and advanced issues pertaining to water supply. The fundamentals of water supply are studied, as is a method for calculating how much water is needed to fight fires. The section on how to make maximum use of the local water supply explains, through real-world applications, a subject that is often not well understood.
- **Chapter 13** provides firefighting personnel with insight into the testing of sprinkler and standpipe systems. The subject of fireground formulas is also discussed, with the intent of steering the serious student of hydraulics away from "guesstimations" and toward more accurate calculations.

Math and Formulas

In many respects, this is a math text. Every chapter contains one to more than a dozen formulas. Many of the formulas are not used every day, but they are needed for a thorough understanding of the subject. Without the ability to use the formulas included in this text, it is impossible to master hydraulics.

To work these formulas, the student of hydraulics must be able to perform basic algebra and have a working knowledge of basic plane and solid geometry. Some of the formulas in this text may look intimidating, but they all are solvable with basic algebra and geometry. Even the formidable Hazen–Williams formula in Chapter 12 is solvable with basic algebra.

Anyone unsure about either basic algebra or geometry should take the time to review the subjects. Appendix A has been included to provide a review of the fundamentals required to understand the examples and to successfully solve the problems included in the text.

To assist in maintaining a high level of accuracy while solving problems, regardless of their complexity, a five-step process is recommended:

1. Read the problem more than once.
2. Draw a schematic.
3. Label what you know.
4. Apply the appropriate formula.
5. Solve for the unknown.

This concept is expanded in Chapter 10.

Also note: Because this text contains a variety of complex problems and examples, some errors may have escaped detection. Should you come across any errors, please bring them to the Senior Acquisitions Editor's attention at: JMaker@jblearning.com.

New to This Edition

There are several important changes in the *Third Edition*:

- An amended basic formula for finding gallons per minute (gpm), Freeman's formula, is introduced, to match the one used in the *National Fire Protection Handbook, 20th Edition*.
- A new constant is featured in the formula used to find nozzle reaction when gpm and pressure are known. In previous editions, separate formulas were used for smooth-bore and fog nozzles. With the newly calculated constant, accurate nozzle reaction numbers can be found for either type of nozzle using only one formula; only the flow and nozzle pressure are needed. (Of course, the original formula for finding nozzle reaction from a smooth-bore nozzle, where tip size and nozzle pressure are known, is still acceptable.)

- There is an expanded discussion of the concepts of enthalpy, specific heat, and latent heat in Chapter 1. The serious fire science student should pay special attention to these topics. Enthalpy is an important subject to the study of thermodynamic systems, of which a fire qualifies, but is not well represented in current fire training literature. In fact, no mainstream fire service publications or texts currently recognize the concept.
- Case studies are given. Each chapter begins with a case study that provides students with a means to test their understanding of the chapter's concepts in the context of a fictional scenario. An additional case study closes each chapter, giving students an opportunity to apply chapter content to a real-life example. Some of the case studies are built around several concepts, so that students can practice applying specific knowledge they have learned. Other case studies are intended to add credibility to some of the more abstract concepts addressed throughout the chapters.
- Updated questions and activities are given at the end of each chapter.
- An expanded appendix featuring answers to all chapter questions and activities (Appendix C) has been provided.

AUTHOR'S ACKNOWLEDGMENTS

Writing a book is far from an individual task. Many people beyond those directly connected with reviewing and publishing this text are deserving of mention.

First, I would like to thank my wife, Karen, for her understanding during the long hours of researching and writing. She has been supportive throughout every endeavor I have undertaken since I met her. I would also like to thank my good friend Bill Whelan for his review of and input on the *First Edition*. I asked Bill to perform an independent review because of my respect for his technical knowledge and his ability to uncover technical glitches. He brought an unbiased perspective to the review process because he is not associated with the fire service.

Others deserving of recognition include Dr. David Thomas, Senior Fire Protection Engineer for the Fairfax County, Virginia, Department of Fire and Rescue. Dave made many suggestions and comments that proved beneficial from both technical and practical perspectives. Thanks to Dr. Douglas Giacoli for supplying an updated description of how a radial flow pump imparts energy to the water. Finally, thanks to Battalion Fire Chief Brett Hartt and Battalion Fire Chief Jeff Morris, both of the Harrisonburg, Virginia, Fire Department, for their assistance with the sections on foam and rural water supply, respectively. These individuals' comments and recommendations laid a solid foundation for the *Third Edition*.

I would be remiss if I did not mention the input of the reviewers assembled by Jones & Bartlett Learning to review the *Third Edition*. Their input keeps works like this grounded in reality, while also meeting the needs of the fire service.

Last, but not least, I want to thank the people at Jones & Bartlett who thought enough of this text to accept the challenge of a third edition and for their patience and dedication to achieving the best possible end result. Hopefully, through the collective work of everyone involved, this text will remain an integral part of fire service training well into the future.

Bill

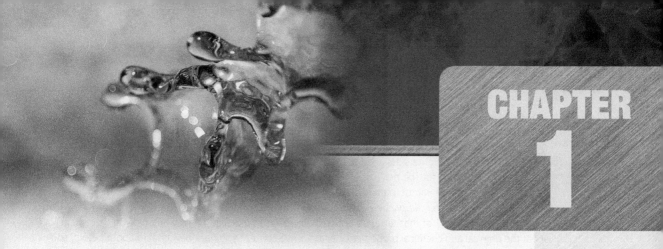

Introduction to Hydraulics

LEARNING OBJECTIVES

Upon completion of this chapter, you should be able to:

- Define hydraulics and explain its origin as a science and its history within the fire service.
- Understand the chemical and physical properties of water.
- Understand the importance, as mechanisms of extinguishment, of enthalpy, latent heat, and specific heat properties of water.
- Calculate needed fire flow (the volume of water needed to extinguish a fire).
- Calculate the volume and weight of water in containers of common shapes.

Case Study

You have recently been assigned to Engine 11. Even though you are one of the youngest firefighters assigned to the company, you have already decided to begin to prepare yourself for the next driver/operator position that becomes available. To learn the hydraulics that will be part of the test, you have selected the text *Fire Protection Hydraulics and Water Supply* by William Crapo.

As you begin going through Chapter 1 of that text, you notice that there isn't much in the way of what you would consider hydraulics. It is all about the physical and chemical properties of water and a concept you have never heard about before, something called *enthalpy*. How is being able to calculate needed fire flow a part of hydraulics, you wonder? But you dive in anyway, eager to learn all the nuances of the subject of hydraulics.

1. Why is it necessary to learn the chemical and physical properties of water?
2. What is enthalpy, and how is it important to the study of hydraulics?
3. How is being able to calculate the needed fire flow useful?

Introduction

The study of the science of hydraulics as we use it in the fire service today began more than 250 years ago **Figure 1-1**. The science began as a study of water flow in relation to water supply, irrigation, river control, and waterpower. Individuals such as Leonhard Euler, Daniel Bernoulli, and Blaise Pascal, to name a few, defined the laws and principles we study today.

Figure 1-1 To be successful at firefighting, we need to get exactly enough water on the fire. To do that requires a thorough knowledge of hydraulics.

Courtesy of J. Arthur Miller.

Hydraulics can be defined as the science of water (or other fluids) at rest and in motion. Hydraulics is an applied science, as opposed to a theoretical science, in that it is a real-world application of well-established laws of physics. A theoretical science, conversely, is just that: only theory. For example, time travel is currently a theoretical science. Until someone actually travels backward or forward in time, it is only a theory.

The science of hydraulics originally began as a study of water alone. It was later that the need to study other liquids and gases borrowed principles and formulas from hydraulics. While the principles and formulas developed for hydraulics served well as a starting point, they were, for some purposes, too inexact. A new, more exact science was necessary. The new science, the mechanics of fluids, includes hydraulics as a specialized phase and only one of its many subspecialties.

The study of hydraulics is divided into two primary areas of study, hydrostatics and hydrodynamics. **Hydrostatics** is the study of water at rest. Its principles deal with water at rest, whether in an open container subject only to atmospheric pressure or in a closed, pressurized container. (Chapter 2 is devoted entirely to hydrostatics.) **Hydrodynamics**, in contrast, is the study of water in motion. Hydrodynamics is more typically what we think of when we consider the fireground application of hydraulics, such as calculating friction loss and gallon-per-minute flow.

Calculations associated with velocity, flow, and nozzle reactions are also dynamic calculations.

Some of the principles studied in hydrostatics also apply to hydrodynamics. With a little careful study and logic, it is easy to identify which ones they are.

History of Hydraulics in the Fire Service

Although the basic principles that apply to hydraulics in the fire service are at least 250 years old, the specific application of hydraulic principles, from engineering to water supply needs, began in earnest around 1889. It was then that engineers such as J. Herbert Shedd, J. T. Fanning, John R. Freeman, and Emil Kuichling started doing serious work on calculating water supply needs for fire protection and associated water distribution systems. In 1910 the National Board of Fire Underwriters (NBFU), known today as the Insurance Services Office Inc. (ISO), published its first fire flow recommendations and requirements.

The person most recognized as the champion of the scientific application of hydraulics to the fire service is Fred Shepherd, considered the father of fire service hydraulics. He wrote his first book about hydraulics for the fire service in 1917. The book *Practical Hydraulics for Firemen* was written for the average firefighter, not engineers.

Water

Firefighters all know how important water is to their job. Without it, firefighting as we know it would be impossible. Water has many characteristics and properties that make it an ideal fire suppression agent. It has been said that if water did not exist naturally, the fire service would have had to invent it.

To begin with, water is relatively abundant (two-thirds of planet Earth is covered with it); can be stored easily, such as in onboard water tanks; and, with the exception of a few substances such as iron and calcium carbide, is relatively nonreactive.

Water itself is a chemical compound made up of two parts of elemental hydrogen and one part of elemental oxygen. The universally recognized compound symbol H_2O is derived from water's composition. Water is chemically stable because of its strong hydrogen–oxygen bond. This bond gives water a polar structure, making it a polar molecule. Because of its strong polar structure, water reacts favorably with many compounds, causing them to go into solution. In fact, water reacts adversely with so few chemicals that in inorganic chemistry it is a universal solvent.

Note

Water is a chemical compound of two parts hydrogen and one part oxygen.

Properties of Water

To understand hydraulics, it is necessary to have a thorough understanding of the physical properties of water. The basic properties of water are summarized in **Table 1-1**. The following list explains the 10 physical properties:

1. In addition to its liquid state or phase, water has a solid state or phase (ice) and a vapor state or phase (humidity).
2. Water, like all substances, gets denser as it approaches its freezing point. However, unlike

Table 1-1 Basic Properties of Water

	1 Gal	1 ft³
Weight, lb	8.34	62.4
Cubic inches	231	1,728
Btu absorbed	9,343	69,886[†]
Cubic feet of steam	227	1,700
Gallons	1	7.48
Specific heat = 1 Btu/lb of water		
Latent heat of vaporization = 970.3 Btu/lb of water		

[†]9,343 × 7.48 = 69,886 Btu.

most other substances, water actually gets less dense just before it freezes. In fact, water is densest at 39.2°F. Because the ice is less dense than the water containing it, the ice floats, forming an ice layer that insulates the rest of the water. Because of the high specific heat of water, the remainder of the body of water will not freeze if there is sufficient volume.

3. At its densest, 1 cubic foot (ft³) of freshwater weighs 62.4 pounds (lb). And 1 ft³ of saltwater weighs 64 lb **Figure 1-2**.

4. Also 1 ft³ of water occupies 1,728 cubic inches (in³). One gallon (gal) of water occupies 231 in³.

Note

Water becomes less dense just before it freezes.

Example 1-1

If there is 1,728 in³ of water in 1 ft³ and there is 231 in³ in 1 gal, how many gallons are in 1 ft³?

Answer

$$1,728 \text{ in}^3/\text{ft}^3 \div 231 \text{ in}^3/\text{gal} = 7.48 \text{ gal/ft}^3$$

5. In addition, 1 ft³ of water contains 7.48 gal **Figure 1-3**.

Example 1-2

If 1 ft³ of water weighs 62.4 lb, what does 1 gal weigh?

Answer

If, at its densest, 1 ft³ of water weighs 62.4 pounds and there are 7.48 gal of water in 1 ft³, then:

$$62.4 \text{ lb/ft}^3 \div 7.48 \text{ gal/ft}^3 = 8.34 \text{ lb/gal}$$

6. At its densest, 1 gal of water weighs 8.34 lb **Figure 1-4**.

7. The boiling point of water is 212°F.

8. The freezing point of water is 32°F.

9. The triple point (a unique temperature and pressure at which the substance exists in all three states at the same time) of water is 32°F at 0.08 pounds per square inch (psi) pressure.

10. Water, for practical purposes, is noncompressible.

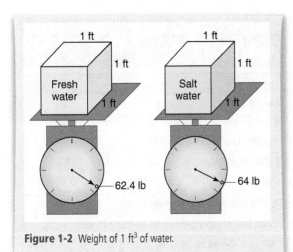

Figure 1-2 Weight of 1 ft³ of water.

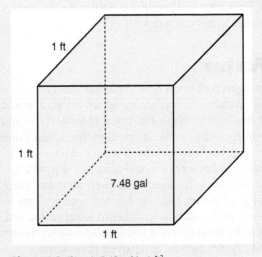

Figure 1-3 There is 7.48 gal in 1 ft³.

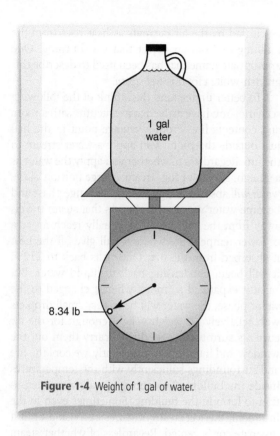

Figure 1-4 Weight of 1 gal of water.

Water as an Extinguishing Agent

In addition to the 10 physical properties of water, it has other properties that make it a nearly ideal extinguishing agent. But first it is necessary to understand what is meant by the term <u>**British thermal unit (Btu)**</u>. A Btu is a unit of energy in the English system, as opposed to the International System (SI) of units. The Btu is the amount of energy needed to raise the temperature of 1 lb of water by 1°F at 60°F. Technically the exact value of 1 Btu will vary marginally with temperature, but for most calculations, 1 Btu = 1°F/lb of water is used.

Conversion of Water to Steam

Another important physical property of water is the ratio of water to steam when water changes phase at 212°F. Water expands to steam at a ratio of 1:1,700. That is, 1 ft³ of water will expand to 1,700 ft³ of steam. In Example 1-3 we will use this expansion ratio to find out how much steam 1 gal of water will generate. If it is taken advantage of, this steam will exclude oxygen from the atmosphere, depriving the fire of oxygen needed for combustion.

Note that this expansion ratio is only valid at 212°F. After water expands to vapor at 212°F, and if it continues to absorb heat from the atmosphere, the steam will continue to expand even further. How much water will ultimately expand depends on the temperature of the atmosphere. The expansion ratios in **Table 1-2** assume atmospheric pressure.

For an example of how Table 1-2 can be used, assume a room flashes over at 1,100°F. What will the expansion ratio of water to steam be at 1,100°F? Table 1-2 shows that at 1,100°F, water will expand 3,997 times when converting to steam. This is 2.35 times greater expansion than at 212°F.

By converting water to steam, two important extinguishing mechanisms are accomplished simultaneously. First, a tremendous amount of energy is absorbed from the fire, taking away the energy that feeds back into the system and causing the chain reaction known as *combustion* to slow down or

Table 1-2 Conversion of Water to Steam above 212°F	
Temperature, °F	**Expansion Ratio**
300	1:1,923
400	1:2,174
500	1:2,429
600	1:2,680
700	1:2,934
800	1:3,190
900	1:3,440
1,000	1:3,696
1,100	1:3,997
1,200	1:4,202

stop completely. Second, the steam generated displaces oxygen-carrying air and further helps extinguish the fire by depriving the fire of the oxygen it needs. This method of depriving the fire of oxygen is not perfect, but it is tremendously effective; and when it is used along with other tactical measures, it can be a key factor in achieving fire control and extinguishment.

A word of caution is appropriate here: *steam can burn.* Suppression personnel should make every effort to wear protective gear so as to prevent burns and should be aware of overproduction of steam so as to prevent it from escaping into areas of the building not involved in the fire.

Example 1-3

If water expands 1,700 times when going from the liquid state to the vapor state, how much steam will be created by 1 gal of water?

Answer

If 1 ft^3 of water expands 1,700 times and there is 7.48 gal of water in 1 ft^3, divide 1,700 ft^3 of steam by the number of gallons in 1 ft^3:

$$1,700 \div 7.48 = 227.27$$

So 1 gal of water will expand into 227 ft^3 of steam.

Before proceeding with our discussion of water as an extinguishing agent, we need to define the term *steam* (or *water vapor*). The nature of the vapor state of water is not well understood by most people, including firefighters. Most people think the grayish-white cloud that rises from a building or object on fire after water has been applied is steam. This understanding is incorrect. Water vapor is invisible; if you can see it, it is *not* steam.

The grayish-white cloud rising from a fire is actually a cloud of water in its liquid state. It might seem counterintuitive, but the grayish-white cloud is a mixture of products of incomplete combustion and water droplets that have formed when steam/water vapor condenses back to its liquid phase. The droplets are so small that they are light enough to

be carried by the air currents as heat rises from the building or from an object that was burning. One appropriate name that has been used to describe this grayish-white cloud is *wet steam*.

To better understand this, think of the following scenario. You have arrived at a structure with a room and contents fire. From a vantage point in the hall just outside the room, you apply a water stream to the fire. Regardless of whether you apply the water as a straight stream or fog stream, a large portion of the water will absorb enough energy to change phase and become water vapor, or steam. As that steam travels away from the fire, it will eventually reach an area of lower temperature where it will give off the heat it absorbed from the fire. Once it is back to 212°F, it will begin condensing back to liquid water. But since it expanded so much when it changed to the vapor phase, the water will be in very small droplets with relatively no weight—light enough for the air currents surrounding the fire to carry them out the window and into the sky. Frequently, we cool the fire and surroundings sufficiently with the temperatures inside the building low enough to allow condensation to form in the building. Sometimes even in the fire room, the wet steam obscures our vision until it dissipates or is vented. Regardless of whether steam is inside or outside the building, the presence of wet steam indicates a temperature of 212°F or lower.

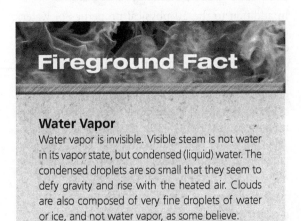

Fireground Fact

Water Vapor

Water vapor is invisible. Visible steam is not water in its vapor state, but condensed (liquid) water. The condensed droplets are so small that they seem to defy gravity and rise with the heated air. Clouds are also composed of very fine droplets of water or ice, and not water vapor, as some believe.

Enthalpy, Specific Heat, and Latent Heat

Enthalpy

The concept of enthalpy is little known to the fire service. It is, however, a key component in the study of any thermodynamic system—and a fire, after all, is a thermodynamic system. **Enthalpy** is the total amount of energy contained in a substance. Regardless of its form—solid, liquid, or gas—all matter has energy in it. This means that even an ice cube on the coldest night in Antarctica has energy in it, that is, until it reaches absolute zero (−459.7°F) where all molecular motion stops. Since our discussion is about thermodynamic systems, we can further define this energy as thermal energy or heat.

As the temperature and volume of a substance change, the enthalpy undergoes a corresponding change. For example, for each degree that 1 gal of water increases, it contains more energy to account for the temperature increase. Additionally, as the temperature goes up, there is technically an increase in volume, although it is extremely small. Together, the increase in energy and the increase in volume caused by the temperature change equate to an increase in enthalpy. In our discussions below of specific heat and latent heat, you will see practical illustrations of how enthalpy change is important to the fire service.

Both specific heat and latent heat are critical concepts to understand if we want to successfully calculate the amount of energy a substance either absorbs or gives off as it changes temperature or phase. Specific heat and latent heat are measures of the change in enthalpy as a substance changes temperature or changes phase. Thus, both specific heat and latent heat are subcategories of enthalpy.

Specific Heat

The **specific heat** of a substance is a measure of how much a substance heats up with the addition of a specified amount of energy or heat. For water,

specific heat is 1 Btu/lb. That is, the enthalpy of water will change by 1 Btu as we raise or lower the temperature of 1 lb of water by 1°F.

For example, how much energy will be necessary to raise the temperature of 1 gal of water by 1°F? We know from Table 1-1 that 1 gal of water weighs 8.34 lb. And we now know that the specific heat of water is 1 Btu. The total amount of heat necessary to raise the temperature of 1 gal of water by 1°F is then

Specific heat = weight of water × change in temperature
= weight of water × change in Btu[†]
= 8.34 lb/gal × 1 Btu/lb
= 8.34 Btu/gal

Specific heat = weight of water × change in temperature

It will take 8.34 Btu to raise the temperature of 1 gal of water 1°F. Or, it can also be said that the enthalpy of 1 gal of water will change by 8.34 Btu for every 1°F that the temperature of the water either increases or decreases.

Note that this value of specific heat only applies to liquid water. For frozen water, or ice, the specific heat value is 0.5 Btu/lb, and for water vapor, or steam, the specific heat is 0.48 Btu/lb. When you are finding the specific heat for steam, calculate it based on the amount of liquid water you started with.

If we want to find how much energy will be absorbed when we apply water to a fire, we simply find the specific heat needed to raise the temperature of water from ambient temperature to the temperature of the fire. But we need to do it in steps. Let us assume that the water starts out at 62°F. Step 1 is to find the specific heat needed to raise water from our starting temperature of 62°F to 212°F. The example given is for 1 gal of water,

---

[†]Note here that when one is dealing with liquid water, the change in Btu per pound of water is exactly the same as the change in temperature since the specific heat of liquid water is 1. When one is dealing with steam or ice, the change in Btu will be less than the change in temperature because of different specific heat factors.

but it can be applied to any quantity as long as you know the weight.

Water has a specific heat of 1 Btu/lb of water.

Example 1-4

How many Btu will it take to raise 1 gal of water to 212°F but not vaporize it?

Answer

To raise 1 pound (weight) of water from 62°F to 212°F requires specific heat of 212°F − 62°F = 150 Btu. However, the problem calls for the specific heat (amount of Btu) for 1 gal (volume) of water. To find the specific heat for 1 gal of water, we need to multiply the 150 Btu for 1 lb of water by the weight of 1 gal of water, which is 8.34 lb. We can then solve this problem using the formula Specific heat = weight of water × change in temperature = 8.34 × 150 = 1,251 Btu. We now know that 1 gal of water will absorb 1,251 Btu when going from ambient temperature (assumed to be 62°F) to 212°F.

Latent Heat

The amount of energy absorbed by water can be measured by calculating its specific heat, but only to a point. A new dynamic comes into play at the point where a substance, such as water, changes phase or state (solid ↔ liquid or liquid ↔ gas). At a phase change, additional energy is needed to overcome the energy that is bonding the substance together. This means that going from water at 212°F to steam at 212°F requires more energy than can be accounted for by specific heat. This additional energy is called the *latent energy* or *latent heat*.

Typically an engineer will calculate energy changes due to temperature changes by looking them up in tables. For example, the table that was referenced while writing this text lists that the enthalpy

of saturated water—water that is already at 212°F and will generate vapor at 212°F with the addition of energy—is 180.16 Btu/lb. The enthalpy of saturated steam—steam at 212°F that can be condensed back into water due to the loss of energy/heat without changing temperature—is 1,150.5 Btu/lb. If we calculate the difference between the two we find that water requires an additional 970.34 Btu/lb, usually rounded to 970.3 Btu/lb, to overcome the bonding forces and change state. This is referred to as the **latent heat of vaporization**, or the **enthalpy of vaporization**. (Recall that specific heat and latent heat are subcategories of enthalpy.) When condensing, once the steam has cooled back to 212°F, the steam must give off 970.3 Btu/lb to become water at 212°F. Note that the temperature at which this all occurs is the boiling point of water, also known as the *saturation temperature*, thus the terms *saturated water* and *saturated steam*. Steam/vapor above this temperature is referred to as *superheated steam* or *superheated vapor*. The latent heat property of water is what makes it such an important tool in fire suppression, absorbing 6.47 times as much heat going from water at 212°F to steam at 212°F as in getting from ambient temperature to 212°F.

The latent heat or enthalpy of vaporization is 970.3 Btu/lb.

The same dynamics apply at the point where water goes from solid to liquid (melts) or goes back to solid from liquid (freezes). Here it is referred to as the **latent heat of fusion** or **enthalpy of fusion**. The latent heat of fusion of water is 143.3 Btu/lb. Typically, this amount will not be included in any calculations necessary to the study of firefighting, but is included here for the sake of being complete.

The four specific properties responsible for water's value as an extinguishing agent are summarized in the following list:

1. Water expands at a ratio of 1:1,700 when converted to steam at 212°F [Figure 1-5]. At higher temperatures the expansion ratio is higher.
2. Water has a specific heat of 1 Btu/lb of water.
3. The latent heat/enthalpy of vaporization (going from liquid to vapor or back) is 970.3 Btu/lb of water.
4. The latent heat/enthalpy of fusion (going from solid to liquid or back) is 143.4 Btu/lb of water.

Just as we calculated the total specific heat of 1 gal of water going from 62°F to 212°F, it is possible to find the total energy/heat absorbed by 1 gal of water going from liquid water at 212°F to steam at 212°F. This is step 2 in calculating how much energy will be absorbed when water is applied to a fire.

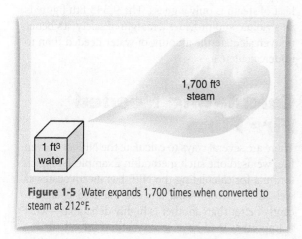

Figure 1-5 Water expands 1,700 times when converted to steam at 212°F.

1,700 ft³ steam

1 ft³ water

Example 1-5

How many Btu will 1 gal of water absorb going from a liquid state at 212°F to steam at 212°F?

Answer

Because we already calculated the specific heat, we only need to determine the latent heat/change in enthalpy needed for 1 gal of water to change state from liquid to vapor. Just as in calculating the specific heat, we simply multiply the latent heat by the weight of water. The formula is Latent heat of vaporization = weight of water × 970.3 Btu. In this example, we find Latent heat = 8.34 × 970.3 Btu = 8,092 Btu. We now know that the latent heat of vaporization of 1 gal of water is 8,092 Btu.

Latent heat of vaporization = weight of water × 970.3

Now we know how much heat is absorbed by 1 gal of water going from ambient temperature to 212°F and how much heat is absorbed when water changes state from a liquid to vapor. To truly appreciate the importance of specific heat and latent heat properties, we now put them together.

Example 1-6

How many Btu are absorbed when 1 gal of water goes from ambient temperature to steam?

Answer

In Example 1-4, we determined that 1 gal of water absorbs 1,251 Btu going from 62°F to 212°F. In Example 1-5, we determined that 1 gal of water absorbs 8,092 Btu changing state from liquid to vapor. To find the answer to this example, we simply add the two values: 1,251 Btu + 8,092 Btu = 9,343 Btu/gal of water [Figure 1-6].

The practical application of Example 1-6 is that it tells us exactly how much heat 1 gal of water will absorb, if completely vaporized, when applied to a fire, which is one method of determining the **needed fire flow (NFF)**. The NFF is an estimate of the amount of

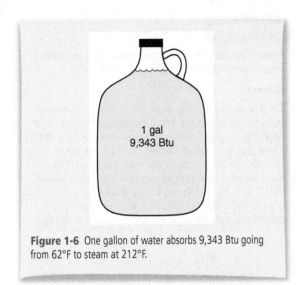

Figure 1-6 One gallon of water absorbs 9,343 Btu going from 62°F to steam at 212°F.

water needed to extinguish a fire in a specific building or involved compartment. While this number is valid for calculating the heat absorption ability of a specific quantity of water, we can be more accurate if we add in the specific heat for steam.

Earlier we mentioned that steam also has a specific heat value of 0.48 Btu/lb. If we know the air temperature of a room on fire, we can actually calculate how much energy the steam will absorb as its temperature goes from 212°F to the temperature of the room. For example, how much additional energy would the water absorb if it were directed into a room at the point of flashover?

If flashover occurs at about 1,100°F, we need to calculate the additional specific heat that the steam will absorb going from 212°F to 1,100°F. The specific heat of steam going from 212°F to 1,100°F is the change in temperature times the specific heat value of steam, or $(1,100 - 212) \times 0.48 = 888 \times 0.48 = 426$ Btu/lb. As the steam increases in temperature from 212°F to 1,100°F, it will absorb an additional 426 Btu for each 1 lb of water we started with. That means that for each 1 gal of water we can absorb an additional

$$\begin{aligned} \text{Specific heat} &= \text{weight of water} \times \text{change in Btu} \\ &= 8.34 \times 426 \\ &= 3{,}552.84 \text{ or } 3{,}553 \text{ Btu} \end{aligned}$$

See Example 1-7 to determine how much heat is absorbed going from 62°F to 1,100°F.

How many Btu are absorbed when 1 gal of water goes from ambient temperature to steam at 1,100°F?

Answer

In Example 1-6 we determined that 1 gal of water absorbs 9,343 Btu when it changes from ambient temperature to steam at 212°F. Also we found that the specific heat of 1 gal of water that has vaporized at 212°F and increased to 1,100°F is 3,553 Btu—that is, the amount of additional heat that it will absorb. To find the answer to Example 1-7, we simply add the two values: 9,343 Btu + 3,553 Btu = 12,896 Btu. So 1 gal of water will absorb 12,896 Btu going from an ambient temperature of 62°F to steam at 1,100°F.

Note from the above calculations that water absorbs 6.47 times as much heat when it turns to steam as when it increases from 62°F to 212°F. Water is most efficient as a cooling agent if it is allowed to change phase and become steam. When we calculate the amount of heat a given quantity of water will absorb, it is best to use 9,343 Btu rather than 12,896 Btu. Unless you know the actual temperature of the atmosphere, any calculation that involves the specific heat of steam is only a guess. The 9,343 Btu figure is an absolute and provides a margin of safety; it's better to overcalculate the amount of water needed than to undercalculate it.

Calculating Needed Fire Flow

There are several ways to calculate the NFF of a given fire; we used one such method in Example 1-6. Each formula for calculating the NFF has its advocates as well as detractors. That any one method is significantly better than another is highly debatable—what is important is to have some method for calculating how much water it might take to put out a given fire.

Having some yardstick for making the necessary calculation is beneficial for preplanning at the very least.

One informal method is simply to calculate the amount of steam needed to fill a room. By excluding the oxygen, the fire cannot burn. This relates to the discussion above about the conversion of water to steam at a ratio of 1:1,700. Example 1-8 walks through the calculations needed to determine the amount of water required to fill a room with steam.

Example 1-8

How much water is required to completely fill a bedroom on fire with steam if the room is 15 ft × 15 ft?

Answer

First determine the volume of the room, and unless otherwise known, assume the height of the room is 8 ft. This requires the formula $V = L \times W \times H$, where V is the volume, L is the length, W is the width, and H is the height. The volume of the room, then, is $V = L \times W \times H = 15 \times 15 \times 8 = 1,800$ ft³. Now, to determine the gallons of water needed, first we divide the volume of the room by 1,700, which is the amount of steam generated by 1 ft³ of water. Using the formula Cubic feet of water (CFW) = $V \div 1,700$, where V is the volume of the room in cubic feet, we find that the amount of water = $V \div 1,700 = 1,800 \div 1,700 = 1.06$ ft³. Finally, to determine how many gallons of water this figure represents, use the formula Gallons = $V \times 7.48$, where V is the volume in cubic feet of water that is needed to generate the appropriate amount of steam. In this particular example, it is the cubic feet of water necessary to make enough steam to fill the room. To convert cubic feet of water into gallons, we use the formula Gallons = $V \times 7.48 = 1.06 \times 7.48 = 7.93$ gal of water.

$$V = L \times W \times H, \text{ where } V \text{ is the volume,}$$
$$W \text{ is the width, and } H \text{ is the height}$$

$$CFW = V \div 1,700$$

$$Gallons = V \times 7.48$$

Once you understand this concept, it is possible to make a direct calculation of the amount of water that will be needed to fill a given volume with steam. Recall from Example 1-3 that 1 gal of water will expand into 227 ft³ of steam. Knowing this, we can develop a new formula—Gallons = $V \div 227$, where V is the volume of the room in cubic feet—that will give us a direct calculation of the gallons of water needed to fill a room with steam.

$$Gallons = V \div 227$$

Example 1-9

We calculated the volume of the bedroom in Example 1-8 to be 1,800 ft³. Use the formula Gallons = $V \div 227$ to find the amount of water needed to fill the room with steam.

Answer

$$Gallons = V \div 227$$
$$= 1,800 \div 227$$
$$= 7.93$$

This amount of water does not seem like much, considering the amount usually put on most fires. For it to take only 7.93 gal of water would require an ideal situation. The room would have to be unvented, and we would have to have perfect conversion of water to steam. Neither one of these conditions occurs at a real fire, but this example still illustrates that a simple room and contents fire can be extinguished with a minimum amount of water.

This method of calculating the volume of steam necessary to extinguish a fire is the basis of one of the oldest formulas for calculating the NFF.

Iowa State Rate-of-Flow Formula

One of the first, if not the first, formal NFFs was developed by Keith Royer and Bill Nelson at Iowa State University in the 1950s. From tests and observation they determined that 1 gal of water could reliably produce 200 ft³ of steam. Royer and Nelson used 200 ft³ of steam per gallon of water instead of

227 to introduce a margin of safety. They also determined that with the proper flow rate, a fire should be knocked down within the first 30 seconds (s). This led them to develop the rate-of-flow formula: Gallons per minute = $V \div 100$, where V is the volume of the room in cubic feet. See Example 1-10 below and the Iowa State rate-of-flow formula case study at the end of this chapter.

> Gallons per minute (gpm) = $V \div 100$, where V is the volume in cubic feet

Fireground Fact

Another Important Iowa State Finding
Among other important findings by the researchers at Iowa State was the fact that using too much water to accomplish extinguishment complicated fire control.

> Btu absorbed = $9{,}343 \times$ gallons

National Fire Academy Fire Flow Formula

The National Fire Academy (NFA) fire flow formula is a more inclusive formula than the Iowa State formula. The major difference is that whereas the Iowa formula is applied for fire control in the single largest open area of a building, the NFA fire flow formula considers the entire building, including exposures.

The NFA fire flow formula is:

$$\text{NFA} = \left(\frac{\text{length} \times \text{width}}{3} + \text{exposure charge} \right) \times \% \text{ involvement.}$$

Practically speaking, this formula needs to be worked in three separate phases:

1. Calculate the area of the building and divide by 3. If it is a multistory building, calculate the area of each floor and add them. This will give you the base fire flow requirement.
2. Take 25 percent of the base flow for each exposure. If there is no exposure, omit this step; if there are two exposures, the exposure charge would be 50 percent (25 percent $\times$ 2),

---

Example 1-10

Applying the Iowa State rate-of-flow formula, the room in Example 1-8 can be filled with steam with just 9 gal of water:

$$\text{Gpm} = V \div 100$$
$$= 1{,}800 \div 100$$
$$= 18$$

Because the Iowa formula expects the water to be applied in the first 30 s, only one-half of the 18 gpm, or 9 gpm of water, is actually required. (The difference between 9 gal of water determined by the Iowa State rate-of-flow formula and the 7.9 gal from Example 1-9 is the margin of safety referred to by Royer and Nelson.) How many Btu will be absorbed?

Answer

In Example 1-6, we determined 1 gal of water would absorb 9,343 Btu going from ambient temperature to steam. Simply multiply 9,343 by the number of gallons needed, using the formula Btu absorbed = $9{,}343 \times$ gallons. Using the Iowa State rate-of-flow formula answer, we can absorb

$$\text{Btu absorbed} = 9{,}343 \times \text{gallons}$$
$$= 9{,}343 \times 9$$
$$= 84{,}087$$

and so on. Then add the basic flow and exposure charge.

3. Multiply the sum of these two numbers (basic flow + exposure charge) by the percentage of involvement. The answer will be the amount of water necessary to extinguish a fire in the given building. This includes water for attack lines, exposure lines, and backup lines. See Example 1-11 for a walk-through of the NFA fire flow formula.

$$NFA = \left(\frac{\text{length} \times \text{width}}{3} + \text{exposure charge} \right) \times \% \text{ involvement}$$

Example 1-11

Use the NFA fire flow formula to calculate the NFF for a residential structure that is 30 ft × 45 ft. This house has a second floor that is one-half the area of the main floor. There is one exposure, and upon arrival you estimate the building to be about 25 percent involved.

Answer

Area = $L \times W$ = 45 × 30 = 1,350 × 1.5 (to include area of second floor)

= 2,025 gpm

Exposure charge = 2,025 × 0.25 = 506.25 gpm

NFF = (base flow + exposure charge) × % involvement

= (2,025 + 506.25) × 0.25

= 632.8 or 633 gpm

Note that no fire flow formula is perfect. For instance, the Iowa State rate-of-flow formula is designed for the single largest compartment involved. It also assumes the water can be delivered within 30 s. The NFA fire flow formula also has its weaknesses. For example, it is designed for interior offensive operations, not defensive surround-and-drown operations. It is also based on the area of the building, assuming a ceiling of no greater than 10 ft. Whichever formula you choose to use, you need to learn it thoroughly so as to be consistent.

Calculating Volume and Weight of Water

At some point, it inevitably becomes necessary to calculate the volume of a container of water, and from there we must calculate the gallons and weight of the water. The shapes that firefighters should be most concerned with are cylinders (hose, water towers, and tankers), spheres (water towers), and rectangles (reservoirs and pools). Most of these are fairly straightforward, but at times calculating the volume calls for a little ingenuity. Appendix A contains formulas for calculating the area and volume of various common geometric shapes.

Most problems call for simple calculations, such as the volume of a water tank on a pumper. These involve the simple formula $V = L \times W \times H$. Once the volume in cubic feet is found, regardless of the shape of the container, finding the volume in gallons is easy. Multiply the cubic feet of the container by the number of gallons in 1 ft³. The formula for finding how many gallons of water are contained in a given size container, once we know the volume in cubic feet, is Gallons = $V \times 7.48$. Example 1-12 demonstrates how to use these calculations to find the volume of a specific water tank.

Example 1-12

What is the volume, in cubic feet and gallons, of a tank 10 ft long, 5 ft wide, and 1.33 ft (16 in) deep? (These measurements represent the approximate shape of a water tank on some fire apparatus.)

Answer

Begin by finding the volume in cubic feet by multiplying length by width by height: 10 ft × 5 ft × 1.33 ft = 66.5 ft³. To convert this figure to gallons, Gallons = $V \times 7.48$, or Gallons = 66.5 × 7.48 = 497.42 gal.

If the need arises to find the volume of a small container, it can sometimes be easier to find the volume in cubic inches and then convert it to gallons. The formula is the same, $V = L \times W \times H$, except in this instance all dimensions must be in inches and the volume will be given in cubic inches. Then, to find the volume in gallons, we must divide by 231, the number of cubic inches in 1 gal. The formula for finding gallons when cubic inches are known is Gallons = $V \div 231$, where Gallons is the number of gallons in the container, V is the volume of the container in cubic inches, and 231 is the number of cubic inches in 1 gal.

$$\text{Gallons} = V \div 231$$

Example 1-13

What is the volume in cubic inches, and how many gallons of water are in a container 18 in × 18 in × 24 in?

Answer

Find the volume in cubic inches by multiplying length by width by height: 18 in × 18 in × 24 in = 7,776 in³. Divide the volume of 7,776 in³ by 231 to get 33.66 gal of water.

One of the more useful shapes whose volume you may need to find is an in-ground swimming pool. In rural areas pools can be useful as supplementary water sources. They are also not the easiest shapes for which to calculate volumes because of their multiple depths. They usually have a shallow end that extends for several feet at one end and a deep end that extends for several feet at the other end. In between, the depth drops off at an angle. These sections can be figured as three separate problems, with the shallow end and deep end separately as rectangles and the sloped area in between as a trapezoid Figure 1-7A .

An easier way is to double the pool back-to-back on itself Figure 1-7B . You now have a rectangle that is easier to work with. Find the volume of the doubled pool and take half of it. It will not be exact, but unless you calculate for all the rounded corners and curved surfaces, any method is only a reasonable approximation. If the pool is composed of

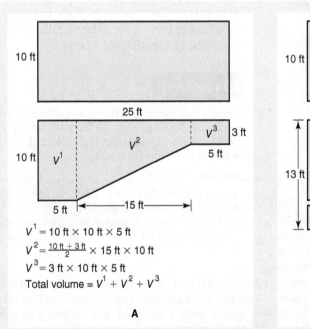

$V^1 = 10\text{ ft} \times 10\text{ ft} \times 5\text{ ft}$
$V^2 = \frac{10\text{ ft} + 3\text{ ft}}{2} \times 15\text{ ft} \times 10\text{ ft}$
$V^3 = 3\text{ ft} \times 10\text{ ft} \times 5\text{ ft}$
Total volume = $V^1 + V^2 + V^3$

A

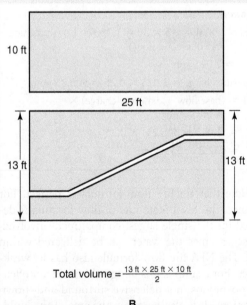

Total volume = $\frac{13\text{ ft} \times 25\text{ ft} \times 10\text{ ft}}{2}$

B

Figure 1-7 Calculating the volume of a pool. **A.** Dimensions of the pool. **B.** The pool doubled back-to-back.

multiple curved or irregularly shaped areas, contact the owner or manufacturer for the volume.

One statistic that always seems to be asked at some point is the volume and weight of charged hose. It has a useful purpose in illustrating to new recruits and interested observers just how much work is really involved in firefighting. The following method can be used to find the volume, in cubic feet and gallons, of any cylinder.

The formula for finding the volume of a cylinder is $V = 0.7854 \times D^2 \times H$ (or L). In this formula V is volume in either cubic feet or cubic inches, 0.7854 is a constant used to find the area of a circle when its diameter is used to solve for area, D is the diameter of the cylinder, and H (or L) is the height (or length) of the cylinder. The units in the answer depend on the units of D and H. If D and H are in inches, the answer will be in cubic inches. If D and H are in feet, the answer will be in cubic feet. There is one stipulation on D and H: they must be in the same units. If one dimension is in feet and the other is in inches, convert the inches to feet by dividing by 12. Finally, if D or H is in inches and you need feet, divide the number of inches by 12. This can then be placed directly into the formula as shown in Example 1-14.

$$V = 0.7854 \times D^2 \times H \text{ (or } L)$$

$$\text{Area} = 0.7854 \times D^2$$

If the formula is written to give the answer in cubic feet, this same formula can be used to find the volume in gallons. Simply multiply the answer by the number of gallons in 1 ft^3. The formula then becomes $V = 7.48 \times 0.7854 \times D^2 \times H$ (or L). Because multiplication can be done in any sequence, we can multiply 7.48 by 0.7854 to get a constant of 5.87. By using this constant we eliminate a step, creating a shortcut, and the formula becomes $V = 5.87 \times D^2 \times H$ (or L).

$$V = 7.48 \times 0.7854 \times D^2 \times H \text{ (or } L)$$

$$V = 5.87 \times D^2 \times H \text{ (or } L)$$

Fireground Fact

Finding the Area of a Circle by Using the Diameter

The classic formula for the area of a circle is Area = $\pi \times r^2$. In the fire service, we use diameter, not radius, for all nozzle openings, so it would be convenient to find a way to use diameter to calculate the area of a circle. Because the radius is one-half the diameter, we can express the radius as $D/2$. Taking this a step further, we can now rewrite the formula as Area = $\pi \times (D/2)^2$. Now insert 3.1416 (the mathematical value of π) and square $D/2$ to get the formula Area = $3.1416 \times (D^2/4)$. Divide the 4 into 3.1416, and the final formula is Area = $0.7854 \times D^2$.

Example 1-14

How much water is in a 50-ft section of charged 2½-in hose?

Answer

$$\begin{aligned}
V &= 5.87 \times D^2 \times L \\
&= 5.87 \times (2.5/12)^2 \times 50 \text{ ft} \\
&= 5.87 \times (0.208)^2 \times 50 \text{ ft} \\
&= 5.87 \times 0.043 \times 50 \text{ ft} \\
&= 12.6 \text{ gal of water}
\end{aligned}$$

If we need to know the weight of a volume of water, we only need to multiply the number of gallons by the weight of 1 gal of water. This requires a formula of $W = V \times 8.34$, where W is the weight of the water and V is the volume of water in gallons.

$$W = V \times 8.34$$

Example 1-15

What is the weight of 12.6 gal of water?

Answer

$$W = V \times 8.34$$
$$= 12.6 \times 8.34$$
$$= 105.08 \text{ lb, or rounded down to } 105 \text{ lb}$$

If we need to know not the volume in gallons but only the weight of water in the cylinder, we can create another shortcut. A single formula to find the weight of water in a cylinder could be written as shown here: Weight $= 8.34 \times 7.49 \times 0.7854 \times D^2 \times H$. Again, since multiplication can be done in any sequence, we can create a constant of 49.06, or 49, simply by multiplying $8.34 \times 7.49 \times 0.7854$. The new formula for finding the weight of the water in a cylinder becomes $W = 49 \times D^2 \times H$ (or L).

$$W = 49 \times D^2 \times H \text{ (or } L)$$

Example 1-16

What is the weight of water in a 50-ft section of 2½-in hose?

Answer

$$W = 49 \times D^2 \times H$$
$$= 49 \times (2.5/12)^2 \times 50 \text{ ft}$$
$$= 49 \times (0.208)^2 \times 50 \text{ ft}$$
$$= 49 \times 0.043 \times 50 \text{ ft}$$
$$= 105.35 \text{ lb, or rounded down to } 105 \text{ lb}$$

Both formulas, while shortcuts, are perfectly legitimate. However, before you use the shortcuts, it is important to fully understand how the formulas work. Also, where and when we round off can affect the answer. In most cases the difference due to rounding will not be significant.

Math and Rounding

Throughout this text, in most instances where calculations render more than two digits to the right of the decimal, the numbers are rounded to two decimal places. Some numbers have traditionally been used with three or four digits to the right of the decimal; in instances where those numbers are used, they will be shown in traditional format. In all instances, answers are rounded to just two decimal places. Where rounding to whole numbers is appropriate, it is noted in the proper section.

The more places to the right of the decimal an answer has, the greater its level of precision. Although precision and accuracy should be the goal of all hydraulics calculations, the fireground does not lend itself to two-decimal-place precision. Practice accuracy in all calculations not associated with the fireground, such as in the classroom and during preplanning. In short, develop habits that result in highly accurate, precise answers. On the fireground, that level of precision will not be possible because of the nature of the beast. However, if you have developed a habit of accuracy, even the fireground can be more than a "best guess" scenario. The chapter on standpipes, sprinklers, and fireground formulas offers recommendations for accuracy on the fire ground while making calculations doable under the chaos of emergency operations.

A Note of Caution

A note of caution concerns some of the figures given as standards in this text. In some instances, these figures will not agree with other references, but for good reason. The figures in this text represent a recent calculation performed for the purpose of checking the validity of the older numbers before including them in this text. In many instances the older figures have been replaced with the results of the recent calculations. An example is the weight of a column of water 1 in × 1 in × 1 ft tall. Older texts give it a weight of 0.434 lb, but it is calculated in the chapter on force and pressure to be 0.433 lb. This difference results because the older figure was calculated using the weight of 1 ft^3 of water as 62.5 lb,

not 62.4 lb. The 62.4-lb figure is the more accurate figure. The 62.5-lb figure, from which 0.434 was derived, was rounded for ease of calculation. In each instance where this text uses a different figure than has been used in the past, the calculations from which the figures were derived are included.

WRAP-UP

Chapter Summary

- Hydrostatics is the study of water at rest.
- Hydrodynamics is the study of water in motion.
- Fred Shepherd wrote the first book on hydraulics for the average firefighter in 1917.
- Freshwater weighs 62.4 lb/ft^3.
- 1 ft^3 of water contains 1,728 in^3.
- 1 gal of water contains 231 in^3.
- 1 ft^3 of water contains 7.48 gal.
- 1 gal of water weighs 8.34 lb.
- Latent heat/entropy of vaporization is 970.3 Btu/lb of water.
- 1 gal of water will absorb 9,343 Btu going from water at 62°F to steam at 212°F.
- NFF is the amount of water needed to extinguish a given fire.
- Water expands 1,700 times going from water at 212°F to steam at 212°F.
- 1 gal of water expands to 227 ft^3 of steam.

Key Terms

British thermal unit (Btu) The amount of heat needed to raise the temperature of 1 lb of water by 1°F at 60°F.

enthalpy The total amount of energy in a substance.

hydraulics The science of water (or other fluids) at rest and in motion.

hydrodynamics The study of water in motion.

hydrostatics The study of water at rest.

latent heat/enthalpy of fusion The amount of heat, in Btu, absorbed or given off when 1 lb of water goes from its solid state to its liquid state or back.

latent heat/enthalpy of vaporization The amount of heat, in Btu, absorbed or given off when 1 lb of water goes from its liquid state to its vapor state or back.

needed fire flow (NFF) An estimate of the amount of water needed to extinguish a fire in a specific building or compartment.

specific heat The amount of heat, in Btu, absorbed or given off as a substance changes temperature 1°F per pound.

Case Study

Iowa State Rate-of-Flow Formula

In the 1950s, Keith Royer and Floyd Nelson from Iowa State University conducted experiments that among other things resulted in the Iowa State rate-of-flow formula. During their many experiments they began to see a pattern of water use while doing research related to uncontrolled fire behavior in structures. This research into a rate-of-flow formula was coordinated with other countries conducting similar research, including Britain, Canada, Germany, Japan, and Australia. The results of their research were published for the first time in 1959 in both *Fire Engineering* magazine and Iowa State University Bulletin no. 18.

The Iowa State rate-of-flow formula is based on several scientific facts:

1. Ordinary combustibles will give off 535 Btu when burned in 1 ft^3 of pure oxygen.
2. Flame production ceases when oxygen falls to 14 percent.
3. Normal air contains 21 percent oxygen.

This means that only 7 percent oxygen is available for combustion in normal air. From this Royer and Nelson were able to calculate that the oxygen in 1 ft^3 of air will produce just 37 Btu. One other important fact they understood was that it is the oxygen in the air, not the fuel, that will determine heat release from a fire.

Additionally, they calculated that 1 gal of water had the capacity to absorb all the heat that would be generated by the oxygen in 200 ft^3 of air, with a margin of safety. This means that if we divide the volume of the room by 200 ft^3, we can determine the amount of water needed to achieve fire control. And dividing the volume of the room by 100, we end up with a flow rate that if applied for 30 s will achieve control.

In his article "Iowa Rate of Flow Formula for Fire Control," Royer points out two limitations of the formula:

1. It is only applied to the single largest open area of the building.

2. It only calculates the amount of water needed to achieve fire control, not fire extinguishment.

Finally, Keith Royer goes on to caution: "If a combination attack is made and the nozzle is not advanced to the proper location and properly manipulated, the fire will not black out. When the fire does black out and the nozzle is not shut down within a reasonable time, the area will be overcooled, and overhaul will be difficult. In multi-storied situations, late shut down may even lead to the loss of the building."

1. How did Nelson and Royer determine that each cubic foot of air had sufficient oxygen to generate 37 Btu during a fire?

   A. Through empirical testing
   B. With a commonly used figure
   C. By calculating 7 percent of 535 Btu
   D. By understanding that only one-third of the oxygen in air goes to combustion

2. Can this formula be used to calculate the amount of water needed to extinguish a fire in a particular building?

   A. No, it is limited to the single largest area of the building.
   B. Yes, as long as the 200 ft^3/gal of water guideline is observed.
   C. Yes, as long as too much water is not used.
   D. Yes, as long as each room is extinguished separately.

3. If the assumption is that 1 gal of water will absorb 200 Btu, why does the formula have the constant 100 and not 200?

A. 1 gal per 200 ft^3 of volume is the amount of water that will be needed to completely extinguish the fire, but the formula is looking for the amount of water to control the fire.

B. 1 gal of water per 200 ft^3 of volume is the amount of water that will be needed without a measure of safety.

C. 1 gal of water per 100 ft^3 of volume is the amount of water needed to achieve extinguishment within 60 s.

D. 1 gal of water per 100 ft^3 of volume is the amount of water needed to achieve fire control within 30 s.

4. Name one other important Iowa State finding.

A. More water can achieve faster control.

B. Using too much water to accomplish extinguishment can complicate fire control.

C. It is not possible to "overcool" a fire.

D. All the above.

Information for this case study came from Keith Royer, "Iowa Rate of Flow Formula for Fire Control," *Fire Engineering*, September 1959; Keith Royer and Floyd Nelson, "Water for Firefighting," *Fire Engineering*, August 1959; Keith Royer and Floyd Nelson, "Water for Firefighting—Rate of Flow Formula," *Iowa State University Bulletin*, no. 18, Ames, Iowa: Iowa State University, 1959.

Review Questions

1. What is the weight of 1 gal of salt water?

2. How much does 1 ft^3 of freshwater weigh?

3. How many gallons are in 1 ft^3 of water?

4. What is the latent heat of vaporization of water?

5. Into how many cubic feet of steam will 1 gal of water expand?

Activities

You are now ready to put your knowledge of water and its properties to work.

Please note that the problems in the Activities and the Challenging Questions sections are designed to challenge your knowledge of the principles learned up to this point. Do not expect these questions to be carbon copies of the examples already given. Be prepared to think creatively in solving these problems throughout this book.

1. How many gallons of water are in a reservoir that measures 100 ft × 50 ft and is 10 ft deep?

2. An office space measures 10,000 ft^3. How many gallons of water will it take to completely fill the space with steam?

3. An Air Crane helicopter, with a special apparatus for dropping water on forest fires, has a capacity of 2,000 gal of water. How much does the water add to the weight of the helicopter?

4. How many Btu will be absorbed by 100 gal of water if the water is completely vaporized?

Challenging Questions

1. A salesperson is selling a tanker that weighs 19,182 lb more when full than when empty.

   A. How many gallons of water does it hold?
   B. What is the volume of the tank in cubic feet?

2. How many gallons will a water tower with a 20-ft sphere at the top store?

3. What is the weight of a tank of water if the tank is 10 ft long and 4½ ft in diameter?

4. What is the weight of the tank in Question 3 if the tank is full of salt water?

5. Two hose streams are operating in the second floor of a small warehouse for 30 minutes (min). If one line is flowing at 210 gpm and the other one is flowing at 265 gpm, how much weight will be added to the weakened floor structure if only one-half of the water is converted to steam? (Water not converted to steam becomes part of the live load of the building.)

6. How many Btu will be absorbed if 2 ft³ of water at 62°F is heated to 200°F?

7. A hose line with a 125-gpm tip can be expected to extinguish a room of what volume, if the line operates for 30 s? Assume 100 percent conversion of water to steam.

8. Find the volume of the swimming pool in Figure 1-7 in

   A. Cubic feet
   B. Gallons

9. How many gallons of water will it take to absorb all the Btu given off in Question 6?

10. How many gallons of water will it take to completely fill a room with steam if the room is 12 ft × 20 ft × 8 ft? Calculate the answer by using the Iowa State rate-of-flow formula and the straight conversion of water-to-steam formula.

11. Explain the difference in the answers for Question 10 due to the method used to calculate each of the answers.

Formulas

To find specific heat:

Specific heat = weight of water ×
change in temperature

To calculate the latent heat of vaporization of a given weight of water:

Latent heat of vaporization = weight of water ×
970.3

To find the volume of a rectangle:

$$V = L \times W \times H,$$ where V is the volume, L is the length, W is the width, and H is the height

To find out how much water is needed to fill a room with steam:

$$CFW = V \div 1,700$$

To calculate how much water is in a container:

$$\text{Gallons} = V \times 7.48$$

To find the gallons of water needed to fill a room with steam:

$$\text{Gallons} = V \div 227$$

The Iowa State rate-of-flow formula:

$$\text{Gallons per minute (gpm)} = V \div 100,$$
where V is the volume in cubic feet

To find out how many Btu will be absorbed by the number of gallons of water:

$$\text{Btu absorbed} = 9,343 \times \text{gallons}$$

The NFA fire flow formula:

$$\text{NFA} = \left(\frac{\text{length} \times \text{width}}{3} + \text{exposure charge} \right) \times \% \text{ involvement}$$

To find the gallons of water when the volume is in cubic inches:

$$\text{Gallons} = V \div 231$$

To find the volume of a cylinder in cubic inches or cubic feet:

$$V = 0.7854 \times D^2 \times H \text{ (or } L)$$

To find the area of a circle, using the diameter:

$$\text{Area} = 0.7854 \times D^2$$

To find the volume of a cylinder in gallons:

$$V = 7.48 \times 0.7854 \times D^2 \times H \text{ (or } L)$$

To perform direct calculation of gallons of water in a cylinder:

$$V = 5.87 \times D^2 \times H \text{ (or } L)$$

To find the weight of water in a container:

$$W = V \times 8.34$$

To preform direct calculation of the weight of water in a cylinder:

$$W = 49 \times D^2 \times H \text{ (or } L)$$

Force and Pressure

LEARNING OBJECTIVES

Upon completion of this chapter, you should be able to:

- Understand the difference between force and pressure.
- Calculate pressure, given force and area.
- Calculate pressure, given elevation.
- Calculate force, given pressure and area.
- Understand the six principles of pressure.
- Understand the difference between absolute pressure and relative pressure.
- Calculate absolute pressure from relative pressure.
- Differentiate among static pressure, residual pressure, and flow pressure.

Case Study

One spring afternoon Engine 11 is out training on pumps. Your company officer knows you are studying hydraulics to prepare for the next driver/operator position and has asked you to help the "probie" learn some of the basic concepts of hydraulics.

The current evolution involves a 200-foot (200-ft) preconnect 1¾-inch (1¾-in) line. You have asked the probie to charge the line at the correct pressure, but she has used an incorrect friction loss for the flow and is pumping at too high a pressure. When the firefighter at the nozzle attempts to open it, the nozzle is extremely difficult to open.

1. Why is the nozzle so difficult to open?
2. Which principles of fluid pressure are at work here before the nozzle is opened?
3. How do you explain to the probie the difference between relative and absolute pressure?

Introduction

The study of hydraulics must begin with a thorough understanding of both force and pressure. Studying force and pressure together is logical, because their definitions are related. Further, it is necessary to understand the six principles of fluid pressure as a basis for our study of hydraulics. Together, these definitions and six principles of fluid pressure lay the foundation for an accurate understanding of hydraulics.

Pressure

Pressure is defined as the *force per unit of area* or, in the form of an algebraic expression, $P = F/A$, where P = pressure, F = force, and A = area. For now let us accept the definition of force, in this equation, simply as the weight of the fluid, in pounds (lb). Units of area can be in square yards (yd^2) or square feet (ft^2), but are almost universally given in square inches (in^2). Therefore, force in pounds, exerted over an area measured in square inches, gives us a reading of pressure in pounds per square inch (psi, or lb/in^2).

The weight of 1 cubic foot (ft^3) of water is 62.4 lb. This figure is also the force, in pounds, of the water on the inside bottom surface of the container. So we can say that the force of 1 ft^3 of water over an area of 12 in × 12 in is 62.4 lb. Note that force is assigned units of pounds. If the container is 12 in × 12 in × 12 in, it can correctly be stated that it represents both a force of 62.4 lb and a pressure of 62.4 lb/ft^2, if we are using square feet as our unit of pressure. But we want our units of pressure to be in pounds per square inch. Since our example is a true cubic foot of water, the surface where the force is acting is 12 in × 12 in, or 144 in^2. Using the formula $P = F/A$, we can calculate the pressure from the force.

$$P = \frac{F}{A}$$

Example 2-1

What is the pressure in pounds per square inch of 1 ft^3 of water if it is exerted over 1 ft^2?

Answer

The pressure is the force of 1 ft^3 of water, or 62.4 lb, divided by 144 in^2:

$$P = F/A$$
$$= 62.4 \div 144$$
$$= 0.433 \text{ psi}$$

We now know that a column of water 1 in × 1 in and 1 ft tall exerts a pressure of 0.433 psi **Figure 2-1**. We have just calculated pressure from force.

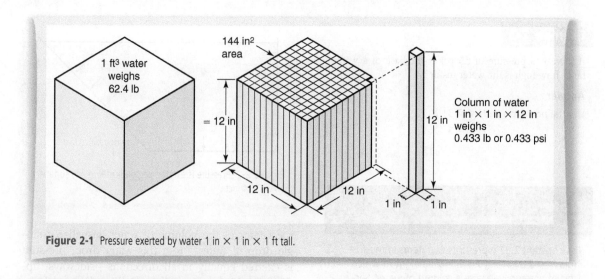

Figure 2-1 Pressure exerted by water 1 in × 1 in × 1 ft tall.

Note

A column of water 1 in × 1 in and 1 ft tall exerts a pressure of 0.433 psi.

$$P = 0.433 \times H$$

In general, in situations where firefighters need to calculate pressure, the force is not known. Therefore, we need another way to calculate pressure. Because we now know that a column of water is capable of exerting a pressure of 0.433 psi for every 1 ft of elevation, we can calculate pressure by using

Example 2-2

What is the pressure at the base of a column of water 10 ft tall?

Answer

Multiply 0.433 by the elevation head.

$$P = 0.433 \times H$$
$$= 0.433 \times 10$$
$$= 4.33 \text{ psi}$$

the equation $P = 0.433 \times H$. In this formula H stands for the elevation or height of water, in feet, over the point where the measurement is taken. It is often referred to as the *elevation head*.

Calculating Elevation from Pressure

Suppose we know the pressure at a given point and want to find the elevation of the surface of the water. We can actually use this same formula by rearranging it to read $H = P/0.433$. Alternatively, we can create a new formula that allows us to use multiplication to find the answer. To create a new formula, first we need to know how far a single pound of pressure will vertically push a column of water. To calculate, simply divide 1 psi of pressure by 0.433, the pressure that will push a column of water 1 ft vertically or, conversely, the pressure exerted by a column of water 1 ft high: 1 psi ÷ 0.433 = 2.309, rounded to 2.31 ft. That is, l psi of pressure will raise a column of water 2.31 ft. Conversely, 2.31 ft of water will exert a pressure of 1 psi at its base. The new formula used to calculate elevation from pressure is $H = 2.31 \times P$, where P is the pressure exerted by the elevation we are trying to find.

$$H = 2.31 \times P$$

If there is a pressure of 65 psi at the base of a water tank, how high is the water inside?

Answer

Use the new formula $H = 2.31 \times P$.

$$H = 2.31 \times 65$$
$$= 150.15 \text{ ft}$$

Note

The constant 2.31 represents two items in the formula $H = 2.31 \times P$. First, it is the height to which 1 psi of pressure will raise a column of water. Second, it is the height of water that will create a pressure of 1 psi at its base. The same formula applies in both instances—you just have to know what you are looking for. In both cases, 2.31 is assigned units of feet per pound (ft/lb). They have been omitted for simplicity but should be understood. H will always be given in feet.

Six Principles of Fluid Pressure

Now that we know what pressure is, the next step is to study the six principles of fluid pressure. These principles explain the behavior of pressure in fluids.

Principle 1

Pressure is exerted perpendicular to any surface on which it acts. Simply stated, pressure must act at right angles to any surface with which it is in contact **Figure 2-2**. If pressure were imposed at other than a right angle, the fluid would tend to flow out of the container or bunch up in the center. But this does not happen, proving that pressure is exerted at right angles to its container.

Principle 2

Pressure in a fluid acts equally in all directions. Understanding this principle requires a bit of imagination. Visualize a bucket full of water. Then, at a point in

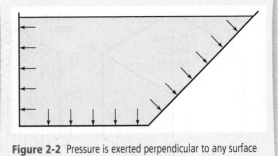

Figure 2-2 Pressure is exerted perpendicular to any surface on which it acts.

the center of the volume of water, visualize a single drop of water. From this water drop, pressure is exerted equally in all directions—sideways, up, down at 45° angles, and so forth with equal force **Figure 2-3A**.

A more familiar example is a fire hose **Figure 2-3B**. Visualize a point of pressure, in the exact middle of the stream of water in the hose. If that pressure is truly equal in all directions, then the hose, which has no intrinsic shape of its own, should be round when charged, and it is.

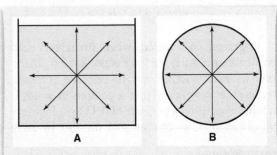

Figure 2-3 Pressure in a fluid acts equally in all directions. **A.** Bucket of water. **B.** Fire hose.

Principle 3

If pressure is applied to a confined liquid, that pressure is transmitted to every point within the liquid without reduction in intensity. This principle, also known as **Pascal's principle**, is named for Blaise Pascal (1623–1662), a French philosopher and scientist who first postulated it.

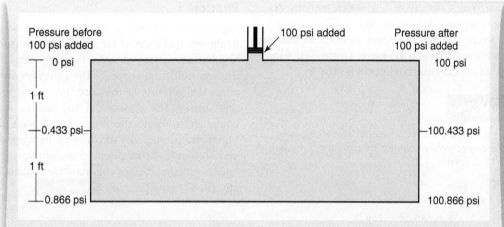

Figure 2-4 If pressure is applied to a confined liquid, that pressure is transmitted to every point within the liquid without reduction in intensity.

Imagine a tank of water 2 ft in diameter and 10 ft long. On the top of this tank is a cylinder with an area of exactly 10 in² (**Figure 2-4**). With no external pressure applied, there would be a pressure of 0.866 psi at the bottom of the tank (0.433 × 2 ft of water), 0.433 psi at midpoint, and 0 at the top. If 100 psi of pressure is applied to the tank by the cylinder, then there is a total pressure of 100.866 psi at the bottom, 100.433 psi at the midpoint, and 100 psi at the top.

Principle 4

The pressure of a liquid in an open vessel is proportional to the depth of the liquid. We use this principle to determine the pressure at the 1- and 2-ft levels in our illustration of Principle 3. Each segment of water 1 in × 1 in × 12 in tall exerts 0.433 psi. Therefore every 1 ft of water above the point where we measure the pressure adds another 0.433 psi (**Figure 2-5**).

Principle 5

The pressure of a liquid in an open container is proportional to the density of the liquid. This principle is the twin of Principle 4, except that in Principle 5 reference is made not to the height of liquid in the

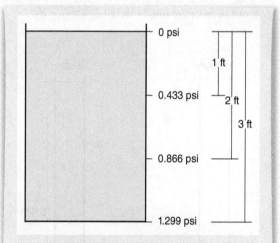

Figure 2-5 The pressure of a liquid in an open container is proportional to the depth of the liquid.

container, but to the density of the liquid. Because different liquids have different densities, the pressures they exert for a given elevation are different. The only liquid, other than water, we would be concerned with in pump operations is mercury. Mercury is used to measure the amount of work a pump does at draft. Let us use mercury to illustrate this principle.

By assigning a density of 1 to water, mercury on the same scale has a relative density of 13.6; in other words, mercury weighs 13.6 times as much as water. This means that if 1 ft³ of water exerts a force of 62.4 lb, then 1 ft³ of mercury exerts a force of 848.64 lb.

Example 2-4

If 1 ft³ of mercury exerts a force of 848.64 lb, how much pressure will a column of mercury 1 in × 1 in × 12 in exert?

Answer

Use the formula $P = F/A$.

$$P = 848.64 \div 144$$
$$= 5.89 \text{ psi}$$

There is another way to compare the density of water to that of mercury. It would take a column of water 13.6 in tall to exert the same pressure as a column of mercury only 1 in tall (Figure 2-6).

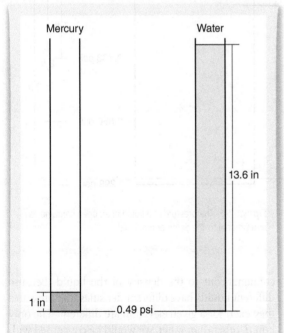

Figure 2-6 The pressure of a liquid in an open container is proportional to the density of the liquid.

Principle 6

Liquid pressure on the bottom of a container is unaffected by the size and shape of the container. This principle, also known as the *hydrostatic paradox*, states that for a given fluid, the shape of the container does not affect the pressure at the base, or at any point for that matter. The pressure of any size or shape of container is always the same at any point along a horizontal plane. Only the distance between the top and bottom (or other points of measurement) of the liquid—not the shape or volume of the container—has any effect on the pressure (Figure 2-7).

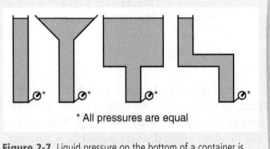

* All pressures are equal

Figure 2-7 Liquid pressure on the bottom of a container is unaffected by the size or shape of the container.

Relative Pressure versus Absolute Pressure

There are two methods for measuring pressure. Just as the weight of a column of water 1 in × 1 in is called *pressure*, and for every 1 ft of water we have 0.433 lb of pressure, the atmosphere (air) around us also has weight. After all, air is a fluid with mass and volume that obeys the laws of pressure we have just outlined, but with some variation due to the fact that gases are compressible while liquids are not. This means that a column of air (gas) is actually denser at the bottom than at the top. The pressure of air at sea level is 14.7 psi. This number represents the weight, at sea level, of a column of air that is 1 in × 1 in. It is referred to as 1 atmosphere (atm) and is the standard used for atmospheric pressure. When we expose a gauge to the atmosphere at sea level, it should naturally read 14.7 psi, but most gauges do not. This brings us to the concept of relative pressure versus absolute pressure.

Technically, all gauges read atmospheric pressure when at rest. However, some gauges read it as relative pressure, whereas a few read it as absolute pressure. **Relative pressure** can be defined as the pressure indicated with 0 psi = atmospheric pressure. This is the most common way of measuring pressure and how the gauges on the pump panel of pumping apparatus measure it. Gauges that read pressure on a relative scale will read 0 psi when the only pressure acting on them is from the weight of air (the atmosphere). **Absolute pressure** can be defined as the method of reading pressure in which 14.7 psi = atmospheric pressure. Absolute pressure gauges differ from relative pressure gauges in that absolute pressure gauges will actually read the exact pressure, including the pressure of the atmosphere.

Gauges used to measure pressure may be marked to indicate whether they measure on an absolute scale, psia, or on a relative scale, psig (relative pressure is also referred to as gauge pressure). Some gauges are not marked, and you have to know the scale to which they correspond. Gauges used on fire apparatus will always read on a relative, psig, scale.

Example 2-5

What formula converts relative pressure to absolute pressure?

Answer

Absolute pressure = relative pressure + atmospheric pressure

Absolute pressure = relative pressure + atmospheric pressure

Distinguishing between absolute and relative pressure might seem trivial, because the pressure is only off by the value of atmospheric pressure, but it is actually a critical distinction. Without the distinction between the two methods of reading pressure, there is no negative pressure, or vacuum. **Figure 2-8** illustrates the difference between absolute and relative pressure.

As far back as Aristotle it has been known that nature abhors a vacuum. In fact, vacuums, in terms of negative pressure, technically do not exist. What we refer to in the fire service as a **vacuum** is just a relative pressure below the surrounding atmospheric pressure. We usually identify this vacuum

Fireground Fact

Relative versus Absolute Pressure Gauges

Relative pressure gauges are by far the most common type of gauge found. Normally, when we want to add pressure to something such as a hose line, we calculate the pressure as what is needed above atmospheric pressure. Although an absolute pressure gauge would work, it would require another step to add atmospheric pressure to the calculated pressure. Using relative pressure simplifies the process by eliminating this step.

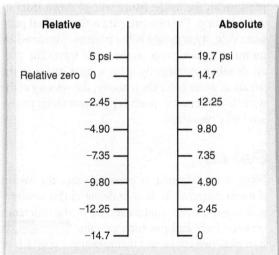

Figure 2-8 Relative versus absolute pressure scales.

in reference to inches of mercury. On an absolute scale, pressure below atmospheric is still a positive pressure. This distinction is important to understand, and its relevance will become more apparent later when we discuss pump theory and drafting. In absolute terms, what is usually referred to as a vacuum is simply the absence of any pressure. The concept of a vacuum is fully explained in Chapter 8.

Types of Pressure

Before leaving our discussion of pressure, we must define a few types of pressure:

- Static pressure
- Residual pressure
- Velocity pressure

These terms' definitions are used throughout the remainder of this text.

Static pressure is the pressure of a fluid at rest. For example, when a pumper is first hooked up to a hydrant and the hydrant is charged (or opened) but no water is flowing, the intake gauge will read static pressure. **Residual pressure** is the pressure remaining when water is flowing. Suppose a pumper is hooked up to a hydrant and the hydrant is charged. After a discharge is opened and water is flowing, the pressure on the intake gauge will go down from its static reading. The new pressure is the residual pressure. **Velocity pressure** is the pressure measured by means of a pitot gauge as the water leaves the nozzle or other opening. By using a pitot gauge in the stream of water as it exits a nozzle, the velocity of the water is converted to pounds per square inch to give a velocity pressure.

Force

Force is best defined as pressure times the area it is exerted against. To best understand this concept, and to get a better understanding of the difference between force and pressure, examine **Figure 2-9**. It depicts two containers that measure 1 ft × 1 ft × 3 ft, each containing 3 ft^3 of water. In Figure 2-9A,

the container is standing on end on a table. Because it is standing on end, it has an area of 144 in^2 in contact with the table. Since the container is 3 ft high, the pressure is 0.433 × 3 psi, or 1.299 psi. Putting this in the form of a formula, we have

$$P = 0.433 \times H$$
$$= 0.433 \times 3$$
$$= 1.299 \text{ psi}$$

To find force once we have found pressure, we need to return to the formula $P = F/A$. By multiplying both sides by A, the left side of the equation becomes $P \times A$ and the right side of the equation becomes F. We now have the formula $F = P \times A$ for finding force when we know area and pressure. With this formula, we can now calculate the force at the base Figure 2-9A.

$$F = P \times A$$
$$= 1.299 \times 144$$
$$= 187.06 \text{ lb}$$

$$F = P \times A$$

In Figure 2-9B, the container has been placed on its side. In this instance the force is exerted over an area of 432 in^2, and the pressure is 0.433 psi because the elevation is only 1 ft. Using the formula for force, we can calculate the force exerted in Figure 2-9B as

$$F = P \times A$$
$$= 0.433 \times 432$$
$$= 187.06 \text{ lb}$$

We can verify that the force in Figure 2-9 is correct because in both cases the force is equal to the weight of water in the container. The force should be 3 ft^3 × 62.4 lb/ft^3, or 187.2 lb. Note that in both instances force stays the same because the weight of the water is the same, but the pressure varies according to the orientation of the container.

Earlier we used a definition of force as the weight of the fluid. However, this definition is not always true. For example, look at **Figure 2-10**. Because

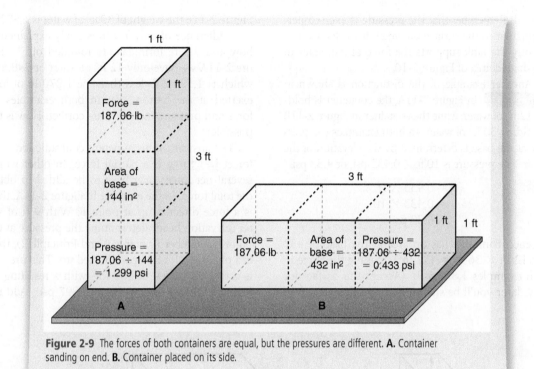

Figure 2-9 The forces of both containers are equal, but the pressures are different. **A.** Container sanding on end. **B.** Container placed on its side.

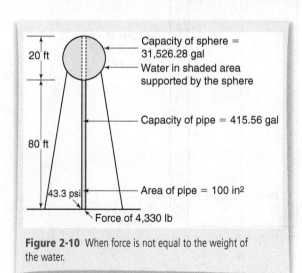

Figure 2-10 When force is not equal to the weight of the water.

Principle 4 and Principle 6 tell us that the pressure at the bottom of the water tank is proportional to the depth of the liquid regardless of shape, the pressure at the base of the water tower in Figure 2-10 is 43.3 psi. If the area of the pipe leading from the tank is 100 in², there should be a force of 4,330 lb (43.3 psi × 100 in²).

$$F = P \times A$$
$$= 43.3 \times 100$$
$$= 4,330 \text{ lb}$$

However, if we use the weight of the fluid (water) to calculate the force, then there should be the force of 31,741.84 gal (total volume of sphere and supply pipe) × 8.34 lb/gal, or 264,726.94 lb. How do we account for the difference?

In this example, the force we are measuring is only the weight exerted by a column of water 100 in² × 100 ft tall (4,332 lb). It is the area represented by the unshaded portion of the tank in Figure 2-10 and the pipe below it. The force (weight) of the water in the shaded area of the tank in the figure is actually bearing against the tank itself, not on the ground. Remember that the pressure is exerted perpendicular to the surface on which it works. Consequently the tank supports the force of the water in the shaded area of Figure 2-10.

Another example of the distinction is shown in **Figure 2-11**. In Figure 2-11A, the container is holding 12 ft³ of water, while the container in Figure 2-11B is holding 30 ft³ of water. In both examples the pressure at the base is determined by the elevation of the water. The pressure is 10 ft × 0.433 psi, or 4.33 psi.

$$P = 0.433 \times H$$
$$= 0.433 \times 10$$
$$= 4.33 \text{ psi}$$

Each container has an area at the base of 432 in² (12 in × 36 in = 432 in²), making the force in both examples 1,870.5 lb. (Accept this as fact for now; later you'll be shown how to calculate it.) In

Figure 2-11B the weight of 30 ft³ of water is 1,872 lb. This difference of only 1.5 lb is easily explained by how and where formulas are rounded off. In Figure 2-11A we have only 12 ft³ of water or 748.8 lb, which is 1,121.2 lb less than the 1,870 lb of force exerted at the bottom. Yet in both examples the force and pressure indicated are correct. How is this possible?

The answer to this apparent contradiction is net force. **Net force** is a partial force. In other words, several net forces may have to be added to obtain the total force. In the example in Figure 2-11A, there is a force directed up at point B. With 9 ft of water (elevation head) determining the pressure at this point (remember Principle 4 and Principle 2), there is a pressure of 3.897 psi directed up. This pressure is acting on a total of 288 in², with a resulting net force of 1,122.3 lb (288 in² × 3.897 psi). Add this

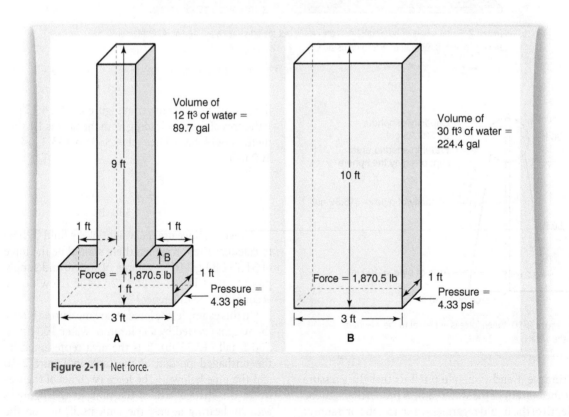

Figure 2-11 Net force.

net force to the weight of 12 gallons (gal) of water (748.8 lb), which in this instance is also a net force, and we have a total force of 1,122.3 lb + 748.8 lb = 1,871.1 lb of force. Figure 2-11A and Figure 2-11B are in agreement. (Again this small difference is due to rounding.)

The purpose of this exercise in net force is to clarify the following explanation of force: Force on a surface is a function of gravity, the density of the fluid, the height of the fluid, and the area it is exerted against. To simplify this, we can say that the influence of gravity and density gives us a constant for water of 0.433 psi per foot (psi/ft) of elevation. This figure was previously shown to be the pressure exerted by a 1-ft column of water. To calculate the force at the bottom of any container regardless of shape or size, simply multiply: 0.433 × height × area. The height referred to here is also called the **piezometric plane**, which is simply the equivalent elevation head H of water at the point where the force is measured Figure 2-12 . By putting all this together, we end up with the following formula for calculating force when height (or elevation) is known: $F = 0.433 \times H \times A$.

$$F = 0.433 \times H \times A$$

Example 2-6

Using both formulas for force, prove that 1,870.5 lb is the correct force in Figure 2-11A and Figure 2-11B.

Answer

In both cases,

$$
\begin{array}{ll}
P = 0.433 \times H & \text{or} \quad F = 0.433 \times H \times A \\
\quad = 0.433 \times 10 \text{ ft} & \qquad\quad = 0.433 \times 10 \times 432 \\
\quad = 4.33 \text{ psi} & \qquad\quad = 1{,}870.56 \text{ lb} \\
F = P \times A & \\
\quad = 4.33 \text{ psi} \times 432 \text{ in}^2 & \\
\quad = 1{,}870.56 \text{ lb} &
\end{array}
$$

This example of net force illustrates the point that the total force at the bottom of a container may not be equivalent to the weight of the liquid. Therefore, whenever you calculate force, always use the formula $F = P \times A$, or $F = 0.433 \times H \times A$. It is accurate 100 percent of the time.

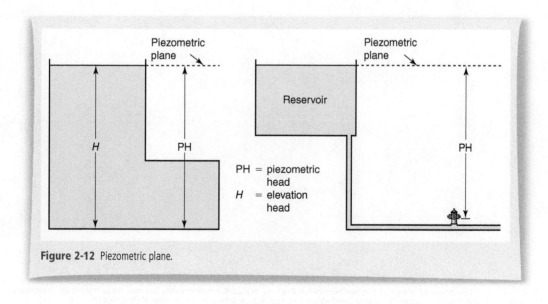

Figure 2-12 Piezometric plane.

Fireground Fact

Force versus Pressure

While we are on the subject of force, note that force, not pressure, is the primary cause of most problems we normally associate with pressure. For example, we always say that too much pressure on a hose line makes it difficult to open the nozzle. However, if you calculate the area over which the pressure is working on the ball valve and multiply that area by the pressure, you will find the force can be quite formidable. It is technically the force on the nozzle that prevents it from opening, not pressure.

By knowing this we can also use force to our advantage. An excellent example reveals how a low pressure of 15 to 20 psi on the dry side of a differential dry pipe sprinkler valve can hold back a water pressure of 60 psi or more.

Chapter Summary

- Pressure is the force per unit of area.
- A column of water 1 ft high will exert a pressure of 0.433 lb.
- There are six principles of fluid pressure:
  1. Pressure works perpendicular to any surface.
  2. Pressure works equally in all directions.
  3. Pressure is proportional to depth.
  4. Pressure applied to a closed container is distributed to all points equally.
  5. Pressure is proportional to density.
  6. Pressure at the base of a container is not affected by the container's size or shape.
- Relative pressure reads 0 in the presence of atmospheric pressure.
- Absolute pressure reads 14.7 in the presence of atmospheric pressure.
- Static pressure is the pressure of a fluid at rest.
- Residual pressure is the pressure remaining when water is flowing.
- Velocity pressure is the pressure measured by a pitot gauge as water leaves the opening.
- Force is pressure times the area it is working against.
- Net force is a partial force.

Key Terms

absolute pressure The pressure indicated when 14.7 psi = atmospheric pressure.

force Pressure times the area it is exerted against.

net force A partial force.

Pascal's principle If pressure is applied to a confined liquid, that pressure is transmitted to every point within the liquid without reduction in intensity.

piezometric plane The equivalent elevation head H of water at the point where the force is measured.

pressure The force per unit of area.

relative pressure The pressure indicated when 0 psi = atmospheric pressure.

residual pressure The pressure remaining when water is flowing.

static pressure The pressure of a fluid at rest.

vacuum A relative pressure below the surrounding atmospheric pressure; or in an absolute scale, the absence of any pressure.

velocity pressure The pressure measured as the water leaves the nozzle or other opening.

Case Study

Pascal's Vases and the Hydrostatic Paradox

The hydrostatic paradox is an assumption that a container of a fluid will exert greater pressure at its base if it has a larger volume. As an example, Figure 2-7 depicts four containers, each of a different shape. Shouldn't it be obvious that the second and third containers would have greater pressure at their base? After all, they contain more liquid.

In the 1660s Blaise Pascal set out to demonstrate that the shape and volume of a container were not factors in determining the pressure at its base. To do this he developed what would later be called *Pascal's vases*.

Pascal built an apparatus that used several glass vases of various shapes and volumes but vases that were the same height and had the same base area (Figure 2-13). Each vase was then placed on a device; it was a pivoted arm that held a cover over the base of the vase at one end and in which weights were suspended from the other.

The downward force of the weights on one end of the arm exerted an upward force on the other end; this held the cover in place, preventing the liquid from leaking out. Weight would then be removed in small increments until the vase began to leak. Regardless of the shape and volume of the vase that was used, as long as the vases were filled to the same height, each one would begin to leak after the same amount of weight was removed. This proved that pressure is independent of volume, but completely dependent on the height of the liquid.

1. Why is the height of the liquid, and not the volume, important in determining pressure?

    A. Pressure is essentially an abstract concept and is simply defined as a function of height.

    B. Pressure is all about net force.

    C. Pressure is force divided by area.

    D. Both B & C are correct.

2. Could atmospheric pressure have an effect on this experiment?

    A. No, atmospheric pressure can vary too much from one geographic area to another.

    B. No, atmospheric pressure is dependent upon weather.

    C. No, atmospheric pressure will have the same effect regardless of shape of the container; that is, the end result is relative, but will be the same regardless of shape.

    D. Yes, the greater the surface area of the container that the atmosphere is working on, the greater the pressure.

3. What would be the result if the size of the bottom opening of each of the various shapes were different for each shape?

    A. The shape with the largest opening at its base would open first.

    B. The shape with the smallest opening at its base would open first.

    C. Since depth determines the pressure at the base, the size of each opening would have no effect.

    D. None of the above.

4. If the hydrostatic principle is correct, then how is it possible for a differential dry pipe valve to work?

    A. The differential dry pipe valve works because of a difference in area on each side of the valve.

    B. In addition to the differential dry pipe valve, these systems employ retarders to make them work.

    C. The pressure on the dry side of the valve is maintained a little higher by a jockey pump that is constantly running to maintain pressure.

    D. The pressure on the dry side of the valve is dynamic because of the jockey pump, while the pressure on the wet side of the valve is static.

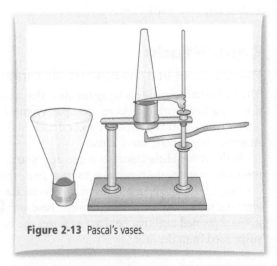

Figure 2-13 Pascal's vases.

Review Questions

1. What is the difference between force and pressure?

2. What is the name for the imaginary line on the same plane as the level of the surface of an open container of water, used as a reference for determining elevation head?

3. What is the pressure reading from water at rest called?

4. If you were to attach a pressure gauge to one of the 2½-in outlets of a hydrant and then open the hydrant, allowing water to flow from one of the remaining outlets, what type of pressure would the gauge read?

Activities

Now you can put your new knowledge of force and pressure to work.

1. If the surface of a reservoir were 123 ft above the only fire hydrant in town, what pressure would you expect at the hydrant?

2. If the pressure at the only hydrant in town is 63 psi, what is the elevation head of the water supply?

3. If a charged fire hydrant has a pressure of only 10 psi on the 4½-in blind cap, what is the force on the cap?

4. What is the pressure at the bottom of a container of water if the force on the bottom of the container is 25 lb and it measures 5 in × 5 in?

5. If atmospheric pressure at sea level is the only pressure in consideration, how high will it raise a column of water?

Challenging Questions

1. If a water tank is 60 ft high, what pressure will it have at its base?

2. What is the force in Question 1 if the base has an area of 2 ft²?

3. How much pressure is needed to lift water 35 ft?

4. How high will a pressure of 70 psi lift water?

5. A water tank is 10 ft in diameter and 40 ft high.

   A. How much force is exerted on the bottom of the tank if the tank is one-half full?

   B. How much pressure is at the base when it is full?

   C. If the tank is three-quarters full and 100 psi of air pressure is maintained in the airspace, what is the force on the bottom of the tank?

6. How much force is there on a 4½-in blind cap of a hydrant if it has a pressure on it of 65 psi?

7. How high is the piezometric plane that represents the water elevation of the reservoir supplying the pressure to the hydrant in Question 6?

8. If an absolute pressure gauge has a reading of 140 psi, what reading will a relative pressure gauge have for the same pressure?

9. If the piezometric plane of a reservoir supplying a fire hydrant is 83 ft above the hydrant, what is the force on the 2½-in cap of the hydrant?

10. A piston at the top of a closed container with a surface area of 10 in² is being driven by a piston rod of just 1-in² area. If a pressure of 100 psi is driving the piston rod, how much pressure will be added to the tank of water?

11. What is the pressure at the 10-ft level of a 20-ft vertical tank if an external pressure source is adding 50 psi to the tank?

Formulas

To find pressure when force and area are known:

$$P = \frac{F}{A}$$

To find pressure when the height (elevation) of water is known:

$$P = 0.433 \times H$$

To find the height of water when the pressure it creates is known:

$$H = 2.31 \times P$$

To find absolute pressure:

$$\text{Absolute pressure} = \text{relative pressure} + \text{atmospheric pressure}$$

To find force when pressure and area are known:

$$F = P \times A$$

To find force when the height of the water and the area it acts on are known:

$$F = 0.433 \times H \times A$$

Bernoulli's Principle

LEARNING OBJECTIVES

Upon completion of this chapter, you should be able to:

- Understand the basic principles of Bernoulli's principle.
- Understand the concept of conservation of energy.
- Distinguish between potential and kinetic energy.
- Understand the application of Bernoulli's principle to hydraulics.
- Apply Bernoulli's equation to solve problems.
- Understand Torricelli's theorem as applied to Bernoulli's principle.
- Apply Torricelli's equation to calculate velocity.

Case Study

You have agreed to give a presentation on Bernoulli's principle to all members of Engine 11. Other than the formulas used to prove Bernoulli's principle, several facts strike you as important. You proceed with your lesson plan, including several examples of real-life application of Bernoulli's principle. You even include Torricelli's theorem in your training.

1. What important concept does Bernoulli's principle teach us?
2. Identify some real-life applications in which Bernoulli's principle is important.
3. How is the formula derived from Torricelli's theorem useful in the study of hydraulics?

Introduction

Bernoulli's principle, perhaps the single most important principle in understanding hydraulics because of its application to such a wide variety of situations, was developed by the Swiss mathematician and physicist Daniel Bernoulli (1700–1782) and first published in 1738. In fact, even beyond our study of hydraulics, everyday life is full of examples of Bernoulli's principle.

Bernoulli's principle is responsible for some everyday phenomena that we normally take for granted, for example, giving lift to wings on aircraft. Or have you ever wondered why the canvas top of a convertible automobile bulges up as it goes down the highway? Or how can a sailboat actually sail into the wind? The answer to each question is Bernoulli's principle.

But since this is a text about hydraulics, what examples apply to hydraulics? The first one that comes to mind is listed on the pump panel of all pumping apparatuses. Note the label that lists specifications for testing the pumps Figure 3-1. As the required test pressure goes up, the rated capacity goes down. Bernoulli's principle is also responsible for explaining how a foam eductor works and how a nozzle converts pressure to velocity.

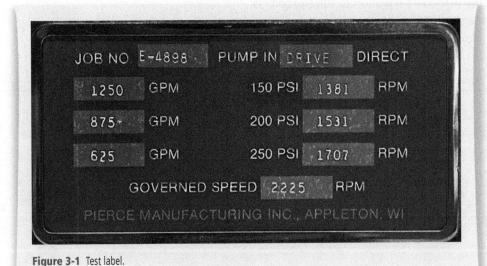

Figure 3-1 Test label.
Courtesy of William F. Crapo.

The study of Bernoulli's principle requires an understanding of what appears to be a complicated algebraic equation. In reality, the equation is not that complicated. In fact, Bernoulli's equation can be, and is, broken down into very simple terms. Remember that the purpose of this chapter is to explain the theory in terms that can be understood.

Conservation of Energy

Bernoulli first postulated his principle to explain the conservation of energy as it pertained to hydraulics. Conservation of energy means that energy can be neither increased nor decreased in any system. It can only change from one form to another or be transferred from one body to another.

To better understand the concept of conservation of energy, let us consider how energy can change forms. A match has energy in the form of a chemical. As it is struck, friction generates heat that initiates an exothermic reaction, and the chemical energy in the head of the match is converted to heat and light energy. (An exothermic reaction is a chemical reaction that gives off heat.) The amount of heat and light the match will give off is proportional to the amount of the chemical on the head of the match. To alter the amount of heat and light, the amount of the chemical on the head of the match must be changed.

Next, we need to understand how energy can be transferred from one object to another. To understand this transfer of energy, consider a batter as he swings at a pitch. If he hits the ball squarely, the energy he has put into the bat will be transferred to the ball. If the batter put enough energy into his swing, the ball will sail over the outfield fence. In short, how far the ball will go depends on how much energy was in the bat and transferred to the ball.

Note

Conservation of energy means that energy can be neither increased nor decreased in any system. It can only change from one form to another or be transferred from one body to another.

Bernoulli's Principle

Bernoulli's principle states that where the velocity of a fluid is high, the pressure is low; and where the velocity is low, the pressure is high. In this statement, Bernoulli is saying that we can have energy, as it pertains to hydraulics, in the form of either high pressure or high velocity, but not both at the same time; one will be dominant. In short, energy can be transferred between pressure and velocity, but if one increases, the other must decrease. Before we continue with Bernoulli's principle, we need to review the concept of energy.

Energy

Energy is the ability to do work. This ability is in the form of either potential energy or kinetic energy. **Potential energy** can be defined as the energy of position or stored energy. There are different kinds of potential energy (PE). For example, a bucket of water held in the air has gravitational potential energy because of its position in relation to the floor Figure 3-2 . A spring has elastic potential energy.

If the bucket in Figure 3-2 is released, it will fall to the floor and cause work to be done. The amount of work (the dent it puts in the floor) is related to the weight of water in the bucket and the distance (position) the bucket is from the ground. This principle can actually be put in the form of an equation: $PE_{gravity} = mgy$. In this equation, $PE_{gravity}$ is the gravitational potential energy, m is mass of the object, g is the acceleration due to gravity, and y is the vertical distance. For our purposes, it is sufficient to use this equation simply to confirm our definition of potential energy as energy of position.

Kinetic energy is defined as energy of motion. If a bucket of water being held in the air is released and allowed to fall, it will eventually hit the floor, but it will take time Figure 3-3 . The distance that an object moves in a measured amount of time is called the **velocity**. Velocity introduces motion to our definition of kinetic energy. Catch the bucket at any point on the way down, and you will feel the energy.

Kinetic energy can also be defined in terms of an equation. Technically this equation refers to translational (straight-line) kinetic energy, as opposed to

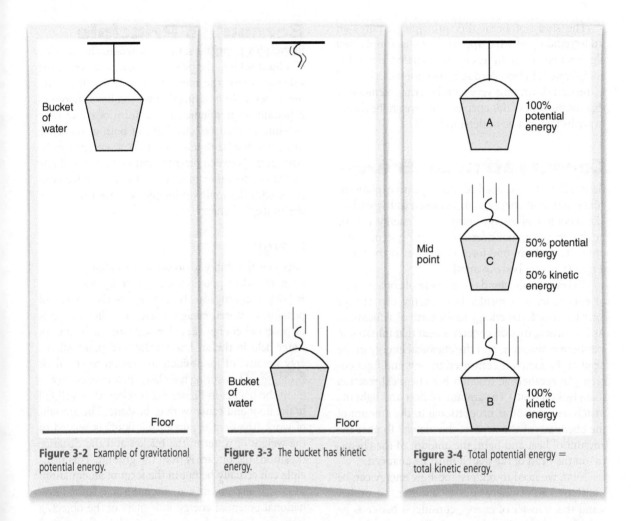

Figure 3-2 Example of gravitational potential energy.

Figure 3-3 The bucket has kinetic energy.

Figure 3-4 Total potential energy = total kinetic energy.

rotational kinetic energy. The equation is KE = ½mv^2, where KE is kinetic energy, m is the mass of the object, and v is the velocity of the object. Velocity in the equation for kinetic energy confirms kinetic energy as energy of motion.

The next logical step is to put both of these formulas together. In a system where energy is conserved, we can say that Total PE = total KE, as illustrated in Figure 3-4 . The bucket at position A represents total PE. The bucket at position B represents total KE. The energy at each position is the same.

What happens if we release the bucket and measure the energy at the halfway point? At point

C in Figure 3-4 we have ½PE (one-half of the PE has been converted to KE) and ½KE (one-half of the PE still exists). Together they equal the total energy in the system. Or ½PE + ½KE = total KE = total PE. In short, energy can exist in the form of either potential energy or kinetic energy, but we cannot have total potential energy and total kinetic energy at the same time. If potential energy and kinetic energy exist at the same time, they must be proportional and add up to the total energy in the system. Energy can be transferred from potential to kinetic and back, but it cannot be increased or decreased.

Total PE = total KE

Understanding the concept of energy is important in the study of hydraulics. Everything done from this point forward is at least indirectly associated with energy in fluids. Pressure is a form of energy, usually a measure of potential energy, but it can also be a measure of kinetic energy. Velocity is a measure of kinetic energy.

Bernoulli's Equation

No good scientific theory is worth its weight without a corresponding mathematical formula to prove it, and Bernoulli's principle is no exception. The purpose of the equation is to prove that energy in (PE) is equal to energy out (KE). Therefore, Bernoulli's equation is $P_1 + \frac{1}{2}v_1^2 d + gh_1 d = P_2 + \frac{1}{2}v_2^2 d + gh_2 d$. The total energy on each side of the equals sign must be the same because each side represents total energy of different points in the same system. Each side of the equals sign has the potential to be a mixture of both potential and kinetic energy. For our purpose, we can redefine the terms of the equation to a simpler form as follows:

$$P = \text{pressure head} = PH$$
$$\tfrac{1}{2}v^2 d = \text{velocity head} = VH$$
$$ghd = \text{elevation head} = EH$$

The term *head* in each of the definitions above refers to an equivalent in feet. For example, 20 pounds per square inch (psi) is equivalent to a pressure head of 46.18 feet (ft). This allows us to restate Bernoulli's equation in the simpler form of $PH_1 + VH_1 + EH_1 = PH_2 + VH_2 + EH_2$. Bernoulli's equation allows us to make comparisons of the total energy between any two points within a system. Our answer is expressed

in feet, instead of some technical scientific jargon, but it is valid just the same.

$$PH_1 + EH_1 + VH_1 = PH_2 + EH_2 + VH_2$$

Example 3-1

A water tank with water to the 20-ft level has a pressure gauge at the base of the tank that reads a pressure of 8.66 psi **Figure 3-5**. Using Bernoulli's equation, compare the total head at Point *1* and Point *2*.

Answer

$PH_1 = 0$	$PH_2 = 8.66 \text{ psi}$
$EH_1 = 20 \text{ ft}$	$EH_2 = 0$
$VH_1 = 0$	$VH_2 = 0$

We need to convert 8.66 psi to feet. The equation is written:

$$PH_1 + EH_1 + VH_1 = PH_2 + EH_2 + VH_2$$
$$0 + 20 \text{ ft} + 0 = 8.66 \text{ psi} + 0 + 0$$
$$0 + 20 \text{ ft} + 0 = (8.66 \text{ psi} \times 2.31) + 0 + 0$$
$$0 + 20 \text{ ft} + 0 = 20 \text{ ft} + 0 + 0$$
$$20 \text{ ft} = 20 \text{ ft}$$

The two sides of the equation are in agreement.

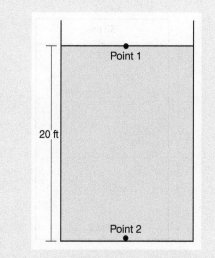

Figure 3-5 Are Point *1* and Point *2* in agreement?

Example 3-2

This example is similar to Example 3-1, except that it is a closed system under 50 psi pressure Figure 3-6.

Answer

$PH_1 = 50$ psi $PH_2 = 58.66$ psi
$EH_1 = 20$ ft $EH_2 = 0$
$VH_1 = 0$ $VH_2 = 0$

Because of Pascal's principle, the pressure at Point 2 will be 58.66 psi.

The equation is written:

$$PH_1 + EH_1 + VH_1 = PH_2 + EH_2 + VH_2$$
$$50 \text{ psi} + 20 \text{ ft} + 0 = (50 \text{ psi} + 8.66 \text{ psi}) + 0 + 0$$
$$(50 \text{ psi} \times 2.31) + 20 \text{ ft} + 0 = (58.66 \text{ psi} \times 2.31) + 0 + 0$$
$$115.50 \text{ ft} + 20 \text{ ft} + 0 = 135.50 \text{ ft} + 0 + 0$$
$$135.50 \text{ ft} = 135.50 \text{ ft}$$

The two sides of the equation are in agreement.

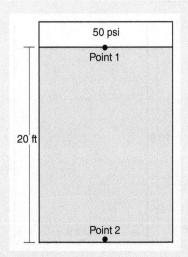

Figure 3-6 Is the total energy at Point *1* equal to the total energy at Point *2*?

Note

Bernoulli's principle assumes a system has no friction loss.

In Examples 3-1 and 3-2, the equations have balanced, which proves that the energy in each system has remained constant or, in other words, has been conserved.

A word of caution is in order here. Bernoulli's principle assumes a system has no friction loss. In real life, however, friction loss occurs and needs to be taken into account. When applying the concepts in Bernoulli's principle to real life, keep the friction loss in mind.

Application of Bernoulli's Principle

Bernoulli's principle has many everyday applications in hydraulics, most notably in the pump itself. As the capacity of the pump increases, the maximum discharge pressure decreases. For example, in order to flow the capacity of a pump, the pump is limited to a maximum of 150 psi discharge pressure. If we limit our discharge to 70 percent of the pump's capacity, we can increase the maximum discharge pressure to 200 psi. If we are content with only 50 percent of the capacity of the pump, we can pump at a maximum of 250 psi discharge pressure.

A **venturi tube** is another excellent example of how Bernoulli's principle is applied in the real world of hydraulics. A venturi tube is a restriction in a conduit intended to increase the velocity with a corresponding reduction in pressure, as depicted in Figure 3-7. Notice that the cross-sectional areas at points *A* and

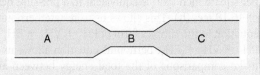

Figure 3-7 The velocity at *B* is greater than at *A* or *C*.

C are the same. However, the cross-sectional area at point *B* is smaller. Water flowing through the conduit at points *A* and *C* will have the same velocity. Because the area in the restricted portion of the conduit is less than that at point *A* or *C*, the water has to flow faster through point *B*. Another way to say this is that the velocity at point *B* will be faster than at *A* or *C*.

Now recall exactly what Bernoulli's principle says: Where the velocity of a fluid is high, the pressure is low; and where the velocity is low, the pressure is high. If we apply this principle to the venturi tube in Figure 3-7, we realize that the pressure at point *B* is less than the pressure at either point *A* or point *C*.

Fireground Fact

Real-Life Application of Bernoulli's Principle

If the venturi tube in **Figure 3-8** is configured with a pickup tube attached at point *B*, and if the velocity of water is high enough, the pressure at point *B* can be reduced below atmospheric pressure. When this happens, the pickup tube can be used to draft a liquid, such as foam concentrate, out of a container. This effect is referred to as a *venturi*. A venturi tube designed to pick up foam concentrate is called a *foam eductor*.

Torricelli's Theorem

A special application of Bernoulli's principle is called Torricelli's theorem, even though the Italian physicist and mathematician Evangelista Torricelli (1608–1647) first proved it about one hundred years before Bernoulli was even born. **Torricelli's theorem** says that the velocity of water escaping from an opening below the surface of a container of water will have the same velocity, minus exit losses, as if it were to fall the same distance. This means that the velocity of water at a discharge point a given distance below the surface of a body of water will be accelerated by the

force of gravity exactly as if the water were to free-fall the same distance **Figure 3-9**. The only thing we need to do to get the actual velocity is to factor in the friction loss at the discharge. This leads to the formula for Torricelli's equation $v = \sqrt{2g(y_2 - y_1)}$, where *v* is velocity in feet per second (fps), *g* is the gravitational acceleration constant of 32.174 feet per second per second (ft/s^2), and $y_2 - y_1$ is the elevation difference between the top of the container and the point where the water is being discharged. Here, the constant 32.174 is one instance in which a number with more than two decimal places is used. Using this number to three decimal places will allow us to later calculate the same formula for finding gallons per minute as is used in the 20th edition of the *Fire Protection Handbook*.

If we rewrite the equation, substituting *H* for $y_2 - y_1$, and simplify it as far as possible, the equation becomes:

$$
\begin{aligned}
v &= \sqrt{2gH} \\
&= \sqrt{2 \times 32.174 \times H} \\
&= \sqrt{64.35 \times H} \\
&= 8.02\sqrt{H}
\end{aligned}
$$

$$v = \sqrt{2g(y_2 - y_1)}$$

or

$$v = \sqrt{2g \times H}$$

The formula for determining the velocity from a discharge when the elevation head is known is now $v = 8.02\sqrt{H}$, where *v* is the velocity in feet per second and *H* is the elevation or elevation head in feet.

$$v = 8.02\sqrt{H}$$

Example 3-3

What is the velocity of a discharge 20 ft below the surface of an open container?

(continues)

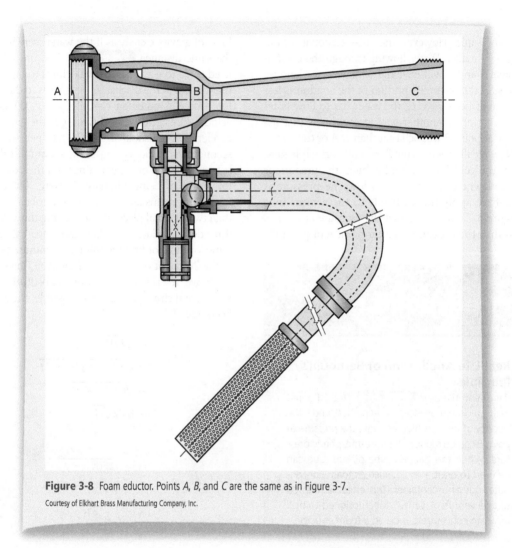

Figure 3-8 Foam eductor. Points *A*, *B*, and *C* are the same as in Figure 3-7.
Courtesy of Elkhart Brass Manufacturing Company, Inc.

(Example 3-3 continued)

Answer

To solve this problem we use the formula:

$$v = 8.02\sqrt{H}$$
$$= 8.02\sqrt{20}$$
$$= 8.02 \times 4.47$$
$$= 35.85 \text{ or } 36 \text{ fps}$$

Revised Bernoulli's Equation

Torricelli's equation can be rewritten as $H = v^2/2g$. If we insert this mathematical expression directly into Bernoulli's equation, we can enter velocity directly into Bernoulli's equation without first converting it to feet (velocity head). If we also insert an expression for pressure so as to get a direct equivalent in feet, then we get a revised Bernoulli's equation. Following are the revised equivalents for PH, EH, and VH:

PH = *P/W*   where *P* = pressure and
             W = weight of water 1 in × 1 in ×
             1 ft tall

EH = *Z*   *H* is replaced with *Z* to indicate
             elevation.

VH = $v^2/2g$ As explained previously.

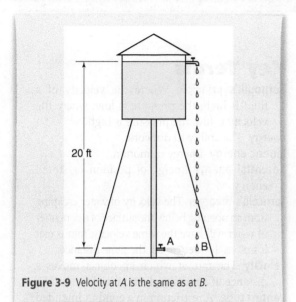

Figure 3-9 Velocity at *A* is the same as at *B*.

The revised form of Bernoulli's equation becomes $P_1/W + Z_1 + v_1^2/2g = P_2/W + Z_2 + v_2^2/2g$. We can also insert the constants we already know, that is, $W = 0.433$ and $2g = 64.35$, and we get $P_1/0.433 + Z_1 + v_1^2/64.35 = P_2/0.433 + Z_2 + v_2^2/64.35$.

This revised form of Bernoulli's equation saves a couple of steps. There is no need for separate calculations, which then have to be put into the equation, to convert pressure to equivalent feet or velocity to equivalent feet. Just plug in the appropriate figures for pressure and/or velocity and do the calculations. The answer will be in feet.

$$\frac{P_1}{0.433} + Z_1 + \frac{v_1^2}{64.35} = \frac{P_2}{0.433} + Z_2 + \frac{v_2^2}{64.35}$$

Example 3-4

A water tower has a water level of 20 ft with 50 psi of pressure added **Figure 3-10**. The discharge has been only partially opened, allowing a discharge velocity of

35.77 fps with a residual pressure of 50 psi. Use the revised Bernoulli's equation to compare Point *1* and Point *2*.

Answer

Simply insert the appropriate numbers in the equation.

$$P_1/0.433 + Z_1 + v_1^2/64.35 = P_2/0.433 + Z_2 + v_2^2/64.35$$

$$50 \text{ psi}/0.433 \text{ psi} + 20 \text{ ft} + 0 = 50 \text{ psi}/0.433 \text{ psi} + 0 + (35.77 \text{ fps})^2/64.35 \text{ fps}^2$$

$$115.47 \text{ ft} + 20 \text{ ft} + 0 = 115.47 \text{ ft} + 1{,}279.49 \text{ fps}^2/64.35 \text{ fps}^2$$

$$135.47 \text{ ft} = 115.47 \text{ ft} + 19.89 \text{ ft}$$

$$135.47 \text{ or } 135 \text{ ft} = 135.36 \text{ or } 135 \text{ ft}$$

The revised formula gives us the answer without having to do separate calculations to find the PH or VH. This form of the equation is also the one found in the 20th edition of the *Fire Protection Handbook*.

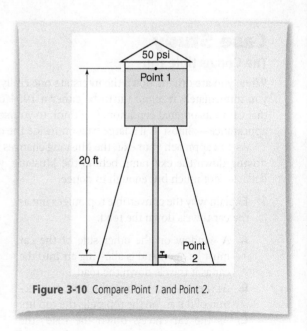

Figure 3-10 Compare Point *1* and Point *2*.

Chapter Summary

- Conservation of energy means that energy can be neither increased nor decreased in any system.
- Energy is the ability to do work.
- Kinetic energy is energy of motion.
- Potential energy is energy at rest.
- Bernoulli's principle assumes a system has no friction loss.
- A venturi tube is a restriction in a conduit intended to increase the velocity with a corresponding decrease in pressure.
- Torricelli's theorem is considered a special application of Bernoulli's principle.

Key Terms

Bernoulli's principle Where the velocity of a fluid is high, the pressure is low; where the velocity is low, the pressure is high.

energy The ability to do work.

kinetic energy Energy of motion.

potential energy Energy of position or stored energy.

Torricelli's theorem The velocity of water escaping from an opening below the surface of a container of water will have the same velocity, minus exit losses, as if it were to fall the same distance.

velocity The rate at which an object moves a distance in a measured amount of time.

venturi tube A restriction in a conduit intended to increase velocity with a corresponding reduction in pressure.

Case Study

The Convertible Top

While you are driving down the interstate one chilly afternoon, a car passes you. It is a convertible, and you immediately recognize it to be either a 1964 or 1965 Ford Mustang. What a great job restoring that car to its original condition, you think to yourself. You also notice that the cloth top has an inflated appearance—almost as if a large balloon inside the car were pushing the top out.

As you approach your exit, the Mustang changes lanes ahead of you and signals to exit as well. While driving down the exit ramp behind the Mustang, you notice that the cloth top of the car appears to deflate—not much but enough to notice.

1. Explain why the convertible top bulges out as the car travels down the road.

   A. A window on the other side of the car must be open and is allowing air into the car as it travels down the road.

   B. As the car travels down the road, the friction of the air on the top pulls the top up.

   C. As the car travels down the road, the higher-velocity air going over the top of the car creates a low-pressure area, allowing the higher static pressure inside the car to push the top out.

   D. The driver has the fan on the heating system so high that it creates a higher-pressure area inside the car.

2. Why does the top deflate when the vehicle stops?

   A. With the car stopped, the pressures inside and outside the car are in equilibrium.

   B. With the car stopped, there is no friction with the air to pull the top up.

   C. With the car stopped, the weight of the top simply holds it down.

   D. None of the above.

3. This case study illustrates which of the following principles?

  A. Torricelli's theorem
  B. Kinetic energy
  C. Conservation of potential energy
  D. Bernoulli's principle

4. Name at least one other circumstance in the fire service where this same principle is in play.

  A. The air pressure in tires will increase when the tires get hot, such as during long responses.
  B. As the output from a pump is increased, the maximum discharge pressure will decrease.
  C. Both A and B are correct.
  D. Neither A nor B is correct.

Review Questions

1. In what year did Bernoulli first publish his principle on the conservation of energy as it pertains to hydraulics?

2. In a hose, how are pressure and velocity related?

3. What is the ability to do work called?

4. Give an example of potential energy not already given in this chapter.

5. If the contents of a bucket of water were to be dropped from an elevation of 10 ft, what would the energy measurement be at the halfway point? (Hint: Define it in terms of potential or kinetic energy.)

6. Are there any situations in which the total energy into a system is not equal to the total energy out?

7. Torricelli's theorem correlates velocity to what factor or "head"?

Activities

You are now ready to put your knowledge of Bernoulli's principle to work.

1. A pumper is hooked up to a hydrant and pumping 100 psi to a line that extends down a hill with an elevation difference of 25 ft **Figure 3-11**. Using Bernoulli's equation, calculate the total head for both Points 1 and 2.

2. **Figure 3-12** depicts a water tank with a pressure at the base of 35 psi. Using Bernoulli's equation, calculate EH_1 for Point 1.

3. Using Torricelli's equation, calculate the velocity at the discharge where the elevation head is 50 ft.

4. A 200-ft-tall water storage tank has developed a ¼-in hole 50 ft off the ground. What is the velocity of the water as it leaves the tank at the point where the water level is 30 ft from the top?

5. A pumper is pumping at 54.33 psi to a smooth bore nozzle 10 ft above the pumper. The nozzle has a discharge velocity of 85.98 fps **Figure 3-13**. Use the revised version of Bernoulli's equation to compare pressure at the engine with pressure at the nozzle. Note: The

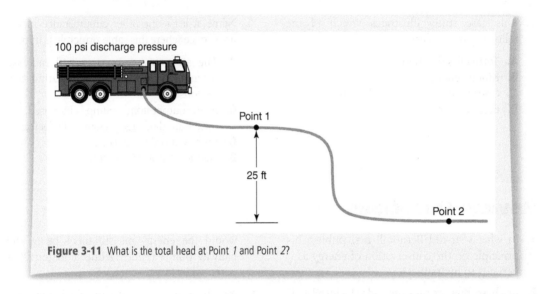

Figure 3-11 What is the total head at Point *1* and Point *2*?

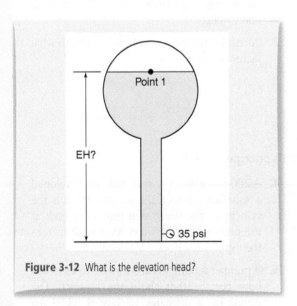

Figure 3-12 What is the elevation head?

engine pressure does not include any friction loss because Bernoulli's equation does not take it into consideration.

Note that in this scenario, because the pumper is pumping uphill, the elevation represents a loss of pressure. It is not a loss in the sense that the energy is lost, but is a loss in the sense that the pump must compensate for this downward pressure or energy. By this we mean that the pressure created by the elevation, 4.33 psi, works against the pressure being created by the pumper, and by the time the pressure reaches the nozzle it is only 50 psi. In this example the pumper has to pump at 54.33 psi to compensate for this downward pressure and still have a nozzle pressure of 50 psi. This pressure is called *back pressure*. We can more precisely define back pressure as the pressure exerted by a column of water against the discharge of a pump.

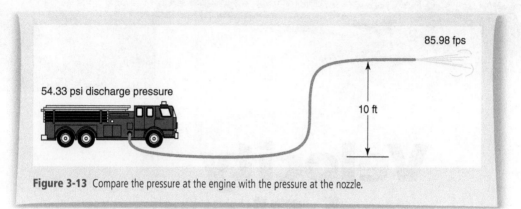

Figure 3-13 Compare the pressure at the engine with the pressure at the nozzle.

Challenging Questions

1. What is the pressure head equivalent to 35 psi?

2. In its original form, what does g stand for in Torricelli's equation?

3. What is the advantage of using the revised form of Bernoulli's equation, $P_1/0.433 + Z_1 + v_1^2/64 = P_2/0.433 + Z_2 + v_2^2/64$, instead of the $PH_1 + EH_1 + VH_1 = PH_2 + EH_2 + VH_2$ version?

4. A gauge reading 6.06 psi is attached to a tank outlet with water discharging at 30 fps. What is the elevation of water in the tank above the level of the outlet?

Formulas

Conservation of energy:

$$\text{Total PE} = \text{total KE}$$

Bernoulli's equation:

$$PH_1 + EH_1 + VH_1 = PH_2 + EH_2 + VH_2$$

Torricelli's equation:

$$v = \sqrt{2g(y_2 - y_1)}$$

To find velocity when height is known with Torricelli's equation simplified:

$$v = 8.02\sqrt{H}$$

Revised Bernoulli equation:

$$\frac{P_1}{0.433} + Z_1 + \frac{v_1^2}{64.35} = \frac{P_2}{0.433} + Z_2 + \frac{v_2^2}{64.35}$$

Reference

Cote, Arthur E., P.E., Editor-in-chief, *Fire Protection Handbook*, 20th ed. Quincy, MA: National Fire Protection Association, 2008.

Velocity and Flow

LEARNING OBJECTIVES

Upon completion of this chapter, you should be able to:

- Calculate velocity, given either height or pressure.
- Calculate the change of velocity corresponding to a change of area.
- Understand the relationship between velocity and flow.
- Understand the relation between velocity and area.
- Calculate flow, given area of discharge and velocity.
- Calculate flow for 1 minute (min).

Case Study

It is late afternoon and Engine 11 has just returned from what will turn out to be the first of two fires you have during your shift. As you and the other members of Engine 11 clean up and prepare for the next fire, members of Truck 6, the company you share quarters with, are doing the same.

During the cleanup, one of the truck personnel asks you how your study of hydraulics is going. He says he knows extra thought needs to be given to such things as calculating flow, but doesn't understand where the formulas come from. He has always heard that such formulas are "empirical" and are simply a "best guess" based on observation and experience. You know different and explain!

1. What role does velocity play in determining flow?
2. What is the relationship between area and velocity?
3. Explain, briefly, how the formula $Q = A \times v$ will ultimately get us the quantity flow for 1 min.

Introduction

This chapter's content actually began in Chapter 3 with Torricelli's theorem, which introduces velocity to our study of hydraulics. Anytime water leaves an opening (e.g., through a nozzle) it has velocity, and being able to calculate the velocity of water flowing from an opening is the first step in calculating how much water is flowing. There is a direct relationship among the velocity of water through an opening, the area of the opening, and the amount of water being discharged. Thus, just by using a simple formula of area times velocity we are able to calculate how much water is flowing.

In this chapter, the quantity of water being discharged is calculated in cubic feet. It is important that close attention be given to the relationship between the area of an opening and the velocity of the water (or other fluid) being discharged in calculating the quantity of water being discharged. In Chapter 5 we take this process one step further by developing a formula that calculates discharge in gallons per minute.

For a fixed flow of water, as the size of the hose changes, so does the velocity. This change is directly proportional to the area of the hose. Here we show how to calculate (1) the change in velocity due to the change in the size of the hose and (2) the change in hose size when the change in velocity is known.

Velocity

Because velocity is critical in calculating flow, it is worth taking time to review how velocity is calculated when elevation is known. Then we alter the velocity formula to obtain a formula to calculate velocity when pressure is known.

The Italian physicist and mathematician Evangelista Torricelli said that water exiting an opening will have the same velocity as if it were to free-fall the same distance. This knowledge allows us to use the formula $v = \sqrt{2gH}$ to calculate velocity. Recall that a simplified version of this formula is $v = 8.02\sqrt{H}$, where v is the velocity in feet per second and H is the elevation or elevation head in feet. A physicist would use this same formula to calculate the velocity of an object in free fall. (In Chapter 3 the concept of velocity was introduced in the form of Torricelli's theorem.)

The fact that the gravitational acceleration constant g is part of this formula should not be taken for granted. Remember that all objects on planet Earth are affected by gravity. When those objects are "falling," they are accelerating at a constant rate determined by the force of gravity, that is, the gravitational acceleration constant. In the study of hydraulics the object "falling" is water.

Example 4-1

Water is flowing from an opening at the base of a water tank **Figure 4-1**. If the water inside is 49 feet (ft) above the point of discharge, what is the velocity?

Answer

$$v = 8.02\sqrt{H}$$
$$= 8.02\sqrt{49}$$
$$= 800 \times 7$$
$$= 56.14 \text{ or } 56 \text{ feet per second (fps)}$$

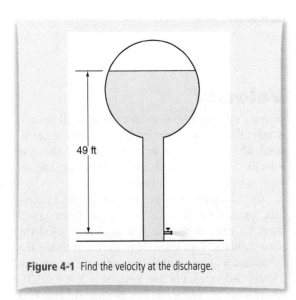

Figure 4-1 Find the velocity at the discharge.

Velocity from Pressure

For our purpose, however, Toricelli's equation does have its limits. We in the fire service usually need to calculate velocity from pressure, not from elevation. The good news is that we can use the same basic formula $v = \sqrt{2gH}$, but to use it, we must replace H in the basic formula with an appropriate expression for pressure. Fortunately, we learned just such an expression in Chapter 2.

That expression (formula) is $H = 2.31 \times P$. The formula for finding velocity from pressure, then, is $v = \sqrt{2 \times g \times 2.31 \times P}$. Recall that g is 32.174 feet per second per second (ft/s^2).

$$v = \sqrt{2 \times g \times 2.31 \times P}$$

Just as we have sought to simplify other formulas, this one is no exception. We can develop the simplified formula as follows:

$$v = \sqrt{2 \times g \times 2.31 \times P}$$
$$= \sqrt{2 \times 32.174 \times 2.31 \times P}$$
$$= \sqrt{148.64 \times P}$$
$$= 12.19\sqrt{P}$$

By multiplying $2 \times 32.174 \times 2.31$ and then finding the square root of the product, we created a constant of 12.19, which is multiplied by the square root of the pressure to obtain velocity. It will work whenever we need to calculate velocity from pressure.

$$v = 12.19\sqrt{P}$$

Example 4-2

What is the discharge velocity of a smooth-bore nozzle with a nozzle pressure of 50 pounds per square inch (psi)?

Answer

$$v = 12.19\sqrt{P}$$
$$= 12.19\sqrt{50}$$
$$= 12.19 \times 7.07$$
$$= 86.18 \text{ or } 86 \text{ fps}$$

Example 4-3

What is the discharge velocity on the 2½-inch (2½-in) discharge of a fire hydrant if it has a pressure of 36 psi?

Answer

$$v = 12.19\sqrt{P}$$
$$= 12.19\sqrt{36}$$
$$= 12.19 \times 6$$
$$= 73.14 \text{ or } 73 \text{ fps}$$

Finding Pressure When Velocity Is Known

Finding the pressure from the velocity has some value in the fire service. We start with the basic formula for finding velocity from pressure, $v = 12.19\sqrt{P}$, and convert it to find pressure. By following the appropriate laws of mathematics, the formula is rearranged to read $P = (v/12.19)^2$. We can use this formula to calculate the pressure whenever we are given velocity at a point of discharge.

$$P = \left(\frac{v}{12.19}\right)^2$$

In Chapter 3, it was said that pressure is a measure of energy, usually potential. By using this formula to convert velocity to pressure, we have an example of using pressure to measure kinetic energy.

Example 4-4

Find the pressure where the discharge velocity is 50 fps.

Answer

Use the formula $P = (v/12.19)^2$.

$$P = (v/12.19)^2$$
$$= (50/12.19)^2$$
$$= (4.1)^2$$
$$= 16.81 \text{ or } 17 \text{ psi}$$

Area and Velocity

The velocity of water in a hose is a function of both the quantity of water flowing and the cross-sectional area of the hose. Just as in a venturi tube, if we reduce the size of the hose, the velocity will increase  . If we know the velocity in either size hose, we can calculate the velocity in the other. To do this, we need to introduce this formula: $Area_1 \times velocity_1 = area_2 \times velocity_2$, or $A_1 \times v_1 = A_2 \times v_2$. Using this formula, as long as we are given any three

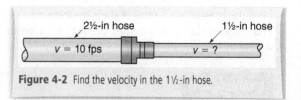

Figure 4-2 Find the velocity in the 1½-in hose.

factors, we can find the fourth. Just as in the other formulas using velocity, v is in feet per second. Here, both A_1 and A_2 are in square inches.

$$A_1 \times v_1 = A_2 \times v_2$$

Fireground Fact

Velocity versus Area

This is actually a simple concept. Basically, as in Figure 4-2, the same amount of water that flows through the 2½-in hose must also flow through the 1½-in hose. For the same amount of water to flow through the smaller hose, the velocity must be proportionally faster. The significance of the proportional relationship between the area and velocity of each hose is evident in the formula $A_1 \times v_1 = A_2 \times v_2$. The formula is a simple ratio and proportion formula.

This formula tells us that velocity of water in a hose, for a given quantity of flow, is inversely proportional to the change in area of the hose. More simply put, for a given flow, if the area of the hose is doubled, the velocity will be reduced by one-half, and if the area of the hose is reduced by one-half, the velocity will double.

A unique feature of this formula is that while velocity is always given in feet per second, area can be in any unit of area. What A_1 and A_2 do, in this formula, is establish a ratio of their respective areas. That ratio is the same whether the area is given in square inches, square feet, or even square yards.

Example 4-5

If the velocity of water in the 2½-in hose is 10 fps, what is the velocity in the 1½-in hose Figure 4-2?

Answer

Use the formula

$$A_1 \times v_1 = A_2 \times v_2$$

where A_1 = area of 2½-in hose

$$v_1 = 10 \text{ fps}$$

A_2 = area of 1½-in hose

and v_2 is what we are seeking.

$$A_1 \times v_1 = A_2 \times v_2$$
$$(0.7854 \times 2.5^2) \times 10 = (0.7854 \times 1.5^2) \times v_2$$
$$4.9 \times 10 = 1.77 \times v_2$$
$$v_2 = 49/1.77$$
$$v_2 = 27.68 \text{ or } 28 \text{ fps}$$

In Example 4-5, the area of the 2½-in hose is 2.768 times larger than the area of the 1½-in hose (4.9/1.77 = 2.768). And the velocity in the 1½-in hose is 2.768 times greater than that in the 2½-in hose (27.68/10 = 2.768). These figures verify that for a given flow, the velocity of the water is inversely proportional to the change in area of the hose.

At times it may be necessary to find the area of a second hose size when the velocity in both hose sizes is known. As already pointed out, we still use the formula $A_1 \times v_1 = A_2 \times v_2$, except in this case we are solving for A_2 instead of v_2.

Example 4-6

The velocity of water in a 3-in hose is 10 fps. Personnel then change to a smaller hose with a velocity of 29.29 fps Figure 4-3 . What is the area, and size, of the smaller hose?

Answer

$$A_1 \times v_1 = A_2 \times v_2$$
$$(0.7854 \times D^2) \times 10 = A_2 \times 29.29$$
$$(0.7854 \times 3^2) \times 10 = A_2 \times 29.29$$
$$7.07 \times 10 = A_2 \times 29.29$$
$$A_2 = 70.7/29.29$$
$$= 2.41 \text{ in}^2$$

Now that we have the area of the hose, we need to find its diameter. To do that we rearrange the formula $A = 0.7854 \times D^2$ to read $D = \sqrt{A/0.7854}$, which gives us the diameter of any circular opening when we know the area.

$$D = \sqrt{A/0.7854}$$
$$= \sqrt{2.41/0.7854}$$
$$= \sqrt{3.07}$$
$$= 1.75\text{-in hose}$$

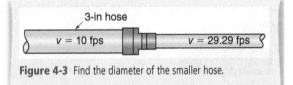

Figure 4-3 Find the diameter of the smaller hose.

$$D = \sqrt{\frac{A}{0.7854}}$$

Fireground Fact

Proof of the Formula $A_1 \times v_1 = A_2 \times v_2$

In Example 4-6, we found that the unknown hose size was 1¾ in. Now that we know the velocity in both sizes of hose and the size of each hose, we can prove the formula is a simple ratio and proportion. The ratio of the velocity of the water in the 1¾-in hose compared to the 3-in hose is 29.9/10 = 2.9 times greater velocity in the 1¾-in hose than the 3-in hose.

The ratio of the area of the 3-in hose to the area of the 1¾-in hose is 7.06/2.41 = 2.9 times more area in the 3-in hose than in the 1¾-in hose. The fact that (1) the velocity of water in the 1¾-in hose is 2.9 times that of the 3-in hose and (2) the area of the two sizes of hose is different by the same 2.9 times proves that this is a simple ratio and proportion problem.

In both Example 4-5 and Example 4-6, we calculated the actual area of the hose involved. When a quick comparison of the areas of two different-size hoses is needed, it is sufficient to simply use D^2 to make the comparison. "Square and compare the diameter of the hose." It will give you the ratio of the velocities, but not the actual velocities, just as if you had taken the extra time to actually find the area of the hose involved.

Note

To make a quick comparison of the areas of two different-size hoses, "Square and compare the diameter of the hose."

Calculating Flow

Because we know how to calculate velocity and we are aware of the relationship of area to velocity, now we can put these two functions together to find the quantity of water flowing. When calculating flow, we use velocity of the water as it exits an opening, such as a nozzle. The velocity of the water is a linear dimension that gives us the distance the water will flow in 1 second (s). When we multiply the velocity by the area of the opening, or nozzle size, the discharge becomes a volume.

This concept is easy to visualize if we think of the water being discharged as a cylinder. The area of the opening is the area of the base of the cylinder, and the velocity, or distance the water will flow in 1 s, is the length. We know that the volume of a cylinder is area times length or height. If we substitute quantity for volume and velocity per second for length, we have the formula $Q = A \times v$, where Q is flow in cubic feet per second, A is area in square feet, and v is velocity in feet per second. Remember, since the quantity of water flowing in 1 s is a volume, the answer must be in cubic feet per second (cfs). While the formula $Q = A \times v$ may not be familiar to most with some prior knowledge of fire service hydraulics, through the remainder of this chapter and the next chapter you will see how this formula morphs into a more familiar formula used to calculate flow in gallons per minute.

$$Q = A \times v$$

Example 4-7

How much water will flow through an opening that is 1 square foot (ft²) in area if it is traveling at a velocity of 89 fps [Figure 4-4]?

Answer

$$Q = A \times v$$
$$= 1 \times 89$$
$$= 89 \text{ cfs}$$

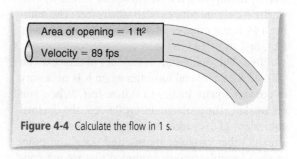

Area of opening = 1 ft²

Velocity = 89 fps

Figure 4-4 Calculate the flow in 1 s.

In the fire service we are rarely, if ever, given the luxury of being given the area in units of square feet. When given the size of openings, we are always given the diameter, never the area. This requires us to insert the formula for calculating area from diameter into the formula for finding quantity flow. The formula then becomes $Q = 0.7854 \times D^2 \times v$.

$$Q = 0.7854 \times D^2 \times v$$

Example 4-8

How much water will an opening discharge if it is 1.128 ft in diameter and has a velocity of 10 fps?

Answer

In this example, the velocity will discharge a cylinder of water 10 ft long in 1 s. Use the formula $Q = 0.7854 \times D^2 \times v$ to solve for volume.

$$Q = 0.7854 \times D^2 \times v$$
$$= 0.7854 \times 1.128^2 \times 10$$
$$= 0.7854 \times 1.272 \times 10$$
$$= 9.99 \text{ or } 10 \text{ cubic feet (ft}^3)$$

In the formula $Q = A \times v$, note that the area and velocity must both be in the same units. Remember that we have already assigned units of feet to velocity, and we usually deal with area of hose and nozzles in units of square inches. To use this formula, when we are given opening sizes in inches, it will be necessary to convert the area of any opening into square feet. By multiplying the area in square inches by $\frac{1}{144}$, square inches are converted to square feet. The number 144 is the number of square inches in a square foot, or $1 \text{ ft}^2 = 144 \text{ in}^2$.

Note: Throughout the remainder of this text you will see $\frac{1}{144}$ in several formulas when it is necessary to convert square inches to square feet. When you are working the problem, insert 0.0069, the decimal equivalent of $\frac{1}{144}$, into the formula in place of $\frac{1}{144}$ after all other variables are inserted. As $\frac{1}{144}$ appears in each formula, we are reminded that we are converting square inches to square feet.

Typically, in the fire service, $Q = A \times v$ is used to find flow through a circular opening of only a couple of inches (at most). If we substitute $0.7854 \times D^2 \times \frac{1}{144}$ for A, we automatically find the area of any

small circular opening in square feet. The formula then becomes $Q = 0.7854 \times D^2 \times \frac{1}{144} \times v$. All we need to do is to plug in the diameter and velocity, and we have the amount of water that will flow in 1 s.

$$Q = 0.7854 \times D^2 \times \frac{1}{144} \times v$$

Fireground Fact

Relationship of Area, Velocity, and Flow

As you saw when calculating the change in the velocity of water in the hose as the diameter changes, the difference in flows from an opening is proportional to both the velocity and the size of the opening. If either the opening size increases or the velocity increases, the flow will increase directly proportional to the change in opening size or velocity. Likewise, if either the opening size or the velocity decreases, the flow will decrease by the same proportion. For example, an increase in opening size by 25 percent will result in a 25 percent increase in flow. Or a decrease of 25 percent in velocity will result in a 25 percent decrease in flow.

Example 4-9

How much water will flow in 1 s through a 1¼-in opening, if it has a velocity of 86 fps **Figure 4-5** ?

Answer

$$Q = 0.7854 \times D^2 \times \frac{1}{144} \times v$$
$$= 0.7854 \times (1.25)^2 \times \frac{1}{144} \times 86$$
$$= 0.7854 \times 1.56 \times 0.0069 \times 86$$
$$= 0.727 \text{ or } 0.73 \text{ cfs}$$

This formula can be used anytime we need to find the flow from any circular opening.

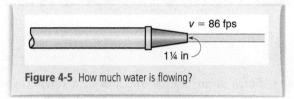

Figure 4-5 How much water is flowing?

Calculating Flow in 1 Minute

The next step in calculating flow is to determine how much water is flowing in units of time traditional to the fire service. We do not normally calculate flow from a nozzle in seconds: we calculate the flow in minutes. To get minutes from the units we have already been working with, we simply multiply the answer by 60, the number of seconds in a minute (1 min = 60 s). The answer will be in cubic feet per minute (cfm).

Example 4-10

In Example 4-9 it was determined that a 1¼-in nozzle would have a discharge of 0.727 cfs if the water were flowing at 86 fps. What would the flow be in 1 min?

Answer

To find the flow in 1 min when we already know that a 1¼-in nozzle will flow 0.727 cfs, we multiply 0.727 by 60.

$$Q = 0.727 \times 60$$
$$= 43.62 \text{ cfm}$$

What would really be helpful is a way to directly calculate quantity flow per minute in a single formula. To do this, we start with the formula we already have for calculating quantity flow, $Q = 0.7854 \times D^2 \times \frac{1}{144} \times v$, and multiply by 60. Our

new formula for directly calculating quantity flow in cubic feet per minute becomes $Q = 0.7854 \times D^2 \times \frac{1}{144} \times v \times 60$. By simply inserting the diameter of the opening in inches and the velocity in feet per second, we get a Q that is in cubic feet per minute.

$$Q = 0.7854 \times D^2 \times \frac{1}{144} \times v \times 60$$

Example 4-11

What is the quantity flow, in cubic feet per minute, from a 1¼-in opening with a velocity of 86 fps?

Answer

$$Q = 0.7854 \times D^2 \times \frac{1}{144} \times v \times 60$$
$$= 0.7854 \times (1.25)^2 \times \frac{1}{144} \times 86 \times 60$$
$$= 0.7854 \times 1.56 \times 0.0069 \times 86 \times 60$$
$$= 43.62 \text{ cfm}$$

We know the answer of 43.62 cfm is correct because between Example 4-9 and Example 4-10 we already calculated the flow to be 43.62 cfm.

Example 4-12

What is the flow for 1 min from a 1-in tip if it has a velocity of 86 fps?

Answer

$$Q = 0.7854 \times D^2 \times \frac{1}{144} \times v \times 60$$
$$= 0.7854 \times 1^2 \times \frac{1}{144} \times 86 \times 60$$
$$= 0.7854 \times 1 \times 0.0069 \times 86 \times 60$$
$$= 27.96 \text{ cfm}$$

WRAP-UP

Chapter Summary

- To find velocity when height is known, we need to know the universal acceleration constant, 32.174 ft/s^2.
- The formula for velocity requires that height be used. Since the fire service uses pressure instead of height, we need to insert $2.31 \times P$ in place of height.

- For a given flow, as the area of a conduit changes, the velocity will change inversely in proportion to the change in area.
- To make a quick comparison of the areas of two different-size hoses, square and compare the diameter.
- To find the flow from a nozzle, we must start with the basic formula, $Q = A \times v$.

Case Study

Force or Pressure

During a training exercise you have been assigned to hook up to and pump the hydrant. You have also been assigned a rookie to assist you. You explain to him that this is a timed exercise and that, in addition to accuracy, speed is important.

When the evolution begins, you pull up to the hydrant after the other company has laid a supply line. While you are setting the hand brake and putting the wheel chocks in position, your helper springs into action. The rookie grabs a hydrant wrench and heads to the hydrant. But instead of first removing the 4½-in cap, he starts to turn on the hydrant. (This is a dry barrel hydrant, so all discharges will receive water at the same time.)

You immediately see what has been done and tell the rookie to remove the 4½-in cap first. Since it has only been turned on a couple of turns, the rookie does not shut off the hydrant before trying to remove the 4½-in cap. After he places the hydrant wrench on the cap, the cap does not budge. The rookie then gets a wrench with a longer handle and tries again, but again with no result.

Up until this point you have been standing back, watching the rookie in his unsuccessful attempt to remove the 4½-in cap from the hydrant. Now you step in. After uttering, "Pay attention and learn," you shut off the hydrant, wait a few seconds for the pressure to drain off, and remove the 4½-in cap with ease. You and the rookie then complete the hookup, turn on the hydrant, and charge the supply line.

After the evolution is over, while you are picking up hose, the rookie asks you, "What did I do wrong? You were able to get that stuck hydrant cap off, but I couldn't."

1. The rookie comments, "The hydrant had only been turned on a couple of turns—surely this would limit enough water from entering the hydrant to prevent the cap from turning!" Which of the following replies provides the best explanation?

   A. Even the slightest volume of water is enough to prevent the cap from turning.

   B. Without water in the barrel of the hydrant there is no water to lubricate the threads on the cap.

   C. Even with the hydrant barely on, there will be sufficient water to fill the barrel and prevent the cap from turning.

   D. None of the above.

2. After answering the rookie in Question 1 above, how do you explain why the cap doesn't turn?

   A. Once the barrel of the hydrant is filled with water, the pressure in the barrel will become static at whatever pressure is in the main. This will create a sufficient force on the cap to prevent it from turning.

   B. Once the barrel of the hydrant is filled with water, there is sufficient flow to prevent the cap from turning.

   C. The flow of water into the hydrant and the presence of pressure prevent the cap from turning.

   D. Most likely, there was enough rust on the threads to prevent the cap from turning.

3. After hearing your explanation and wishing to further understand what happened, the rookie goes back to the hydrant and turns it on a couple of turns again. Again the 4½-in cap won't budge, but on a whim the rookie tries and is able to turn the 2½-in cap. Explain why.

   A. The 2½-in cap was not put on all the way after the last evolution.

   B. At whatever pressure is present in the barrel of the hydrant, the force on the 2½-in cap will only be about 30 percent of what it is on the 4½-in cap.

   C. The 4½-in caps are actually designed not to be turned, even under the slightest pressure, as a safety precaution.

   D. All the above.

4. Which of the following best explains to the rookie how to avoid the issue in the future?

   A. It is sufficient to just barely open the cap prior to "charging" the hydrant.

   B. Back the desired cap off until it is just barely on and charge the hydrant.

   C. Crack open the hydrant, then quickly remove the desired cap and attach the hose or appliance, and finally finish opening up the hydrant.

   D. Remove the desired cap, attach the hose or appliance, and then charge the hydrant.

WRAP-UP

Review Questions

1. What is the gravitational acceleration constant?

2. Why is the gravitational acceleration constant important in determining velocity?

3. How did we arrive at the constant 8.02 in the formula $v = 8.02\sqrt{H}$?

4. In the formula $Q = 0.7854 \times D^2 \times \frac{1}{144} \times v$, what is the purpose of $\frac{1}{144}$?

Activities

Put your knowledge of velocity and flow to use by solving the following problems.

1. What is the velocity of discharge at an opening at the base of a tank if the water in the tank is 125 ft above the point of discharge and the tank holds 100,000 gallons (gal) of water **Figure 4-6** ?

2. In the fire service normally we calculate discharge by using pressure—not elevation of water—because we can impart pressure to water by means of pumps, even where there is no elevation. In Question 1, what would be the pressure at the point of discharge; and using the formula for calculating the velocity from pressure, what is the velocity of discharge?

3. What is the velocity of water in a 2½-in hose if it is flowing water from a 1⅛-in tip at 50 psi?

4. Verify the answer to Question 3 is correct by comparing the area of the two different-size hoses and the velocity of water in the two different-size hoses.

5. What is the quantity of water flowing if the opening is 1.5 ft² and the water is flowing at 3 fps?

6. What is the quantity of water flowing if the opening has a 2-in diameter and the water has a velocity of 108.76 fps?

7. How much water will flow in 1 min from an opening at the base of a tank if the water level is 85 ft above the opening and the opening has a 2½-in diameter?

   Note: As the tank empties, the elevation of the water changes in relation to the discharge, which changes the velocity and flow of water. Problems like this one are meant to be approximate, because in real life you would have to recalculate the velocity and flow every few seconds to account for the elevation change as the tank emptied.

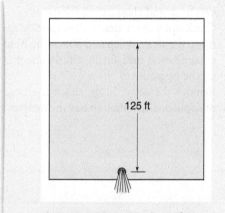

Figure 4-6 What is the velocity of the water at the discharge?

125 ft

Challenging Questions

1. Calculate the velocity of water discharging from an open pipe if it is 37 ft below the water level.

2. If the area of hose 1 is 1.5 times greater than the area of hose 2, what will the velocity of water be in hose 1 compared to that in hose 2?

3. Find the area of hose 2 if hose 1 has a 2½-in diameter and the velocity of water is 10 fps. The velocity of water in hose 2 is 20.42 fps.

4. How much water will flow in 1 s from a 1¾-in opening if it has a velocity of 109 fps?

5. How many cubic feet of water will flow from a 1⅝-in nozzle at 80 psi nozzle pressure in 1 min?

6. How much water will flow from an opening with a 3-in diameter if the water has a velocity of 35 fps?

7. What is the velocity if the pressure is 80 psi?

8. What is the flow per minute from a 1½-in opening with a velocity of 108.7 fps?

Formulas

To find velocity when pressure is known:

$$v = \sqrt{2 \times g \times 2.31 \times P}$$

To find velocity when pressure is known, simplified:

$$v = 12.19\sqrt{P}$$

To find pressure when velocity is known:

$$P = \left(\frac{v}{12.19}\right)^2$$

To compare the area and velocity in two sizes of hose for the same flow:

$$A_1 \times v_1 = A_2 \times v_2$$

To find the diameter of a circle when area is known:

$$D = \sqrt{\frac{A}{0.7854}}$$

Basic quantity flow formula:

$$Q = A \times v$$

To find flow from a circular opening:

$$Q = 0.7854 \times D^2 \times v$$

To find flow from a circular opening when the diameter is in inches:

$$Q = 0.7854 \times D^2 \times \frac{1}{144} \times v$$

To calculate flow in 1 min:

$$Q = 0.7854 \times D^2 \times \frac{1}{144} \times v \times 60$$

GPM

LEARNING OBJECTIVES

Upon completion of this chapter, you should be able to:

- Understand the origins of the formula used to calculate gallons per minute (gpm).
- Calculate the gpm.
- Calculate the nozzle diameter, where gpm and nozzle pressure are known.
- Calculate the nozzle pressure, where gpm and nozzle diameter are known.
- Calculate velocity, where gpm and nozzle diameter are known.
- Calculate the flow from sprinkler heads.

Case Study

It has been a slow afternoon. Engine 11 has had only two runs so far since you got to the firehouse. You use this time to continue studying your hydraulics book. You find yourself beginning to understand how the laws of science have shaped hydraulics as we know it today. You have come to the conclusion that calculating the flow from a tip in gallons per minute isn't just some haphazard solution that just happens to come "close enough." You are really starting to understand hydraulics.

1. How is the constant 29.84 developed?
2. Why is it important to use a discharge coefficient to calculate flow?
3. Explain how to find velocity, when pressure and flow are known.

NFPA 1002 Standard for Fire Apparatus Driver/Operator Professional Qualifications, 2014 Edition

This chapter addresses the following requisite knowledge elements within sections
5.2.1 and **5.2.2:** hydraulic calculations for friction loss and flow using both written formulas and estimation methods.

Introduction

Earlier we began to develop the formula that will ultimately give us the flow from a nozzle in gallons per minute (gpm). The origin of the formula has already been explained, from the basic formula $Q = A \times v$ to $Q = 0.7854 \times D^2 \times \frac{1}{144} \times v \times 60$, which gives the quantity flow in cubic feet per minute (cfm). The next logical step is to find gpm and then develop a simplified formula for calculating it. (See Chapter 4 for more about the origin of the formula.)

This chapter also discusses the use of velocity in the formula for calculating gpm. It is far more convenient for us to calculate gpm from pressure, because it is a standard fire service unit of measurement. But the fact remains that the formula for calculating gpm requires an expression of velocity, so we find a way to introduce pressure into the formula.

After the gpm formula has been developed, we use a variation of the formula to find the diameter of the nozzle when gpm and nozzle pressure are known and to calculate the nozzle pressure when

gpm and nozzle diameter are known. The formula for finding nozzle pressure and nozzle diameter can be handy when you are testing new equipment or finding equivalent diameters when the flow from a fog nozzle is known.

As a bonus, we learn how to calculate the velocity of water being discharged when gpm and nozzle diameter are known. Finally, we learn how to calculate the flow from sprinkler heads.

Calculating Gallons per Minute

We have the formula $Q = 0.7854 \times D^2 \times \frac{1}{144} \times v \times 60$ to make a direct calculation of the quantity of water flowing in cubic feet per minute. The fire service, however, uses quantity flow in units of gallons per minute (gal/min). Fortunately, because there is 7.48 gal in 1 cubic foot (ft^3), the conversion of cubic feet per minute to gallons per minute is just a matter of multiplying the cubic feet per minute by 7.48.

Example 5-1

If a 1-in nozzle operating at 86.18 feet per second (fps) flows 27.96 cfm, how many gallons per minute is it flowing?

Answer

Multiply the cubic feet of water that flows for 1 min by 7.48.

$$\text{gpm} = 27.96 \times 7.48$$
$$= 209.14 \text{ or } 209 \text{ gpm}$$

This entire process can be combined into one formula, eliminating the need to do independent calculations. In addition, by putting all these factors into one formula, we will eventually develop a gpm formula.

The new formula for directly calculating gpm where velocity and nozzle size are known is gpm = $0.7854 \times D^2 \times \frac{1}{144} \times v \times 60 \times 7.48$. We need only to insert the diameter and velocity into the formula, and the answer will be in gallons per minute. But before we do that, since this formula finally has all the factors needed to calculate gpm, we can simplify it to a single constant and two variables, which makes the formula much easier to use. To do that, we simply multiply $0.7854 \times \frac{1}{144} \times 60 \times 7.48$ to arrive at a constant of 2.448. The new formula for calculating gpm when nozzle diameter and velocity are known is gpm = $2.448 \times D^2 \times v$.

$$\text{gpm} = 2.448 \times D^2 \times v$$

Example 5-2

What is the flow, in gallons per minute, for a 1¼-in tip, if the velocity is 86.18 fps?

Answer

Insert 1¼-in for D and 86.18 for v.

$$\text{gpm} = 2.448 \times D^2 \times v$$
$$= 2.448 \times (1.25)^2 \times 86.18$$
$$= 2.448 \times 1.56 \times 86.18$$
$$= 329.11 \text{ or } 329 \text{ gpm}$$

Calculating Gallons per Minute When Pressure Is Known

The formula given for finding gpm encompasses all the factors we need to calculate the flow from a nozzle. The one disadvantage of the formula is that we need to know the velocity of the water exiting the discharge. The easiest way to deal with this shortcoming is to insert an expression in the formula that converts pressure to velocity. Fortunately, we have a formula that converts pressure to velocity, $v = 12.19\sqrt{P}$. By inserting $12.19\sqrt{P}$ into the formula in place of v, we make a direct calculation of gpm without having to first convert pressure to velocity. The formula for calculating gpm when the pressure is given is gpm = $0.7854 \times D^2 \times \frac{1}{144} \times 12.19 \times \sqrt{P} \times 60 \times 7.48$. Or, we can just insert $12.19\sqrt{P}$ into the already abbreviated formula gpm = $2.448 \times D^2 \times v$ and get gpm = $2.448 \times D^2 \times 12.19 \times \sqrt{P}$. See Chapter 4 for a discussion of the formula that converts pressure to velocity.

$$\text{gpm} = 0.7854 \times D^2 \times \frac{1}{144} \times 12.19 \times \sqrt{P} \times 60 \times 7.48$$

$$\text{gpm} = 2.448 \times D^2 \times 12.19 \times \sqrt{P}$$

Example 5-3

Calculate the gpm from a 1¼-in tip with a 50 psi nozzle pressure Figure 5-1 .

Answer

Substitute 1¼ in for D and 50 for P.

$$\text{gpm} = 2.448 \times D^2 \times 12.19 \times \sqrt{P}$$
$$= 2.448 \times (1.25)^2 \times 12.19 \times \sqrt{50}$$
$$= 2.448 \times 1.56 \times 12.19 \times 7.07$$
$$= 329.12 \text{ or } 329 \text{ gpm}$$

Example 5-2 and Example 5-3 are the same problem, with Example 5-2 using velocity to calculate gpm and Example 5-3 using pressure. You can easily verify that 86.18 fps is equivalent to 50 psi by

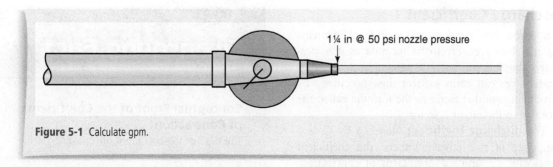

1¼ in @ 50 psi nozzle pressure

Figure 5-1 Calculate gpm.

using the formula to calculate velocity when given pressure. In fact, this was already done in Example 4-2 in Chapter 4.

Freeman's Formula

As already mentioned, the formula used to calculate gpm from pressure in Example 5-3 is a short version of gpm = $0.7854 \times D^2 \times \frac{1}{144} \times 12.19 \times \sqrt{P} \times 60 \times 7.48$. Starting with the entire formula, we create the formula used today to calculate flow from a nozzle. By multiplying $0.7854 \times \frac{1}{144} \times 12.19 \times 60 \times 7.48$ we get a constant of 29.839, which is 29.84 when rounded to two decimal places. The new, more concise formula for calculating gpm becomes gpm = $29.84 \times D^2 \times \sqrt{P}$.

Example 5-4

Calculate the gpm from a 1¼-in tip if it has a 50 psi nozzle pressure.

Answer

This problem is the same as in Example 5-3.

$$\text{gpm} = 29.84 \times D^2 \times \sqrt{P}$$
$$= 29.84 \times (1.25)^2 \times \sqrt{50}$$
$$= 29.84 \times 1.56 \times 7.07$$
$$= 329.11 \text{ rounded to } 329 \text{ gpm}$$

Once again we have arrived at a flow of 329 gpm by using a nozzle pressure. Because the discharge in gallons per minute is the same as the figure we calculate in Example 5-3 by using a longer version of

the gpm formula, we have just verified the validity and origin of the formula gpm = $29.84 \times D^2 \times \sqrt{P}$.

This simplified formula is called *Freeman's formula*, after John Freeman, who first developed it in 1888. Technically, Freeman's formula is slightly different from the one we derived in that he arrived at a constant of 29.71. Freeman's original formula is then gpm = $29.71 \times D^2 \times \sqrt{P}$.

The difference between Freeman's gpm = $29.71 \times D^2 \times \sqrt{P}$ and our gpm = $29.84 \times D^2 \times \sqrt{P}$ is extremely small. The primary reason for the difference lies in the constant for calculating velocity from pressure. Freeman probably rounded off the weight of 1 ft³ of water to 62.5 pounds (lb), instead of using 62.4 lb/ft³, making the formula for converting velocity to pressure slightly different. If Freeman had used 62.5 lb as the weight of 1 ft³ of water, his formula for calculating velocity from pressure would have been $v = 12.14 \times \sqrt{P}$ instead of our $v = 12.19 \times \sqrt{P}$; this difference would account for a constant of 29.71.

Example 5-5

Using Freeman's original formula, find the gpm flow from a 1¼-in tip at 50 psi nozzle pressure.

Answer

Freeman's original formula is gpm = $29.71 \times D^2 \times \sqrt{P}$.

$$\text{gpm} = 29.71 \times D^2 \times \sqrt{P}$$
$$= 29.71 \times (1.25)^2 \times \sqrt{50}$$
$$= 29.71 \times 1.56 \times 7.07$$
$$= 327.67 \text{ rounded to } 328 \text{ gpm}$$

Now go back and look at the answer to Example 5-4; the answers differ by less than 1½ gal before rounding.

Discharge Coefficient

By now you should understand the origin of Freeman's formula. However, Freeman's formula as it is presented above is still not 100 percent accurate. Fortunately, we can easily correct this inaccuracy by introducing another factor to the formula called the *discharge coefficient*, or *C factor*.

The **discharge coefficient**, denoted by *C*, is a composite of two other numbers, the coefficient of velocity, C_v, and the coefficient of contraction, C_c, which have been mathematically manipulated to create a single *C* factor. The first element of the *C* factor, the **coefficient of velocity**, introduces a concept called *laminar flow*. In **laminar flow**, water moves faster at the center of the stream than it does at the edges. Consequently, the velocity of flow is not constant over a cross section of the discharge stream. In the original formula for calculating quantity flow, $Q = A \times V$, it is assumed that the velocity is constant over the entire cross section of the opening. Because we now know that velocity is not constant over the cross section of the opening, we need to introduce the coefficient of velocity to correct for this. Essentially, the coefficient of velocity averages the velocity across the entire area of the opening. (The concept of laminar flow is explained more thoroughly in Chapter 6.)

The second factor that makes up the *C* factor is a **coefficient of contraction**. Again, with water as an example, at the molecular level water wants to cling to the material that the discharge device is made of, as well as to itself. When the water leaves the discharge, it no longer has the material the discharge is made of to cling to, so it clings only to itself. This makes the stream of water contract very slightly, but it is enough to make the cross section of the water stream smaller than the diameter of the discharge. Because the diameter of the stream is smaller than the diameter of the discharge, if this contraction is not factored in, we will be calculating the discharge for a larger stream. The coefficient of contraction factors in the necessary adjustment.

Not every nozzle or discharge device will have a coefficient of contraction. Depending on the

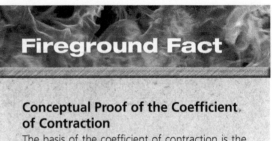

Fireground Fact

Conceptual Proof of the Coefficient of Contraction

The basis of the coefficient of contraction is the molecular attraction of water, in this example, for the material that makes up the discharge device or nozzle. Dipping an object into a glass of water can prove the concept of molecular attraction. When you withdraw the object from the glass, water will adhere to it (the object will be wet), and a small drop of water will form at the bottom.

efficiency of the design, the coefficient of contraction can range from nonexistent to fairly significant. This makes it convenient that both the coefficient of velocity and the coefficient of contraction have been combined into one number called the *C* factor.

The *C* factors for nozzles are usually very large, because they are designed for efficiency. For instance, the *C* factor for a tip for a monitor gun can be 0.997, which means that 99.7 percent of the calculated flow from a monitor nozzle will actually be delivered. **Table 5-1** shows a sample list of discharge coefficients for various types of solid stream devices. For devices not on this list,

Table 5-1 Discharge Coefficients	
Device	**C Factor**
Sprinkler, ½-in	0.75
Sprinkler, ¹⁷⁄₃₂-in	0.95
Smooth-bore nozzle (hand line)	0.96–0.98
Underwriter's playpipe	0.97
Deluge or monitor nozzle	0.997

consult the *Fire Protection Handbook* or check with the manufacturer.

When you are calculating gpm by using Freeman's formula, the *C* factor should always be included. With the *C* factor included, Freeman's formula becomes gpm $= 29.84 \times D^2 \times C \times \sqrt{P}$, where *C* is the discharge coefficient. This is the same formula that appears in Section 15, Chapter 3, "Hydraulics for Fire Protection," of the 20th edition of the *Fire Protection Handbook*.

$$gpm = 29.84 \times D^2 \times C \times \sqrt{P}$$

Example 5-6

Calculate the discharge from a 1¼-in nozzle, at 50 psi nozzle pressure, with a *C* factor of 0.97.

Answer

The *C* factor was obtained from Table 5-1.
$$\begin{aligned} gpm &= 29.84 \times D^2 \times C \times \sqrt{P} \\ &= 29.84 \times (1.25)^2 \times 0.97 \times \sqrt{50} \\ &= 29.84 \times 1.56 \times 0.97 \times 7.07 \\ &= 319.24 \text{ or } 319 \text{ gpm} \end{aligned}$$

In Example 5-4 we calculated the flow from a 1¼-in tip at 50 psi nozzle pressure to obtain an answer of 329 gpm, which is only a 10 gpm difference. Or another way to look at it is that the flow calculated by using the correction factor is only about 3 percent off from the flow found without using the correction factor. This difference is so small for smooth-bore nozzles that the *C* factor is often erroneously omitted. The real value of the *C* factor will become evident when you calculate friction loss for these two flows. You will find the difference in friction loss is about 10 percent.

A word of caution is in order if you choose to omit the *C* factor from your calculations when working with smooth-bore nozzles. When you calculate the discharge from other types of devices, the *C* factor is critical and cannot be omitted. It makes more sense to include the *C* factor as a matter of habit than to forget

it when it is needed. Where the *C* factor is not known, use a factor from Table 5-1. For smooth-bore hand lines, simply split the difference and use 0.97 as above.

Variations of Freeman's Formula

Above you were shown how Freeman's formula was developed. This version is slightly more precise than the original one John Freeman developed in 1888. There are also other versions of Freeman's formula. The biggest discrepancy with other formulas is that they usually omit the coefficient of discharge. With a coefficient of 0.997 the answer will not be significantly in error; with other coefficients, however, the error can be significant. It is good practice to factor in a coefficient of discharge, and it is especially easy to do when you are using a handheld calculator or personal computer. Remember what Albert Einstein once said about taking shortcuts: "Simplify as far as possible, but no further."

Another variation of Freeman's formula was developed many years ago by the American Insurance Association (AIA) for calculating flows from hydrants. The **AIA formula** is gpm $= 29.83 \times D^2 \times C \times \sqrt{P}$. This formula differs from the one developed above only in how and where numbers were rounded off. The formulas are so close that in this text when you are doing calculations for determining hydrant flow, use this formula: gpm $= 29.84 \times D^2 \times C \times \sqrt{P}$.

$$gpm = 29.83 \times D^2 \times C \times \sqrt{P}$$

Example 5-7

What is the flow from a 1½-in monitor nozzle at 80 psi nozzle pressure?

Answer

From Table 5-1 the *C* factor is 0.997.
$$\begin{aligned} gpm &= 29.84 \times D^2 \times C \times \sqrt{P} \\ &= 29.84 \times (1.5)^2 \times 0.997 \times \sqrt{80} \\ &= 29.84 \times 2.25 \times 0.997 \times 8.94 \\ &= 598.43 \text{ or } 598 \text{ gpm} \end{aligned}$$

Calculating Nozzle Diameter When Gallons per Minute and Pressure Are Known

By using Freeman's formula it is possible to calculate the nozzle diameter when we know gpm and pressure. This formula is sometimes used to calculate an "equivalent" nozzle diameter when working with fog nozzles. To find the nozzle diameter from Freeman's formula, the formula must be rearranged to read $D = \sqrt{\text{gpm}/(29.84 \times C \times \sqrt{P})}$.

$$D = \sqrt{\frac{\text{gpm}}{29.84 \times C \times \sqrt{P}}}$$

Example 5-8

What size nozzle delivers 812 gpm at 80 psi nozzle pressure if the C factor is 0.997 Figure 5-2 ?

Answer

Hint: With a C factor of 0.997 we are looking for a monitor nozzle tip.

$$D = \sqrt{\text{gpm}/(29.84 \times C \times \sqrt{P})}$$
$$= \sqrt{812/(29.84 \times 0.997 \times \sqrt{80})}$$
$$= \sqrt{812/(29.84 \times 0.997 \times 8.94)}$$
$$= \sqrt{812/265.97}$$
$$= \sqrt{3.05}$$
$$= 1.746 \text{ or } 1.75 = 1\frac{3}{4} \text{ in}$$

Example 5-8 results in an easily recognized tip size for D. This is not always the case, not because the formula is not good or precise, but because of how we round off numbers. As much as we always want to be as precise as possible, some judgment must be used with this formula to determine the closest actual tip size to our answer. Also if we are trying to find an equivalent tip size for a fog nozzle,

there may be no common tip size equivalent. Finally, the answer is always given in decimals and must be converted to a fraction. Table 5-2 gives the decimal equivalent of various tip sizes.

As mentioned previously, this formula can be used to calculate an equivalent smooth-bore tip size to correspond with the flow from a fog nozzle.

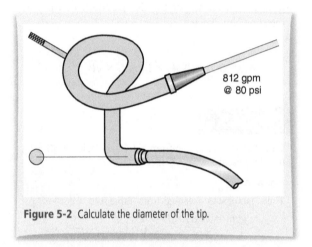

Figure 5-2 Calculate the diameter of the tip.

Table 5-2	Decimal Equivalents
Tip Size, in	**Decimal Equivalent**
$\frac{1}{2}$	0.5
$\frac{17}{32}$	0.53
$\frac{9}{16}$	0.563
$\frac{3}{4}$	0.75
$\frac{15}{16}$	0.938
$1\frac{1}{8}$	1.125
$1\frac{1}{4}$	1.25
$1\frac{3}{8}$	1.375
$1\frac{1}{2}$	1.5
$1\frac{5}{8}$	1.625
$1\frac{3}{4}$	1.75
$1\frac{7}{8}$	1.875

With this knowledge, we can change a fog tip to an equivalent smooth-bore tip without changing the flow. Some people advocate the use of this formula to determine an equivalent tip size, as a means to determine what the flow from the fog nozzle will be if the nozzle pressure is varied. This, however, is not a good practice. The reach and effective pattern of the fog nozzle will be changed if the nozzle pressure and subsequent flow are altered. In short, the nozzle pressure on a fog nozzle should never be altered from what is recommended by the manufacturer.

Example 5-9

If a 2½-in fog nozzle delivers 225 gpm at 100 psi nozzle pressure, what size smooth-bore tip will deliver a similar flow?

Answer

Assume a nozzle pressure of 50 psi and C factor of 0.97.

$$D = \sqrt{gpm/(29.84 \times C \times \sqrt{P})}$$
$$= \sqrt{225/(29.84 \times 0.97 \times \sqrt{50})}$$
$$= \sqrt{225/(29.84 \times 0.97 \times 7.07)}$$
$$= \sqrt{225/204.64}$$
$$= \sqrt{1.1}$$
$$= 1.048 \text{ or } 1.05 \text{ in}$$

Because this formula is used to calculate an equivalent tip size in inches, an exact fractional equivalent is not necessary. This formula has a very practical use when we calculate nozzle reaction for fog nozzles. See Chapter 9 for the practical use of this formula.

Calculating Nozzle Pressure When Nozzle Diameter and Gallons per Minute Are Known

In Example 5-9 we were able to determine that it would take a 1.05-in tip at 50 psi nozzle pressure to flow 225 gpm. The only problem is there is no 1.05-in

tip. Smooth-bore hand line tips are usually sized in ⅛-in increments. We still need to be able to find out how much pressure is needed on a 1-in tip to flow 225 gpm. By rearranging Freeman's formula to find pressure, we get the formula $P = [gpm/(29.84 \times D^2 \times C)]^2$. Now we can find what pressure is needed to deliver a flow of 225 gpm from a 1-in tip.

$$P = \left(\frac{gpm}{29.84 \times D^2 \times C}\right)^2$$

Example 5-10

How much nozzle pressure is needed to flow 225 gpm from a 1-in tip Figure 5-3 ?

Answer

Use a C factor of 0.97.

$$P = [gpm/(29.84 \times D^2 \times C)]^2$$
$$= (225/[29.84 \times (1)^2 \times 0.97])^2$$
$$= (225/28.95)^2$$
$$= (7.77)^2$$
$$= 60.37 \text{ or } 60 \text{ psi}$$

Example 5-11

Use Freeman's formula to verify the results of Example 5-9 and Example 5-10 Figure 5-4 .

Answer

$$gpm = 29.84 \times D^2 \times C \times \sqrt{P}$$
$$= 29.84 \times (1)^2 \times 0.97 \times \sqrt{60}$$
$$= 29.84 \times 1 \times 0.97 \times 7.75$$
$$= 224.32 \text{ rounded to } 224 \text{ gpm}$$

Once again the answer is off only a very small amount due to rounding.

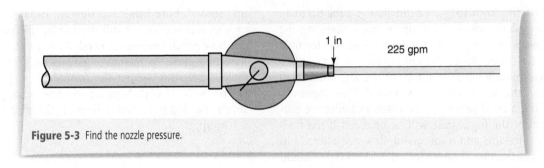

Figure 5-3 Find the nozzle pressure.

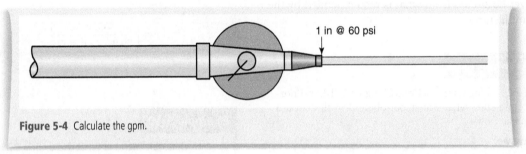

Figure 5-4 Calculate the gpm.

Calculating Velocity When gpm and Nozzle Diameter Are Known

You have learned to use variations of Freeman's formula to calculate the nozzle diameter and nozzle pressure when gpm is known. It is also possible to calculate the velocity of water as it exits a nozzle when the gpm is known, but to do so, we need to go back one step to our original gpm formula, gpm = 2.448 × D^2 × v × C. This formula has an expression for velocity built in, v, and enables us to get an answer directly as velocity. Just as with Freeman's formula, we need to manipulate the formula a little to make velocity the unknown.

When we rearrange the formula to solve for v, the resulting formula becomes v = gpm/(2.448 × D^2 × C). With this formula, we can calculate the velocity of water as it leaves the nozzle, as long as we know the gpm and diameter of the nozzle.

$$v = \frac{\text{gpm}}{2.448 \times D^2 \times C}$$

Example 5-12

If a flow meter indicates a flow of 250 gpm and the tip size is 1¼-in, what is the velocity of the water as it leaves the nozzle if C is 0.97 **Figure 5-5** ?

Answer

$$v = \text{gpm}/(2.448 \times D^2 \times C)$$
$$= 250/[2.448 \times (1.25)^2 \times 0.97]$$
$$= 250/(2.448 \times 1.56 \times 0.97)$$
$$= 250/3.7$$
$$= 67.57 \text{ fps}$$

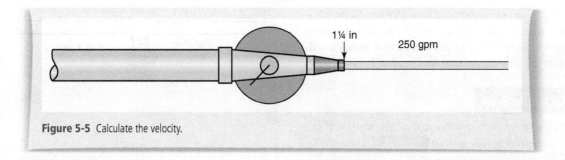

Figure 5-5 Calculate the velocity.

Example 5-13

Verify Example 5-12.

Answer

We verify our result by converting the velocity to pressure and then plugging it into Freeman's formula.

$$P = (v/12.19)^2$$
$$= (67.57/12.19)^2$$
$$= (5.54)^2$$
$$= 30.69 \text{ psi}$$

Now use Freeman's formula to find gpm.

$$\text{gpm} = 29.77 \times D^2 \times C \times \sqrt{P}$$
$$= 29.84 \times (1.25)^2 \times 0.97 \times \sqrt{30.69}$$
$$= 29.84 \times 1.56 \times 0.97 \times 5.54$$
$$= 250.15 \text{ or } 250 \text{ gpm}$$

Once again, the answer is off by only a very slight amount due to rounding.

Calculating Gallons per Minute from Sprinkler Heads

The invention of the sprinkler system is probably the most important discovery, in terms of property protection, in the history of the fire service. For sprinkler systems to be effective, it is necessary that firefighters understand them so that sprinkler systems can be utilized to their fullest. The first step in understanding sprinkler systems is to be aware that

Fireground Fact

Use of Discharge Coefficient

You may have noticed that other texts on hydraulics do not use the discharge coefficient. However, the discharge coefficient is a critical part of any attempt to find an accurate flow. In many cases, the change in flow found by using the discharge coefficient and not using it will not be very large. Just the same, not using it will make your answers less accurate. In this text, the discharge coefficients are given in each problem where they are needed. In real life, you should make every attempt to find the discharge coefficients for the tips you are using. The C factor is important enough that National Fire Protection Association (NFPA) technical standards that include charts of flows from various smooth-bore tips include the C factor(s) used to calculate the listed gpm flows. Also, any NFPA standard that includes the formula for calculating gpm flow from smooth-bore tips includes the C factor.

they are fairly simple in design and operate according to the laws of hydraulics (physics) contained in this book.

Sprinkler heads are designed to operate at a minimum of 7 psi pressure. At that pressure, a sprinkler with a ½-in opening should flow 15 gpm. As the pressure increases, so will the flow, just as with any other nozzle. For now we can use Freeman's formula to calculate the flow from a sprinkler head at 7 psi pressure.

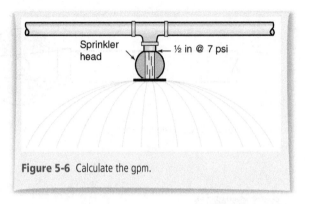

Figure 5-6 Calculate the gpm.

Example 5-14

Apply Freeman's formula to verify a 15 gpm flow from a ½-in sprinkler head at 7 psi **Figure 5-6** .

Answer

As listed in Table 5-1, the C factor for a ½-in sprinkler is 0.75.

$$\text{gpm} = 29.84 \times D^2 \times C \times \sqrt{P}$$
$$= 29.84 \times (0.5)^2 \times 0.75 \times \sqrt{7}$$
$$= 29.84 \times 0.25 \times 0.75 \times 2.65$$
$$= 14.83 \text{ rounded to } 15 \text{ gpm}$$

Remember, different size sprinkler heads have different C factors. It is important to use the correct C factor in order to calculate the correct flow. While it is valuable to be able to calculate the correct flow

from a sprinkler head by using Freeman's formula, if you have to verify flow from dozens or hundreds of heads on a system, this task can become tedious, to say the least. To simplify the process, the concept of the K factor was developed for sprinkler heads. The K factor is simply the constant 29.84 multiplied by the D^2 of the head and the appropriate C factor, based on the head size. The product of these three factors is the K factor. We discuss the K factor further in Chapter 13.

Chapter Summary

- Freeman's formula is used to calculate gpm from a circular, solid stream opening.
- For Freeman's formula to be most accurate, we must factor in the discharge coefficient.
- The discharge coefficient is a composite number made up of a coefficient of contraction and a coefficient of friction.
- By rearranging the terms in Freeman's formula, we can solve for pressure or nozzle diameter.
- By using the proper discharge coefficient, Freeman's formula can be used to calculate flow from a sprinkler head.

Key Terms

AIA formula The formula for determining flow from hydrants.

coefficient of contraction The coefficient that makes a stream of water contract very slightly.

coefficient of velocity The coefficient that averages the velocity of flow across the entire area of an opening.

discharge coefficient The C factor; a composite of two other numbers, the coefficient of velocity and coefficient of contraction.

laminar flow Flow of water that is smooth and orderly, with layers, or cores, of water effortlessly gliding over the next layer of water.

Case Study

Lloyd Layman

During World War II, Chief Lloyd Layman was commander of the U.S. Coast Guard firefighting school at Fort McHenry, Maryland. During that time he conducted research on how to apply water to compartment fires on ships and the best form for applying that water, thereby developing the indirect method of fire attack.

Chief Layman theorized that if water could be broken down into small drops, thus greatly increasing the surface-to-volume ratios of the water, then a greater heat-absorbing capacity would be realized. By applying the water to a compartment in the form of small droplets, it would be converted to steam more efficiently. This steam would in turn displace the air that contained the oxygen necessary for combustion. Displacing the air would deprive the fire of the oxygen it needs to burn.

In his experiments Chief Layman applied water to compartments in the form of fog with low-velocity fog applicators. He found that his theory worked brilliantly as long as all air intake openings were closed, so as to severely limit the amount of oxygen that could feed the fire. He even estimated that he was getting as great as 90 percent conversion of water to steam.

After the war, Chief Layman returned to the Parkersburg, West Virginia, Fire Department where he had been chief since 1931. He continued his experiment with his "indirect" method of fire attack, successfully applying the same principles to structural fire firefighting.

In the 1980s, Swedish fire researchers refined Chief Layman's research even further, noting that, for maximum efficiency, the ideal water droplet size cannot exceed 0.3 millimeter (mm). In his book, *Water and Other Extinguishing Agents*, Stefan Särdqvist emphasizes the importance of the water droplet size and water fog in optimizing water's efficiency as an extinguishing agent.

From Chief Layman's experiments in the 1940s to research nearly 70 years later, the laws of physics still apply: a greater surface area-to-volume ratio equals greater cooling efficiency.

1. What is the basic premise behind Chief Layman's theory?

   A. Water needs to be applied as rapidly as possible.

   B. Water must be applied from as many openings as possible.

   C. Water must be applied in the form of small droplets.

   D. Solid streams of water are most effective.

2. What percentage conversion of water to steam did Chief Layman estimate he was achieving?

   A. 90 percent.

   B. 100 percent.

   C. 75 percent.

   D. None of the above.

3. What was Chief Layman's primary reason for converting the water into steam?

   A. Converting water into steam reduced water damage.

   B. The steam more readily flowed from one part of the compartment to another.

   C. The steam excluded air, which in turn deprived the fire of oxygen.

   D. Moist air will not support combustion.

4. In the 1980s, Swedish fire researches determined the ideal water droplet size to be:

   A. 0.5 mm.

   B. 1 mm.

   C. 0.35 mm.

   D. 0.3 mm.

Review Questions

1. Give the specific reason why each of the following is included in the formula for gpm: $0.7854 \times D^2$, $\frac{1}{144}$, v, 60, 7.48.

2. In the formula gpm $= 2.448 \times D^2 \times v$, from where was the constant 2.448 derived?

3. In the formula gpm $= 2.448 \times D^2 \times v$, how is it possible to introduce pressure in place of velocity?

4. Why did John Freeman's original formula have a constant of 29.71 instead of 29.84?

Activities

Use your knowledge of Freeman's formula and its variations to solve the following problems.

1. How much water will flow, in gallons per minute, from a 1¾-in monitor nozzle at a velocity of 109 fps **Figure 5-7** ?

2. How many gpm will a ¹⅝₆-in tip flow at 50 psi nozzle pressure? $C = 0.97$

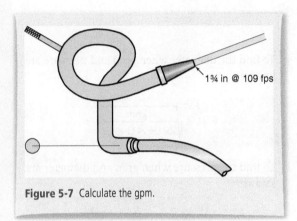

1¾ in @ 109 fps

Figure 5-7 Calculate the gpm.

3. What size tip is needed to flow 650 gpm from a monitor nozzle at 80 psi nozzle pressure?

4. What nozzle pressure is needed to flow 300 gpm from a 1¼-in tip? $C = 0.97$

5. How much velocity is needed to flow 600 gpm through a 1¼-in tip? $C = 0.97$

6. How much water will a ¹⁷⁄₃₂-in sprinkler head flow at 15 psi pressure?

Challenging Questions

1. If a 1⅛-in smoothbore tip is used on a hand line and has a velocity at the tip of 81.5 fps, what is the gpm flow? (Assume a C factor of 0.97.)

2. How many gallons per minute will flow from a 1⅛-in tip on a monitor nozzle if the nozzle pressure is 70 psi? (Assume a C factor of 0.997.)

3. What size tip will flow 247.69 gpm at 45 psi? (Do not forget to include the C factor. Assume a C factor of 0.97.)

4. At what pressure will a ¹⁵⁄₁₆-in tip flow 181.5 gpm? (Assume a C factor of 0.97.)

5. If a 1-in tip is flowing 210 gpm, what is the velocity of the water as it leaves the tip? (Assume a C factor of 0.97.)

6. If a ½-in sprinkler head is flowing 20 gpm, what is the pressure at the sprinkler head?

7. How many gallons per minute will flow from a 1½-in tip if it has a velocity at the tip of 85.98 fps and a C of 0.997?

8. How many gallons per minute will flow from a 1½-in tip if it has a nozzle pressure of 50 psi and a C of 0.997?

9. What tip velocity is necessary to achieve a flow of 325 gpm from a 1¼-in tip? (Assume a C factor of 0.997.)

10. What is the flow from a sprinkler head with a ½-in opening at 15 psi?

11. What is the flow from a 1⅜-in tip at 80 psi?

Formulas

To calculate gpm when diameter and velocity are known:

$$\text{gpm} = 2.448 \times D^2 \times v$$

Basic gpm formula:

$$\text{gpm} = 0.7854 \times D^2 \times \frac{1}{144} \times 12.19 \times \sqrt{P} \times 60 \times 7.48$$

To calculate gpm when diameter and pressure are known:

$$\text{gpm} = 2.448 \times D^2 \times 12.19 \times \sqrt{P}$$

Freeman's formula with the coefficient of discharge:

$$\text{gpm} = 29.84 \times D^2 \times C \times \sqrt{P}$$

AIA formula:

$$\text{gpm} = 29.83 \times D^2 \times C \times \sqrt{P}$$

To find the diameter when gpm and pressure are known:

$$D = \sqrt{\frac{\text{gpm}}{29.84 \times C \times \sqrt{P}}}$$

To find the pressure when gpm and diameter are known:

$$P = \left(\frac{\text{gpm}}{29.84 \times D^2 \times C}\right)^2$$

To find the velocity when gpm and diameter are known:

$$v = \frac{\text{gpm}}{2.448 \times D^2 \times C}$$

Reference

Cote, Arthur E., P.E., Editor-in-Chief, *Fire Protection Handbook*, 20th ed. Quincy, MA: National Fire Protection Association, 2008.

Särdqvist, Stefan. *Water and Other Extinguishing Agents.* Karlstad, Swelen: Swedish Rescue Service Agency, 2002.

Friction Loss

LEARNING OBJECTIVES

Upon completion of this chapter, you should be able to:

- Understand the causes of friction loss.
- Understand the four laws of hydraulics governing friction loss.
- Given the gallons per minute (gpm), calculate friction loss for various sizes of hose.
- Given friction loss and gpm, calculate the friction loss conversion factor.
- Given the friction loss conversion factor and length of any size hose, calculate an equivalent length of hose for the same gpm in any other size hose.
- Given the friction loss in any size hose, calculate the friction loss in any other size hose for the same gpm.
- Given the friction loss and conversion factor, calculate the gpm for a specified hose size.

Case Study

The big day has come and you are driving for the first time. The regular driver has taken off to take his family to the Outer Banks of North Carolina for a week's vacation at the beach. You are a little nervous and apprehensive about finally having the responsibility required of the position, but you also feel ready.

At 12:30 a.m., Engine 11 is dispatched as the second-due engine on a second alarm. You find your in way to your assigned position in the rear of the building with little trouble. After you have charged the hand line your company is using, charged a second line supplying a ladder tower, and ensured you have sufficient water supply yourself, you take a moment to look around. As you do, you see lots of charged hose lines. You can think only about water flowing through the hose, how fast it is moving, friction loss, and whether you have gotten it right.

1. What are the laws governing friction loss?
2. What conversion factors does your department use for various sizes of hose and parallel lines?
3. Give an example of when it might be useful to calculate gpm if the friction loss is known.

NFPA 1002 Standard for Fire Apparatus Driver/Operator Professional Qualifications, 2014 Edition

This chapter addresses the following requisite knowledge elements within sections
5.2.1 and **5.2.2:** hydraulic calculations for friction loss and flow, using both written formulas and estimation methods.

Introduction

Earlier we talked about energy as it pertains to hydraulics. Although energy cannot be lost or destroyed, it does often change into less useful forms. One of the most prevalent and unforgiving causes of energy conversion is friction. (See Chapter 3 for a discussion of energy and hydraulics.)

Friction loss plays a major role in hydraulics. We must always compensate for it when calculating pump discharge pressure, figuring hose lays, and determining how much water we can flow. Fortunately, it is easy to calculate, and charts can be made with friction loss at various flows and in different sizes of hose.

In this chapter we learn the principles of friction loss, how to calculate friction loss, and how to adjust hose size and length based on a thorough understanding of friction loss. In addition, we learn how to calculate the gpm when the hose size and friction loss are known. Finally, we learn to find equivalent length of different sizes of hose, or equivalent friction losses for a given gpm, by using conversion factors.

The formula used in this chapter to find friction loss can be used to find a friction loss for virtually any amount of gpm through any size hose. Just because you can calculate a friction loss does not mean you should actually attempt to pump that much water through the hose. Practically speaking, when the flow in a given size hose reaches a friction loss of 50 pounds per square inch (psi) per 100 feet (ft), the capacity of the hose has been reached. At pressures higher than this, even small-diameter hose can become difficult to maneuver. Excessively high friction loss also leads to excessive engine pressure, which can reduce the capacity of the pump. Pump capacity is covered further in Chapter 7.

While you are reading this chapter, keep in mind that the formulas identified for calculating friction loss can provide very close estimates, but are not 100 percent accurate. This is so because as the velocity of a fluid increases, the fluid reacts to friction differently, so no single formula can precisely calculate friction loss over a wide range of pressure. With the proper attention to detail, however, the friction loss formula presented in this text can be extremely accurate over the range of pressure needed for fire suppression.

> **Note**
>
> When the flow in a given size hose reaches a friction loss of 50 psi per 100 ft, the capacity of the hose has been reached.

Factors Affecting Friction Loss

Friction can be defined as the resistance to movement of two surfaces in contact. For example, if you rub your hand across the surface of any solid object, you will feel a slight resistance. That resistance is friction trying to keep your hand from sliding across the surface of the object. The conversion of useful energy into nonuseful energy (heat) due to friction is called *friction loss*.

In hydraulics, friction is created when water (or another fluid) rubs against the inside of hose or pipe and against itself. Friction loss in hose is usually expressed in terms of loss of pressure per 100 ft. Several factors affect friction loss, including the roughness of the hose, roughness of hose appliances, and restrictions to the flow of the water. Restrictions can include protruding gaskets, sharp bends in hose, and excessive or incorrect use of hose appliances.

Viscosity

Another factor affecting friction loss is the viscosity of the liquid. Viscosity is the resistance to flow of a liquid. The viscosity of a liquid determines how much friction loss is created inside the liquid itself as it tries to flow. In general, the thicker the liquid, the greater the viscosity or resistance to flow; this means that as viscosity increases, so does friction loss within the liquid.

To better understand viscosity, think of the comparison of maple syrup to corn syrup. Maple syrup is very watery and easily flows off your pancakes; it has a very low viscosity. However, corn syrup is thick, and when it is put on pancakes, mostly it stays put; corn syrup has a very high viscosity.

From the previous discussion of friction loss, clearly friction loss in hose has multiple causes: the condition of the hose, the condition of the appliances, the hose lay itself, and the quantity of water flowing. Along with these, two other conditions affect friction loss and need to be taken into consideration whenever friction loss is addressed: laminar flow and turbulent flow.

Laminar Flow

Laminar flow represents a best-case scenario with the least amount of friction loss. In laminar flow, the flow of the water is smooth and orderly, with layers, or cores, of water effortlessly gliding over the next layer of water and velocity gradually increasing from edge to center, much as depicted in **Figure 6-1**. The figure depicts the inner core of water, *A*, smoothly passing through core *B*, which in turn is smoothly passing through core *C*. Each core moves progressively faster as it gets farther away from the walls of the hose; that is, core *A* is moving the fastest.

Laminar flow is found where hose has a relatively smooth lining, the hose is laid straight, and hose appliances are in good condition. Most of the friction loss is created by the friction of the water against the walls of the hose. In laminar flow, because the flow of water is orderly and in layers or cores, there is little friction of water against itself.

Turbulent Flow

Turbulent flow results when water flows in a disorganized, random manner. **Figure 6-2** represents the flow of water under turbulent conditions, where a great deal of the friction loss is due to water rubbing against itself.

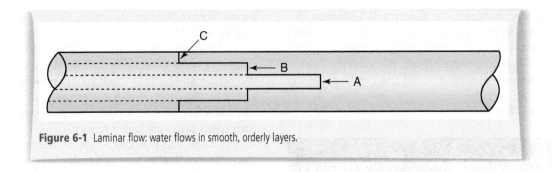

Figure 6-1 Laminar flow: water flows in smooth, orderly layers.

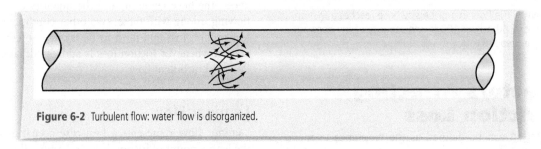

Figure 6-2 Turbulent flow: water flow is disorganized.

Rough hose, excessive pressure, protruding gaskets, excessive bending, and kinking of hose all contribute to turbulence in the hose. In fact, it can logically be assumed that in most situations familiar to the fire service, turbulent flow is present to some degree. To reduce the effects of turbulent flow, deliberate efforts need to be made in laying hose to prevent kinking and excessive bends. Reasonable efforts must also be made to keep pump pressures as low as possible. For this reason, this introduction recommends considering hose capacity to be reached once the friction loss per 100 ft of hose reaches 50 psi. Above this pressure, friction loss goes up at a disproportionally quick rate compared to the increase in flow.

Four Laws Governing Friction Loss

Within the discipline of hydraulics we have encountered principles that govern the behavior of pressure. Hydraulics has four laws governing how friction behaves. By knowing these laws, we are able to calculate and understand friction loss. There is

one difference between the principles of pressure we learned and these four laws of friction loss. The principles of pressure applied to all applications of hydraulics, whereas some of these laws of friction loss apply only to hose.

Law 1

Friction loss varies directly as the length of the hose, provided all other conditions are equal. Law 1 tells us that as the length of the hose line increases, so does the total friction loss **Figure 6-3**. The friction loss increases proportionally to the change in the length of hose. If the length of the line is doubled, the friction loss also doubles. If the length of the line is increased by a factor of 3, the friction loss increases by a factor of 3.

Law 2

Friction loss varies as the square of the change in velocity. The fact that friction loss changes exponentially is a direct result of the flow of water becoming more turbulent as velocity is increased. This principle is

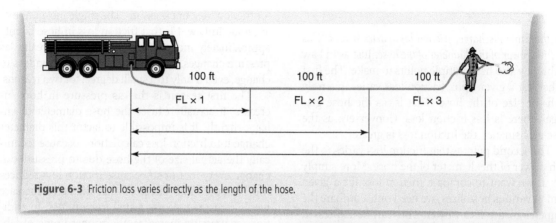

Figure 6-3 Friction loss varies directly as the length of the hose.

a perfect illustration of the discussion of laminar versus turbulent flow. As velocity is increased, the turbulence of water increases, disproportionately faster than the flow increases, causing what can be described as an exponential increase in friction loss. In short, for a given size hose, friction loss will increase much faster than the flow causing the friction loss, as illustrated in Example 6-1.

$$F_m = \left(\frac{v_1}{v_2}\right)^2$$

This change in friction loss, due to the change in velocity, can be expressed as a mathematical formula, $F_m = (v_2/v_1)^2$, where F_m is the friction loss multiplier, v_1 is the reference or original velocity, and v_2 is the new velocity.

Velocity versus Flow

Friction loss is a function of velocity, not gpm. This is an important technicality to understand because gpm and velocity are directly related in a given size conduit. If velocity is increased by a factor of 2, the gpm is also increased by a factor of 2. But because friction is the result of movement of the water against the hose and against the water itself, velocity is technically the cause of friction loss. This is evident by the fact that friction loss varies exponentially with the change in velocity.

Example 6-1

If the friction loss in 100 ft of hose is 5 psi, what will the new friction loss be if the velocity is doubled?

Answer

In this example, v_1 is 1 because this is the reference velocity, and v_2 is 2 because the velocity is doubled.

$$F_m = (v_2 / v_1)^2$$
$$= (2/1)^2$$
$$= (2)^2$$
$$= 4$$

This means the friction loss will be 4 times as great after the velocity (flow) is doubled, or equal to the square of the change in velocity.

The new friction loss is 5 psi × 4, or 20 psi.

Law 3

For the same discharge, friction loss varies inversely as the fifth power of the diameter of the hose. Just as in Law 2, this law also has multiple points to make. The first is that, for a given gpm, friction loss varies inversely with the size of the hose. That is, as the hose gets larger, there is less friction loss. Conversely, as the hose gets smaller, the friction loss is greater.

The second point is that friction loss varies to the fifth power of the diameter of the hose. More simply put, if we want to compare friction loss for a given gpm in various hose sizes, we need only compare the diameter to the fifth power. This can be expressed in the formula $CF = D_1^5 / D_2^5$, where CF is the conversion factor, D_1 is the diameter of the hose with which you are making the comparison, and D_2 is the hose for which you are trying to find the conversion factor. This law, however, has one serious limitation: it calculates a conversion factor only for rubber-lined hose.

$$CF = \frac{D_1^5}{D_2^5}$$

Example 6-2

If a given gpm in 2½-inch (2½-in) hose has a friction loss of 10 psi, what will be the friction loss in 3-in hose for the same gpm?

Answer

$$CF = D_1^5 / D_2^5$$
$$= (2.5)^5 / (3)^5$$
$$= 97.66 / 243$$
$$= 0.4$$

The friction loss for the same gpm in 3-in hose is 10 psi × 0.4, or 4 psi.

Law 4

For a given velocity, the friction loss in hose is approximately the same no matter what the pressure may be. Remember, Law 2 states that velocity is responsible for friction loss, not pressure. The single most important issue in Law 4 is that friction loss in hose is only approximately independent of pressure. In reality, as pressure changes, it is possible for the friction loss to change, even if only to a small degree, for two reasons.

The first reason is that as pressure in hose increases, it actually causes the hose diameter to enlarge slightly. It is impossible to factor this diameter change into friction loss calculations because technically the actual size of the hose due to pressure can change every foot or so because friction loss reduces the pressure. The effect of this diameter increase would be to reduce the actual friction loss. In addition, as pressure increases, the rubber lining of fabric-covered hose can assume the texture of the fabric, causing a rougher surface and increased friction loss.

The second reason why that friction loss is *approximately* independent of pressure is that as hose is pressurized, it elongates. After the hose is charged, there is technically more hose to generate greater friction loss. Again, this condition is impossible to factor into friction loss calculations because not all hose behave similarly, but the serious student of hydraulics should be aware of this behavior.

Even if we consider the change in friction loss due to these factors, the resulting numbers can be highly accurate if proper care is given to making accurate calculations.

Calculating Friction Loss

The National Board of Fire Underwriters, now known as the Insurance Services Office Inc., developed the formula used today to calculate friction loss in hose. Originally the formula was $FL = 2Q^2 + Q$, where FL is friction loss and Q is gpm flow in the hundreds. In the late 1960s the formula was altered slightly to account for improved hose manufacturing processes that reduced the friction loss. Today we use the formula $FL = 2Q^2$, known as the *Underwriters' formula*. The Underwriters' formula is used to find friction loss in *only* 2½-in hose and is calculated for 100 ft of hose. The answer obtained by use of the Underwriters' formula is in units of pounds per square inch.

To use the Underwriters' formula, we need to know only the gpm in hundreds. The gpm in hundreds is easily calculated from gpm by moving the decimal point two places to the left. For example, 250 gpm in hundreds is 2.5; 2.5 would then replace Q in the equation.

Before we begin doing friction loss calculations, we are going to make a slight change to the formula for friction loss. Instead of using the formula $FL = 2Q^2$, we are going to modify it slightly to read $FL\ 100 = 2Q^2$. The notation FL 100 reminds us that this friction loss is only for 100 ft of hose. By using FL 100 here we can keep the two friction loss figures separate later. Also in Chapter 10 the pump discharge pressure formula uses FL for the total friction loss in a line.

$$FL\ 100 = 2Q^2$$

Example 6-3

What is the friction loss in 2½-in hose for 375 gallons per minute (gpm)?

Answer

The 375 gpm in hundreds is 3.75.

$$
\begin{aligned}
FL\ 100 &= 2Q^2 \\
&= 2 \times (3.75)^2 \\
&= 2 \times 14.06 \\
&= 28.12 \text{ or } 28 \text{ psi}
\end{aligned}
$$

Today's fire service uses a variety of hose sizes, and we need a means of calculating friction loss for hose other than 2½-in hose. To calculate the friction loss for hose other than 2½-in hose, we begin by calculating the friction loss for 2½-in hose and then converting it to the friction loss for the hose size in question. At this point Law 3 becomes important. By comparing other hose to 2½-in hose, conversion factors can be generated for any size rubber-lined hose. The formula

for calculating friction loss for hose other than 2½-in hose then becomes FL 100 = CF × $2Q^2$, where CF is the conversion factor. In short, this formula calculates friction loss for 2½-in hose ($2Q^2$) and then multiplies it by a conversion factor to find the friction loss for the same gpm in another size hose.

$$FL\ 100 = CF \times 2Q^2$$

Example 6-4

What is the friction loss in 3-in hose for 375 gpm?

Answer

In Example 6-2, the conversion factor for 3-in hose was calculated to be 0.4.

$$
\begin{aligned}
FL\ 100 &= CF \times 2Q^2 \\
&= 0.4 \times 2 \times (3.75)^2 \\
&= 0.4 \times 2 \times 14.06 \\
&= 11.25 \text{ or } 11 \text{ psi}
\end{aligned}
$$

By using Law 3 we can calculate the conversion factor to compare friction loss for any two hose sizes, as long as they are rubber-lined hose. Because the fire service today uses more than just rubber-lined hose, we also need conversion factors for the other hoses.

Recall that this text emphasizes the need to use C factors to obtain correct flows from smooth-bore tips. Here is where it begins to matter. In the chapter on gallons per minute, we calculated the flow from a 1½-in tip without use of the C factor and got a flow of 329 gpm. In that same chapter, we also calculated a flow from a 1½-in tip using the C factor and got a flow of 319 gpm. Using the above formula for calculating friction loss, we get a FL 100 of 22 for 329 gpm but a FL 100 of 20 for 319 gpm—a 10 percent increase in friction loss for 10 gallons (gal) more of flow. Here is an instance in which calculating the gpm without use of the C factor will result in overpumping by 10 percent. See Chapter 5, and Example 5-4 and Example 5-6 in particular, for a discussion of using C factors to obtain correct flows from smooth-bore tips.

Empirical Method for Calculating Conversion Factors

Because Law 3 directly pertains to only rubber-lined hose, we need a method to calculate the conversion factor for all hoses. For that we look to the formula we just developed for calculating friction loss, FL 100 = $CF \times 2Q^2$. Just as we have done several times already in this text, we simply rearrange the formula to find CF. The empirical formula for calculating the conversion factor then becomes CF = FL $100/2Q^2$.

$$CF = \frac{FL\ 100}{2Q^2}$$

Because friction loss for a given size hose may or may not be constant from manufacturer to manufacturer, it is sensible to verify the conversion factor by doing random sampling. The conversion factor formula allows us to find conversion factors for hose that does not fit Law 3 as well as for hose that does. The good thing about the conversion factor formula is that we need to develop the conversion factor for only a single flow in the test hose. The derived conversion factor will then be good for all flows in that size hose.

To use the formula CF = FL $100/2Q^2$, we need to know two things: (1) the gpm and (2) the friction loss for the gpm. This process may seem a little backward, but if the following procedure is followed, the formula will work with a high degree of accuracy.

This procedure is fairly simple.

1. Start by laying a single line of the hose that you are going to test.
2. Make the line several hundred feet (at least 300 ft) long so you will end up with an averaged friction loss.
3. Flow enough water to truly test the hose. For hand line hose, test at the midpoint of the advertised flow range. For a supply line, test at 150 gpm per 1 in of hose diameter at a minimum.

4. Calculate the friction loss by subtracting the nozzle pressure from the pump discharge pressure. (When you test the supply line, subtract the intake pressure of pumper 2 from the discharge pressure of pumper 1 to obtain the friction loss.)
5. Divide the calculated total friction loss by the amount of hose, in hundreds, to get the friction loss per 100 ft.
6. Next calculate the gpm flow based on the nozzle pressure and tip size.
7. Finally, plug these figures into the formula CF = FL $100/2Q^2$ and calculate the conversion factor.

When you are conducting a test to determine the conversion factor, use a smooth-bore tip so you can take a pitot reading. The pitot reading serves two purposes: (1) it gives a more accurate nozzle

Fireground Fact

Correct Hose Size?

How is it possible that the conversion factor for hose of a given size may vary from manufacturer to manufacturer? Several years ago the author discussed this point with a firefighter who was involved in testing hose in his department. While attempting to find accurate correction factors for their hose, firefighters discovered that one manufacturer's 1¾-in hose had significantly less friction loss than the others. In their attempt to find out why the hose had less friction loss, they decided to actually measure the inside diameter of the hose. They found that the hose with the significantly lower friction loss, while being sold as 1¾-in hose, had a diameter actually just a shade less than 2 in. Of course it would have less friction loss. The takeaway is to conduct your own test to determine the correct conversion factor for your hose.

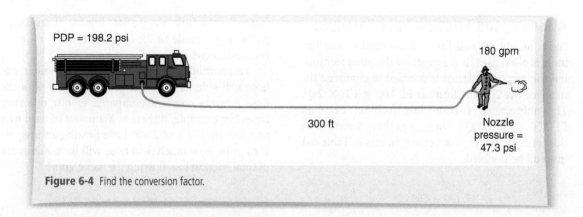

Figure 6-4 Find the conversion factor.

pressure than using a gauge at the base of the noz-zle, and (2) it can be used to calculate the exact gpm flow during the test. Use Freeman's formula, gpm = $29.84 \times D^2 \times C \times \sqrt{P}$, to calculate the exact amount of water flowing. However, do not round it down to a whole number; two decimal places should be sufficient to maintain as much accuracy

as is reasonably possible. (Freeman's formula can be found in Chapter 5.)

Table 6-1 contains the conversion factors for the most common sizes of hose. All hose, unless otherwise noted, is rubber-lined. For hose of a type or size not contained in Table 6-1, you should run your own test to determine the correct conversion factor.

Example 6-5

You have recently conducted a test of 1¾-in hose. You flowed a ¹⁵⁄₁₆-in tip at 50 psi per your pitot gauge. Your pump discharge pressure (PDP) was 200 psi while using 300 ft of hose. Find the proper correction factor for your hose Figure 6-4.

Answer

First, find the exact flow from the ¹⁵⁄₁₆-in tip. You should calculate a flow of 179.88 gpm (assuming a C factor of 0.97). Next subtract the nozzle pressure from the PDP and divide by 3 to obtain the friction loss per 100 ft: $200 - 50 \div 3 = 50$ psi. Now find the correct correction factor.

$$CF = FL\ 100/2Q^2$$
$$= 50/[2 \times (1.7988)^2]$$
$$= 50/(2 \times 3.24)$$
$$= 50/6.48$$
$$= 7.716\ \text{or}\ 7.72$$

The correct conversion factor for your hose was found to be 7.72.

Table 6-1 Conversion Factors for Common Hose Sizes

Hose Diameter, in	Conversion Factor
¾	500
1	75
1½	12
1½ (linen)	25.6
1¾ with 1½-in couplings	7.76
2	4
2 (linen)	6.25
2½ (linen)	2.13
3 with 2½-in couplings	0.4
3½	0.17
4	0.1
5	0.04
6	0.025

Abbreviated Friction Loss Formula

Today an abbreviated form of the friction loss formula is often used. It is essentially the same formula, but the correction factor is doubled to eliminate the need for the 2 in the formula $FL_{100} = CF \times 2Q^2$. The abbreviated friction loss formula becomes $FL_{100} = CF \times Q^2$. There is nothing wrong with this formula, but the conversion factors in Table 6-1 must all be doubled.

$$FL_{100} = CF \times Q^2$$

One important warning is necessary at this point if you choose to use the abbreviated formula. Recall that the original friction loss formula only pertains to 2½-in hose. This does not change with the new formula. Therefore, there is no conversion factor for 2½-in hose in the original formula. However, friction loss coefficient charts for use with the abbreviated formula show a factor of 2 for 2½-in hose. Remember, this only gets 2½-in hose back to its original formula. It is not an actual conversion factor.

The other point to remember about the abbreviated formula is that the conversion factors are incompatible with friction loss Law 3. If a conversion factor is found using Law 3, it must first be multiplied by 2 to be compatible with the abbreviated formula.

Finding Equivalent Hose Length

At times it is helpful to determine how much of one size hose is equivalent to another size hose. For example, how much 3-in hose is equivalent to 300 ft of 2½-in hose? To make these comparisons, it is helpful to have a formula that will directly compare any given size hose to any other size hose. That formula is $L_2 = L_1 \times (CF_1/CF_2)$, where L_1 is the known length of hose and L_2 is the length of the hose we are looking for. The conversion factors used in this formula are the same as those in Table 6-1, with CF_1 the conversion factor of the length of the hose we already know and CF_2 the conversion

factor for the hose we are looking for. If any comparisons are made to 2½-in hose, use 1 for CF for the 2½-in hose.

This formula tells us how much of another size hose will give us the same total friction loss, if the flow remains constant compared to our reference hose. For example, if there is X amount of total friction loss in 300 ft of 2½-in hose flowing an unspecified gpm, how much 3-in hose will have X amount of total friction loss flowing the same gpm?

$$L_2 = L_1 \times \left(\frac{CF_1}{CF_2} \right)$$

Example 6-6

How much 3-in hose is equivalent to 300 ft of 2½-in hose?

Answer

Because there is no CF for 2½-in hose, use 1; the CF for 3-in hose is 0.4.

$$L_2 = L_1 \times (CF_1/CF_2)$$
$$= 300 \times (1/0.4)$$
$$= 300 \times 2.5$$
$$= 750 \text{ ft}$$

So 750 ft of 3-in hose is equivalent to 300 ft of 2½-in hose.

Example 6-7

How much 1¾-in hose is equivalent to 150 ft of 1½-in hose?

Answer

From Table 6-1, CF_1 is 12 and CF_2 is 7.76.

$$L_2 = L_1 \times (CF_1/CF_2)$$
$$= 150 \times (12/7.76)$$
$$= 150 \times 1.55$$
$$= 232.5 \text{ ft}$$

Since hose this size must be in 50-ft lengths, this answer would need to be rounded down to 200 ft.

Finding Equivalent Friction Loss

Being able to compare the friction losses between two different-size hose is another method of comparing hose efficiency. To do this, we need to know the friction loss for the reference hose. Then we need to find the friction loss for the comparison hose. The formula to make the comparison is $FL_2\ 100 = FL_1\ 100 \times (CF_2/CF_1)$. In this equation, FL_1 is the friction loss we already know, FL_2 is the friction loss we are looking for, CF_1 is for the hose whose friction loss we know, and CF_2 is for the size hose for which we are calculating friction loss.

$$FL_2\ 100 = FL_1\ 100 \times \left(\frac{CF_2}{CF_1}\right)$$

If we were comparing the friction loss for a given gpm, we could easily make the comparison by using the Underwriters' formula. The formula $FL_2\ 100 = FL_1\ 100 \times (CF_2/CF_1)$ allows us to make friction loss comparisons even when we do not know the gpm. This formula can be helpful when doing maximum flow problems comparing various hose sizes. This is done in Chapter 11.

Example 6-8

If a 4-in supply line has a friction loss of 15 psi, how much friction loss will there be in 3-in hose for the same gpm?

Answer

From Table 6-1, CF_1 is 0.1 and CF_2 is 0.4.

$$FL_2\ 100 = FL_1\ 100 \times (CF_2/CF_1)$$
$$= 15 \times (0.4/0.1)$$
$$= 15 \times 4$$
$$= 60\ \text{psi}$$

In another useful application of this formula we find a factor to compare the friction loss of any two sizes of hose. This formula is useful in making quick comparisons without having to go through the entire process for each friction loss adjustment. To calculate

the comparison factor, simply use a friction loss of 1 for $FL_1\ 100$. Rather than create a whole new formula just to calculate the comparison factor, we will use $FL_2\ 100 = FL_1\ 100 \times (CF_2/CF_1)$. Just remember the answer is actually a comparison factor and not a friction loss.

Example 6-9

Calculate the friction loss conversion factor to compare friction loss in 1½-in hose to that of 1¾-in hose.

Answer

From Table 6-1, CF_1 is 12 and CF_2 is 7.76.

$$FL_2\ 100 = FL_1\ 100 \times (CF_2/CF_1)$$
$$= 1 \times (7.76/12)$$
$$= 1 \times 0.65$$
$$= 0.65$$

Just multiply the known friction loss for the 1½-in hose by 0.65, and you have the friction loss for the same gpm in 1¾-in hose. If you happen to have the friction loss in 1¾-in hose, divide it by 0.65 to get friction loss for the same flow in 1½-in hose. In other words, friction loss for a given gpm in 1¾-in hose is 0.65 times the friction loss in 1½-in hose.

Example 6-10

If 1½-in hose has an FL 100 of 37.5 psi for 125 gpm, what would be the friction loss in 1¾-in hose for the same flow?

Answer

Simply multiply FL 100 for the 1½-in hose by 0.65.

$$37.5 \times 0.65 = 24.38\ \text{or}\ 24\ \text{psi}$$

Thus FL 100 for 1¾-in hose flowing 125 gpm is 24 psi.

Calculating Gallons per Minute from Friction Loss

To say we truly understand hydraulics, we have to be able to perform calculations that are not routine, such as determining how much water is flowing if all we know is the friction loss of a particular size hose. Under these conditions, it is possible to find the gpm by using the Underwriters' formula, FL 100 = CF × $2Q^2$. By rearranging the formula to solve for Q, the formula becomes $Q = \sqrt{FL\ 100/(CF \times 2)}$. The conversion factor in this formula corresponds to the size hose being used, as listed in Table 6-1.

$$Q = \sqrt{\frac{FL\ 100}{CF \times 2}}$$

Example 6-11

If 5-in hose has a friction loss of 15 psi, how much water is it flowing?

Answer

The conversion factor for 5-in hose from Table 6-1 is 0.04.

$$Q = \sqrt{FL\ 100/(CF \times 2)}$$
$$= \sqrt{15/(0.04 \times 2)}$$
$$= \sqrt{15/0.08}$$
$$= \sqrt{187.5}$$
$$= 13.69 \text{ or } 1,369 \text{ gpm (remember, } Q \text{ is in hundreds)}$$

Conversion Factors for Parallel Hose Layouts

There may be times when pumpers, master stream devices, or sprinkler and standpipe systems are supplied by more than one hose from the same pumper, a setup that is called *parallel lines*. If the hose used is all the same size, this situation does not present a problem, because each line will be supplying an equal share of the water, assuming the lines are of equal length. The friction loss for the appropriate share of water will be the same for all the lines. For example, if two 2½-in lines are supplying a monitor nozzle from the same pumper, then each line will be delivering exactly one-half of the water. If the monitor is flowing 600 gpm, then each line is flowing 300 gpm. When the friction loss is calculated for this setup, it will be found for a flow of 300 gpm.

At times hoses of different diameters are used to supply the same device **Figure 6-5**. During pumping from one pumper through parallel lines to a device, to another pumper, or to a system of some kind, the pressures in the parallel lines will equalize at the point where both lines have the same friction loss, assuming they are both being pumped at the same pressure. Under these circumstances, the only way to find the correct friction loss is to calculate every possible combination of flows until you find a flow in each hose that has the same friction loss at the correct total flow. For practical purposes, this is not something you want to do.

The solution to this problem is to develop conversion factors for possible multiple-line evolutions. This can be done in the same way as we figured the conversion factor in the section "Empirical Method for Calculating Conversion Factors." Instead of using a single line, use two equal-length lines of two different diameters. Pump from one pumper to a second pumper that is delivering a precise amount of water. Then you can calculate the friction loss per 100 ft for the gpm being supplied. Finally, use the formula CF = FL $100/2Q^2$ to determine the conversion factor for the multiple-line layout as done previously for the single line.

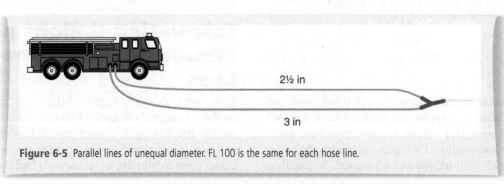

Figure 6-5 Parallel lines of unequal diameter. FL 100 is the same for each hose line.

There is also another method that does not require an actual flow test to determine the conversion factor. Start by selecting a friction loss at random, and then find the flow for the other size hose at that friction loss. You now have the gpm and friction loss to plug into the conversion factor formula.

Example 6-12

What is the conversion factor for parallel lines, one 2½-in and the other 3-in?

Answer

In Example 6-4, we determined that 3-in hose has a friction loss of 11.25 psi when flowing 375 gpm. This gives us a convenient place to start. Now let us find how much water is flowing in 2½-in hose for the same friction loss. Remember, substitute 1 for CF for the 2½-in hose.

$$Q = \sqrt{FL\ 100/(CF \times 2)}$$
$$= \sqrt{11.25/(1 \times 2)}$$
$$= \sqrt{11.25/2}$$
$$= \sqrt{5.625}$$
$$= 2.37 \text{ or } 237 \text{ gpm}$$

The flow will be 375 gpm for the 3-in hose and 237 gpm for the 2½-in hose for a total flow of 612 gpm. To calculate the new conversion factor, FL 100 will be the same 11.25, and Q will be the total of both the 2½-in hose and the 3-in hose, or 612 gpm.

$$CF = FL\ 100/2Q^2$$
$$= 11.25/[2 \times (6.12)^2]$$
$$= 11.25/(2 \times 37.45)$$
$$= 11.25/74.9$$
$$= 0.15$$

The conversion factor for parallel lines of 1–2½ in and 1–3 in at any flow is 0.15.

By inserting a conversion factor of 0.15 into the formula FL 100 = CF × $2Q^2$, the friction loss can be found for any flow through parallel lines of one 2½-in hose and another 3-in hose. This procedure can be applied to any parallel-line situation to determine the conversion factor. Just remember to total all the flows at the same friction loss for each size hose, even if some of the lines are of the same diameter.

Fireground Fact

Parallel Lines of Equal Diameter

A convenient rule of thumb applies to finding FL 100 when you are dealing with parallel lines of equal diameter and length. Calculate the friction loss for one line flowing a proportional share of the water. That is, if you are using two lines, then divide the total flow in half, and so forth. Then find the friction loss for that flow, and you will have your FL 100 for the parallel-hose layout.

Chapter Summary

- When the flow in a given size hose reaches 50 psi per 100 ft, the capacity of the hose has been reached.
- Friction is the resistance to movement of two surfaces in contact.
- Viscosity is the resistance to flow of a liquid.
- Laminar flow occurs when water flows in smooth, orderly layers.
- Turbulent flow occurs when the flow of water is disorganized and random.
- Friction loss varies directly with respect to the length of the hose, provided all other conditions are equal.
- In the same size hose, friction loss varies approximately as the square of the velocity.
- For the same discharge, friction loss varies inversely as the fifth power of the diameter of the hose.
- For a given velocity, the friction loss in hose is approximately the same, no matter what the pressure may be.
- The formula for friction loss calculates the friction loss in 2½-in hose.
- To find friction loss in hose other than 2½-in hose requires the addition of a correction factor to the friction loss formula.
- Conversion factors can be developed to calculate friction loss in parallel-hose layouts.

Key Terms

friction The resistance to movement of two surfaces in contact.

laminar flow The smooth and orderly flow of water, with layers, or cores, of water gliding effortlessly over the next layer of water and velocity gradually increasing from edge to center.

turbulent flow The flow of water that is disorganized and random.

viscosity Resistance to flow of a liquid.

Case Study

Thornton's Rule

In 1917 British college professor W. M. Thornton wrote an article about the heat release rate of burning hydrocarbons entitled "The Relation of Oxygen to the Heat of Combustion of Organic Compounds." In his article Dr. Thornton identifies the heat of combustion of both carbon and hydrogen and links the energy (heat) release to the internal energy of the carbon–hydrogen bonds. More importantly, Dr. Thornton notes that this energy is not released; that is, the carbon–hydrogen bonds will not break and release their energy without the presence of oxygen.

All firefighters are aware that oxygen is needed for combustion, but Dr. Thornton's article tells us for the first time that the oxygen available for combustion determines the energy (heat) release rate during combustion, not what is burning. In fact, in his article Dr. Thornton writes, "The heat of the reaction, being all translational energy, is the result of the speed with which the oxygen atoms rush into combination, and is therefore proportional only to their number."

Dr. Thornton's findings were independently verified in the late 1970s by a researcher at the Center for Fire Research, National Bureau of Standards (now the National Institute of Standards and Technology). The researcher, Clayton Huggett, used a modern research tool, oxygen consumption calorimetry, to determine the exact energy release for a large number of ordinary combustibles. Huggett, in conducting his research, determined the heats of combustion per gram of oxygen for "typical organic liquids and gases," "typical synthetic polymers," and "selected natural fuels." What Huggett found is that the heat release rate for all materials he tested is 13.1 kilojoules per gram (kJ/g) of oxygen with a variance of only ±5 percent or less.

1. What are the implications of Dr. Thornton's statement that oxygen is necessary for carbon–hydrogen bonds to break?

   A. Without oxygen present, there can be no breakdown of the fuel, that is, no energy (heat) release.
   B. Oxygen in any form or concentration is adequate to cause combustion.
   C. Oxygen is a catalyst (a substance necessary for a reaction but not consumed in the reaction) and so does not get consumed in combustion.
   D. The heat of combustion will vary from one fuel to the next, regardless of the oxygen content.

2. Which of the following best characterizes the relationship between Clayton Huggett's research and Thornton's findings?

   A. While Huggett's research is interesting, it has no direct relationship to Thornton's primary findings.
   B. The primary point of agreement between Thornton and Huggett is that when things burn, they give off heat.
   C. Huggett quantified Thornton's findings to the point of establishing an energy release rate per gram of oxygen for common hydrocarbons.
   D. Neither Thornton's nor Huggett's article has any real bearing on heat release rates from real-life fires.

3. Based on Thorton's rule and Huggett's research, what assumption can now be made about heat release from fire?

A. More complex molecules that involve more hydrogen–carbon bonds will release more energy.

B. While the more complex molecules with more carbon–hydrogen bonds release more energy, they require proportionally more oxygen to release the energy.

C. Both A and B.

D. Neither A nor B.

4. Based on Huggett's findings that all materials release 13.1 kJ/g of oxygen, what conclusion can be drawn about heat release from a typical fire?

A. No conclusion can be drawn because there is no direct correlation between Thornton's and Huggett's work.

B. Huggett's research proves that heat is a by-product of combustion.

C. Greater energy (heat) will be released from fuels with more hydrogen–carbon bonds.

D. Regardless of fuel, the release of energy (heat) ±5 percent is dependent on one thing: the concentration of oxygen available for combustion.

Information for this case study came from: Clayton Huggett, "Estimation of Rate of Heat Release by Means of Oxygen Consumption Measurements," *Fire and Materials*, vol. 4, no. 2 (1980): 61–65; and W. M. Thornton, "The Relationship of Oxygen to the Heat of Combustion of Organic Compounds," *Philosophical Magazine Series 6*, vol. 33, no. 194 (1917): 196–203.

Review Questions

1. Define friction loss.

2. Define viscosity.

3. What is laminar flow?

4. What is turbulent flow?

5. While testing 2½-in hose with a flow of 150 gpm and a friction loss of 4.5 psi, you increase the flow to 300 gpm.

A. How can you determine the new friction loss without using the friction loss formula?

B. What is the new friction loss?

6. Assume hose size does not vary according to pressure. Why is friction loss not affected by pressure?

7. What is the formula for calculating friction loss for hose other than 2½-in hose?

Activities

1. What is the friction loss for 200 gpm in 2-in, rubber-lined hose?

2. While testing a sample of hose to verify the conversion factor, you determine the hose has a friction loss of 18 psi at 300 gpm.

   A. What is the conversion factor?
   B. What size is the hose?

3. Find the length of 1¾-in hose that is equivalent to 500 ft of 2½-in hose Figure 6-6 .

4. If 3-in hose has a friction loss of 20 psi for a given flow, how much friction loss will 3½-in hose have for the same flow?

5. Calculate a friction loss conversion factor to compare the friction loss in 4-in hose to friction loss in 3-in hose.

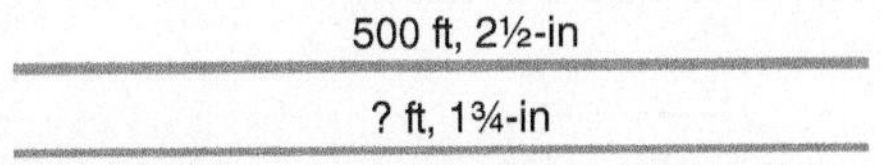

500 ft, 2½-in
? ft, 1¾-in

Figure 6-6 Find the equivalent length of 1¾-in hose.

6. Calculate the amount of water flowing in 3½-in hose if it has a friction loss of 45 psi.

7. Return to Example 6-7 and prove that 231.9 ft of 1¾-in hose is equivalent to 150 ft of 1½-in hose. The flow is 125 gpm.

8. Calculate a conversion factor for a parallel layout of one 3-in hose and another 3½-in hose.

Challenging Questions

1. Using Law 3, how much less friction loss will you have if the hose size is doubled but the flow stays the same?

2. What is the friction loss for 150 gpm in 1¾-in hose?

3. Return to Review Question 5 and use the friction loss formula to verify your answer to part B of that question.

4. You have been given the task of testing new hose for your department. During the test you have 500 ft of supply line laid from the hydrant. The discharge pressure at the pumper on the hydrant is 125 psi, and the intake at the second pumper is 89 psi. During the test,

the second pumper is flowing a 1½-in wagon pipe ($C = 0.997$) at 80 psi tip pressure.

   A. What is the CF?
   B. What size hose is it?

5. During your testing of the hose in the previous question, you are asked to consider the possibility of changing to 2-in hose. If you replaced your 300-ft, 1¾-in attack line and ran a test with 2-in hose at the same flow, how much more 2-in hose could you carry in place of the 1¾-in hose?

6. As you test the hose, the question of using 1¾-in hose in place of 2½-in hose is investigated. You know that some gpm flows in

2½-in hose have friction losses as high as 20 and 25 psi. How much friction loss would there be in the 1¾-in hose to compare with 20 psi in the 2½-in hose?

7. After calculating the friction loss in Question 6, you have decided that the friction in the 1¾-in hose is prohibitive. What was the gpm?

8. What is the maximum flow in the 1¾-in hose?

Formulas

To find the friction loss multiplier:

$$F_m = \left(\frac{v_2}{v_1}\right)^2$$

To find the conversion factor:

$$CF = \frac{D_1^5}{D_2^5}$$

To find the friction loss per 100 ft of 2½-in hose:

$$FL\ 100 = 2Q^2$$

To find the friction loss per 100 ft of hose other than 2½-in hose:

$$FL\ 100 = CF \times 2Q^2$$

To find the conversion factor when the friction loss and quantity of flow are known:

$$CF = \frac{FL\ 100}{2Q^2}$$

To use the abbreviated friction loss formula:

$$FL\ 100 = CF \times Q^2$$

To use the equivalent-length formula:

$$L_2 = L_1 \times \left(\frac{CF_1}{CF_2}\right)$$

To use the equivalent friction loss formula:

$$FL_2\ 100 = FL_1\ 100 \times \left(\frac{CF_2}{CF_1}\right)$$

To find the quantity flow when the friction loss and conversion factor are known:

$$Q = \sqrt{\frac{FL\ 100}{CF \times 2}}$$

Pump Theory and Operation

LEARNING OBJECTIVES

Upon completion of this chapter, you should be able to:

- Identify the types of pumps familiar to the fire service.
- Understand the operation of positive displacement pumps.
- Understand the operation of nonpositive displacement pumps.
- Identify the parts and function of nonpositive displacement pumps.
- Identify the operational difference between multistage and single-stage pumps.
- Explain the steps involved in the efficient operation of the pump.

You and the other members of Engine 11 have now been on the scene for about 1½ hours. The two-story building has been completely gutted, and all hand lines near the base of the building have been pulled back and shut down. Right now you are supplying a ladder tower and a monitor that sits on the roof of the building next door. Your primary responsibility is to monitor the pump, a two-stage radial flow pump, and be alert for anything that might be a problem.

1. Why is the radial flow pump so popular within the fire service?
2. What are the limits of the radial flow pump?
3. How is a radial flow pump made to draft?

Introduction

No course on hydraulics is complete without a discussion of pump theory. The pump generates the pressure (energy) we have been talking about up to this point. In this chapter, the operation of pumps is examined in detail. Differences between various types of pumps are examined, and the advantages and disadvantages of each are explored.

Pumps can be generally divided into two different types: positive displacement and nonpositive displacement. Each has its role in today's fire service, but they are generally not interchangeable. In addition to these two larger types, positive and nonpositive displacement pumps, each can be further divided into subtypes of pumps.

The method of operation of all positive displacement pumps is similar. However, subtypes of nonpositive displacement pumps have different means of creating pressure.

Because the nonpositive displacement pump is the primary pump used in the fire service today, the inner workings of both single-stage and multistage pumps are explained. Specific terminology concerning the naming of pump components and how these components are essential to the creation of pressure is discussed in detail.

In conclusion, the basic steps of pump operation are addressed.

A Brief History of Fire Pumps

The first fire pumps were hand-operated units that were drawn to the fire by the firefighters. Once at the fire, the firefighters who had just pulled the unit to the fire then became the motive power to operate the pump. As few as 2 or as many as 15 firefighters would line up on each side of the pumper and alternately pump, discharging water to a hose or monitor device. While some of the firefighters pumped, others would form an old-fashioned bucket brigade to fill the water tub on some models. Eventually, the idea of using horses to pull the fire engine caught on, and the apparatus became horse-drawn. In the early 1800s, the motive power began to switch from human to steam **Figure 7-1**. By the mid-1800s, after the invention of the gasoline engine, gasoline was first used to supply power to operate pumps that were still drawn to the fire by horses. Eventually as gasoline engines improved, they replaced the horses, and the rest, as they say, is history.

The early fire engines used piston pumps, the leading edge of technology at the time. However, as time passed and apparatus eventually became fully mechanized (run exclusively by gasoline engines), rotary gear pumps began to show up on fire engines. And still later, rotary gear and piston pumps gave way to what are known as centrifugal pumps, which

Figure 7-1 An American LaFrance steam pumper delivered to the Harrisonburg, Virginia, Fire Department in 1911.
Courtesy of City of Harrisonburg Virginia Fire Department.

of only 300 or 400 gallons per minute (gpm) were not unusual. Today it is common to find pumps that have a capacity of as great as 2,000 gpm or more.

Types of Pumps

Pumps are divided into two primary categories: positive and nonpositive displacement pumps. Each has specific advantages and disadvantages **Table 7-1**. For that reason they both fill a specific niche in the fire service today. To understand how each category of pump serves the needs of the fire service, it is important that the operating principles of each be thoroughly understood.

Positive Displacement Pumps

The **positive displacement pump** operates on the principle of discharging a fixed quantity of fluid with each pump cycle. The quantity discharged per cycle is the same, regardless of how fast or slow the pump is operating. The faster the pump operates, the larger the total volume of fluid discharged. Since these pumps will always discharge the same volume of fluid with each cycle and since water is, for practical purposes, incompressible, it is necessary to have some sort of pressure-regulating device to prevent overpressurizing. If the pump is operating and the discharge is closed off, without a pressure-regulating

should more accurately be referred to as radial flow pumps. It is impossible to identify exactly when the transition from piston and rotary pumps to centrifugal pumps took place. As early as 1912, Seagrave was using centrifugal pumps almost exclusively, and Ahrens-Fox did not stop producing piston pumps until 1952. Today, except for special applications, the centrifugal pump is the pump of choice for the fire service.

It is interesting to note that on some of the early mechanized fire apparatus, pumps with a capacity

Table 7-1 Comparison of Pump Types				
Type of Pump	**Operating Principle**	**Examples**	**Advantages**	**Disadvantages**
Positive displacement	Discharges a fixed volume with each cycle	Piston pump	Can pump air Takes advantage of incoming pressure	Needs pressure manifold to reduce pulsation Slippage can occur
		Rotary vein pump Rotary gear pump Rotary lobe pump	Can pump air	Slippage can occur Cannot take advantage of incoming pressure
Nonpositive displacement	Volume of discharge is dependent on resistance to movement of liquid	Radial flow pump Mixed flow pump Axial flow pump	Takes advantage of incoming pressure	Cannot prime without assistance

device, either something will break or at a minimum the drive motor will stall. For this reason, it is necessary to always have an open discharge when operating a positive displacement pump.

The amount of fluid discharged per cycle is determined by the volume of the pump itself. Factored into this is the loss of volume due to slippage. **Slippage** is the tendency of water to slip past the pump mechanism. Fluid lost by slippage will either go back to the intake side of the pump or be lost, depending on the type of pump. This effectively reduces the capacity of the pump and must be taken into account in designing positive displacement pumps.

Most positive displacement pumps are also incapable of taking advantage of any incoming pressure. The only pressure that is discharged from one of these pumps is what the pump itself generates. It is also impossible for water to flow through rotary-type positive displacement pumps if the pump is not engaged.

There are also some advantages to positive displacement pumps. They have the ability to develop higher pressure than their nonpositive displacement counterparts. In fact, where high pressures are needed for special applications, positive displacement pumps are most often used. Additionally, because there is no continuous, unobstructed pathway through the pump, they also have the ability to pump air. This

is critically important when drafting. We will learn more about this in Chapter 8.

The positive displacement pump is not just a single kind of pump, but a class of pumps. Positive displacement pumps can be piston pumps, rotary gear pumps, rotary lobe, or rotary vane pumps. They all operate by the same general principle of discharging a fixed amount of fluid with each cycle.

Piston Pumps

The piston pump is the oldest type of pump. Its principles of operation are very simple and readily illustrate the general operation principles of positive displacement pumps. Essentially, the piston pump operates much as a two-cycle piston engine does. On the intake stroke, the piston draws fluid into the cylinder. Then on the discharge stroke the fluid is discharged.

If you examine **Figure 7-2** closely, it is easy to understand in greater detail how the pump works. As the piston retreats from the top of the cylinder, the piston draws in a fixed volume of fluid, causing the intake valve to open and the discharge valve to close. After the piston has retreated enough to allow the cylinder to completely fill with fluid, it changes direction and begins to push the fluid out. This movement

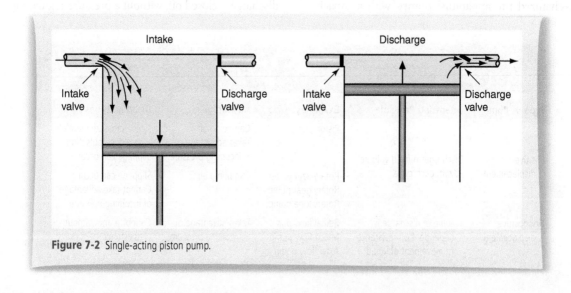

Figure 7-2 Single-acting piston pump.

causes the intake valve to close and the discharge valve to open, allowing the fluid to discharge.

Previously we said that most positive displacement pumps cannot take advantage of incoming pressure or allow water to run through the pump unless the pump is operating. The piston pump is the exception to both these conditions. The valves on the piston pump, which are only held in the closed position by springs, are activated by a differential in pressure. At draft, the pressure on the piston side of the intake valve is reduced as the piston retreats, allowing the atmospheric pressure to push the valve open and admit water and at the same time close the discharge valve. As the piston advances, the pressure on the piston side of the intake valve is greater, and the piston closes the intake valve and opens the discharge valve. If the water on the intake side of the pump has any pressure on it, such as from a fire hydrant, the pressure of the water coming in is sufficient to open the intake valve and admit water into the piston, even if the pump is not operating. Once the piston fills with water, the pressure opens the discharge valve and discharges water to the discharge manifold. When the pump is operating with a positive intake pressure, this condition is met each time the piston retreats. As the piston begins to advance, the intake pressure is already present so that any pressure the piston generates is in addition to the intake pressure.

Piston pumps can be built as single-acting pumps or double-acting pumps. The pump illustrated in Figure 7-2 is a single-acting pump. That is, it pumps fluid on only one stroke of the piston. But if a second set of valves were placed at the other end of the cylinder, a double-acting pump would be created Figure 7-3 . A single cylinder could then discharge almost twice as much fluid in a single pump cycle.

Because the piston pump discharges fluid on only one-half of the pump cycle, the fluid comes out in surges, similar to how water discharges from a water pistol Figure 7-4 . For firefighting this action is unacceptable. To remedy this, piston pumps have used a combination of solutions. First, there were multiple cylinders. To reduce the surging, the cylinders all discharged at different times, equally spaced apart; the more cylinders, the less the effect of the

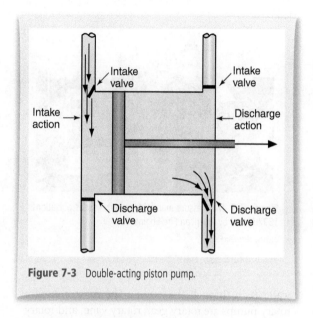

Figure 7-3   Double-acting piston pump.

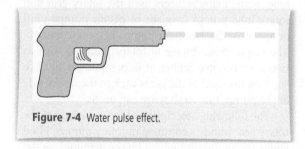

Figure 7-4   Water pulse effect.

surges. However, the surges were not completely eliminated. Another strategy to reduce the effects of the surge is to discharge all the fluid into a common discharge manifold. There the surges are further dampened before the fluid is finally discharged. This surge-dampening manifold is the shiny brass or chrome pressure dome evident on so many old pumpers Figure 7-5 .

Rotary Pumps

Piston pumps are no longer common in the fire service, but a close cousin—the rotary pump— is very much in use. The rotary pump, like the piston pump, is a positive displacement pump. It discharges a fixed volume of fluid with each cycle, and there is no clear

Figure 7-5 The pressure manifold (arrow) reduces pulsations. 1897 Model by American Fire Engine Co.

Courtesy of William F. Crapo.

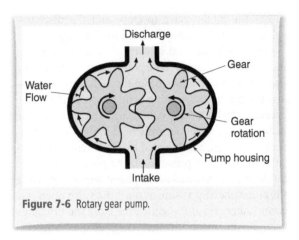

Figure 7-6 Rotary gear pump.

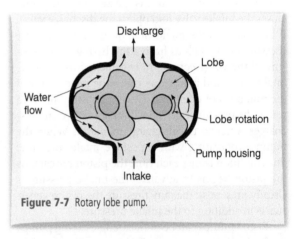

Figure 7-7 Rotary lobe pump.

water path through the pump. The most common rotary pumps are rotary gear, rotary vane, and rotary lobe. Some manufacturers used the rotary gear and rotary lobe pumps in place of piston pumps until nonpositive displacement pumps became common. The single disadvantage of rotary pumps is that as pressures become higher, it is possible for fluid to slip past the sides of the gears back to the intake side of the pump; that is, slippage can occur.

The following description of the operation of the rotary gear pump is common in rotary pumps: Fluid enters the pump through an intake manifold, as depicted in **Figure 7-6**, and is then captured by the gears of the pump. The gear teeth, which are turning away from the center of the pump at the intake, pick up a fixed volume of fluid between them. As the teeth approach the discharge, they are turning toward the center of the pump and the teeth engage, forcing the fluid out the discharge. This same mechanism is how the rotary lobe pump operates, only with two or three lobes instead of multiple gear teeth **Figure 7-7**.

Rotary vane pumps operate much as the rotary gear pumps do, except instead of the fluid being forced out by gears meshing, the fluid is forced out as the vanes retract. The vanes retract because the hub with the vanes is placed **eccentric**, or off-center, to the pump casing. As the hub rotates, the vanes

extend, trapping fluid and carrying it to the discharge side of the pump. On the discharge side, the vanes retract and the fluid is forced out the discharge by the next volume of fluid being brought around by the vanes, as illustrated in **Figure 7-8**.

The rotary pumps still in use in the fire service are used primarily as priming pumps. Because they are positive displacement, they can pump air, an impossible feat for nonpositive displacement pumps.

Nonpositive Displacement Pumps

The second type of pump, the nonpositive displacement pump, is the type primarily used in the fire service today. It is used not only on fire apparatus, but also where pumps are needed to provide pressure for

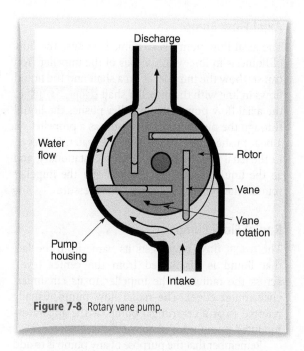

Figure 7-8 Rotary vane pump.

water-based fire protection systems. In short, wherever pumps are mentioned in the fire service, 99.9 percent of the time it is nonpositive displacement pumps being discussed.

Unlike their positive displacement counterparts, **nonpositive displacement pumps** do not pump a fixed volume of liquid with each cycle of operation. Instead, the volume of liquid discharged is dependent on the resistance offered to the movement of the liquid. The pump exerts a force on the liquid that is constant for any given speed of the pump. If the force of resistance equals the force being created by the pump, the liquid will reach a state of equilibrium and the liquid will not flow. The pump will then churn the liquid, and heat will be generated. However, if the force being generated is greater than the force of resistance, the water flows.

The most common nonpositive displacement pumps are referred to as *centrifugal pumps*, a name given to them a long time ago when it was thought that there was a force called *centrifugal force*. We now know that centrifugal force does not exist, but the term is still used. Later the section "Radial Flow

Pump" explains how the nonpositive displacement pump imparts energy to water.

Just as there were several types of positive displacement pump, several types of nonpositive displacement pumps are broadly classified as kinetic energy pumps. In the fire service, we use only the subcategory of kinetic energy pumps classified as centrifugal pumps. These centrifugal pumps come in three types: radial flow, mixed flow, and axial flow. Of the three types, the radial flow pump is the workhorse of the fire service; however, the other pumps deserve mention.

All nonpositive displacement centrifugal pumps have one thing in common: the part of the pump that actually imparts the energy to the liquid is called the **impeller** Figure 7-9 . The impellers for each type of centrifugal pump are different because they use different means of imparting energy to the liquid. If the three types of centrifugal pumps were placed on a continuum, the radial flow pump would come first, operating at the lowest revolutions per minute (rpm), and the axial flow pump would be on the other end, operating at the highest rpm. The mixed flow pump would be in the middle.

Also unlike their positive displacement counterparts, nonpositive displacement pumps do not have any valves or other restriction in the pumps that

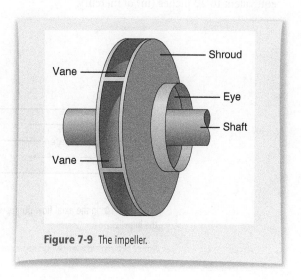

Figure 7-9 The impeller.

would prevent the free flow of liquid. It is possible to open an intake and discharge on a nonpositive displacement pump and let liquid flow through the pumps without the pumps running, enabling the nonpositive displacement pump to take advantage of the intake pressure. You can even **backflush** a nonpositive displacement pump by pumping water into the discharge of a pump and letting it come out the intake. Backflushing is commonly used to clean out pumps after drafting.

The primary disadvantage of the nonpositive displacement pump is that it cannot pump air. This disadvantage is serious where liquid has to be drafted (drawn) from a static supply, such as a pond or river. Once all the air is exhausted from the pump, the nonpositive displacement pump can pump liquid, even from a draft source, as long as air does not reenter the system. To overcome this inability to pump air, nonpositive displacement pumps that are required to draft are fitted with a small positive displacement pump for priming the main pump. The sole purpose of this priming pump is to exhaust air from the main pump when operating from a draft. The National Fire Protection Association (NFPA) has a standard that addresses this need. NFPA 1901, *Standard for Automotive Fire Apparatus*, requires that all apparatus-mounted fire pumps include a permanently mounted priming pump capable of developing a vacuum equivalent to 22 inches (in) of mercury.

Axial Flow Pump

The axial flow pump is so named because the flow of liquid is in line with the axis of the impeller. You can see how the impeller is on a shaft and the liquid flows in line with the impeller shaft Figure 7-10 . In the axial flow pump, the impeller pushes the liquid through the pump, much the same as a propeller on a boat pushes the boat through the water. However, in the axial flow pump, the pump is stationary, and so the liquid has to move. The faster the impeller rotates, the greater the flow and/or pressure.

Radial Flow Pump

The **radial flow pump** gets its name from the fact that liquid is discharged from the center (eye), across the radius of the impeller to its circumference (outer edge). The radial flow pump, usually referred to as a centrifugal pump, is the apparatus-mounted pump most often used today.

Remember that the purpose of any pump is to add energy to the fluid it is pumping. That energy is in the form of pressure, velocity (volume), or some combination of both. The moving part of the pump is called the *impeller*. Liquid enters the impeller at the eye and flies to the outer circumference of the impeller as it rotates. This impeller is enclosed in a pump housing that has a volute shape with the impeller placed eccentric (off-center) to the pump housing Figure 7-11 . The

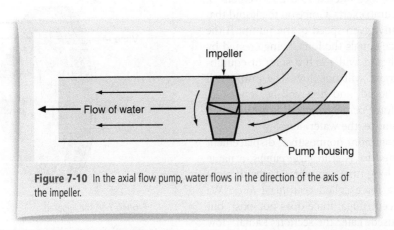

Figure 7-10 In the axial flow pump, water flows in the direction of the axis of the impeller.

volute serves two purposes. First, it accommodates an increasing volume of liquid as the impeller rotates from point *A* in Figure 7-12 to point *B*. Second, after the impeller has imparted velocity to the liquid, the volute confines the accelerated liquid, converting velocity to pressure. Impellers come in one of two types: single-suction `Figure 7-12` or double-suction `Figure 7-13`.

Between the walls of the impeller there are structures in the impeller called *vanes*. The purpose of the vanes is to maintain an orderly flow of liquid through the impeller by preventing eddying of the water. **Eddying** is water running contrary to the direction of the main flow, creating friction loss. The vanes also allow partial conversion of velocity to pressure. Note that the vanes are curved away from the direction of rotation of the impeller so as to allow a natural movement of liquid through the impeller.

The nonpositive displacement pump operates by allowing liquid to enter the pump by the force of an external pressure. That pressure comes from either

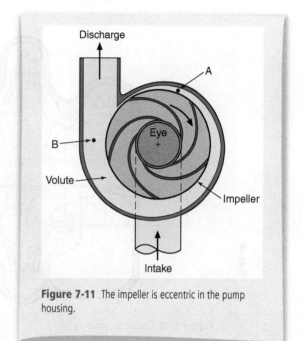

Figure 7-11 The impeller is eccentric in the pump housing.

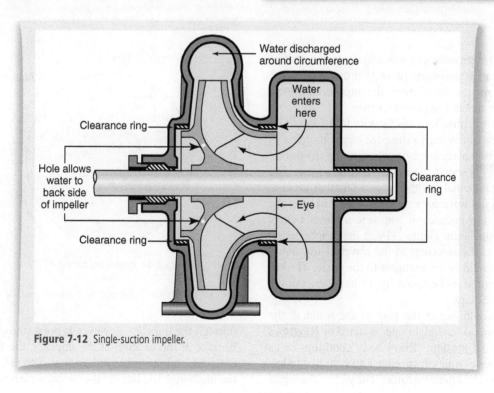

Figure 7-12 Single-suction impeller.

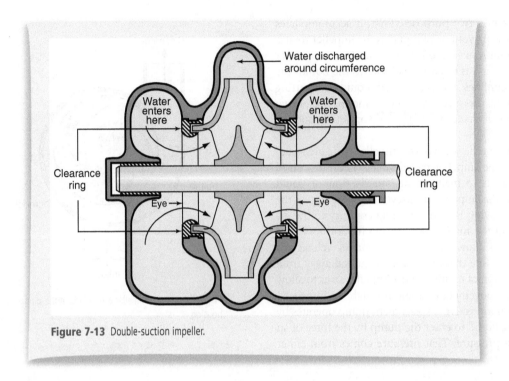

Figure 7-13 Double-suction impeller.

the positive pressure of a municipal water system or the force of atmospheric pressure if drafting. As the impeller rotates, liquid enters the impeller at the eye where an area of low pressure exists.

Next, the impeller adds velocity to the liquid by "throwing" the water to the edge of the impeller in the direction of the force being created by the pump. **Figure 7-14** shows the direction of that force. The force is tangential (a **tangent** is a line that touches but does not intersect a circle) to the rotation of the impeller. For instance, if you were to draw a circle on the impeller in Figure 7-14 so that it intersected droplet A, the direction of the travel of the liquid droplet would be on a tangent to the circle. The energy imparted to the liquid by the impeller is in the form of velocity.

We would expect the path of the water in the impeller to take a straight line, as stated by **Newton's first law of motion**: "Every body continues in its state of rest or uniform speed in a straight line unless acted upon by a nonzero force." The path of the liquid

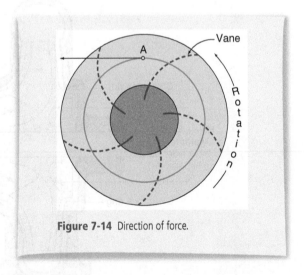

Figure 7-14 Direction of force.

through the impeller, however, is not a straight line. Because of the rotation of the impeller, the volume of liquid, and the fact that the volute directs water to the discharge, the liquid takes a spiral path through the impeller. Because the outer edge of the impeller is traveling at a much faster rate than the edge of the inlet, additional velocity is imparted to the liquid as it approaches the outer edge. This results in a tangential velocity increase because the radius of the spiral increases as the liquid passes through the impeller. In short, liquid passing through an impeller changes direction numerous times, each time in the direction of the force, resulting in the liquid spiraling through the impeller and gaining velocity Figure 7-15.

Finally, after the liquid has been accelerated (energy added), the volute of the pump confines the liquid. By confining the liquid that has just been accelerated, the velocity of the water is reduced with a corresponding increase in pressure.

In most instances dealing with pumps, energy simultaneously exists as pressure and velocity. However, because pressure is the more easily measured energy form, and the one that needs to be compensated for because of the friction loss created by velocity, we usually refer to pressure only. Be aware, however, that there is velocity energy, even if only a minimal amount, anytime water is flowing.

Thus far we have referred to the pressure added by the pumps as energy. Before any action can take place, that is, before liquid can be moved through a distance, energy must be transformed into work. Because work is defined as force moving an object through some distance, the **work done by the pump** is a combination of the pressure being developed and how much liquid is being discharged. An expression for the work done by the pumps would look like this: 400 gpm at 150 pounds per square in (psi).

Since the nonpositive displacement pump is capable of taking advantage of incoming pressure, where there is a positive intake pressure, the pumps do not work as hard. For example, if the pump is hooked up to a hydrant that has a residual pressure of 50 psi and the desired discharge pressure is 150 psi, the pump only needs to work hard enough to develop 100 psi of discharge pressure. As the intake pressure goes up in relation to the discharge

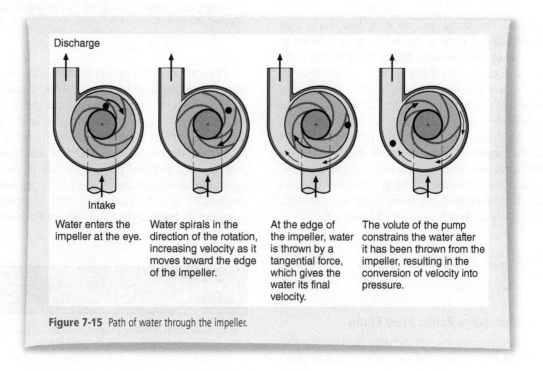

Discharge

Intake

Water enters the impeller at the eye.

Water spirals in the direction of the rotation, increasing velocity as it moves toward the edge of the impeller.

At the edge of the impeller, water is thrown by a tangential force, which gives the water its final velocity.

The volute of the pump constrains the water after it has been thrown from the impeller, resulting in the conversion of velocity into pressure.

Figure 7-15 Path of water through the impeller.

Fireground Fact

Bernoulli's Principle Applied

According to Bernoulli's principle, energy can change form but cannot be created or destroyed. The method by which the radial flow impeller imparts energy to water is an example of the application of Bernoulli's principle. The rotating impeller, taking energy from the driving force of the engine, transfers that energy to the water in the form of velocity by throwing the water toward the circumference of the impeller into the volute. When the water under velocity gets to the volute, it is confined. By confining the liquid at this point, velocity is reduced, with a corresponding increase in pressure, just as Bernoulli's principle would predict. (See Chapter 3 for an earlier discussion of Bernoulli's principle.)

pressure, the pump will do even less work. This pressure, or work done by the pump, is referred to as the net pump pressure. **Net pump pressure** is the pressure generated by the pump. From a hydrant or another positive pressure source, this can be expressed mathematically by this formula: discharge pressure − intake pressure = net pump pressure. The **total pump pressure** is the pressure indicated on the master discharge gauge of the pump. From draft not only must the pumps do enough work to account for all the discharge pressure, but also they have to do work necessary to get water into the pump. We discuss this further in Chapter 8.

> Discharge pressure − intake pressure = net pump pressure

The Multistage Radial Flow Pump

The multistage pump gets its name from the fact that it has more than one impeller, where each impeller

is a separate stage. There are two different configurations of a two-stage radial flow pump **Figure 7-16**. In Figure 7-16A, liquid enters the first stage from the intake, and pressure is added by the first-stage impeller and is then discharged to the second stage. The liquid is further pressurized by the second-stage impeller before being discharged.

When a multistage pump is configured to route liquid first through one impeller and then the other impeller, it is capable of obtaining its highest operational pressure. This arrangement is called the *series* or *pressure* configuration. To better remember how the two-stage pump is configured in series, remember that in series, each impeller pumps all the liquid and generates one-half of the added pressure. The disadvantage to the series position is that it limits the pump to a maximum of about 70 percent of its rated capacity.

Note

In series, each impeller pumps all the liquid and generates one-half of the added pressure.

The second configuration possible with the two-stage pump, depicted in Figure 7-16B, is called the *parallel* or *volume* configuration. In the parallel configuration, water enters each impeller at the same time directly from the intake manifold and is then discharged directly to the discharge manifold. In parallel, each impeller pumps one-half of the total liquid but must generate all the added pressure. In the parallel configuration the pump is capable of pumping its rated capacity, but cannot do so at high pressure.

Note

In parallel, each impeller pumps one-half of the total liquid but must generate all the added pressure.

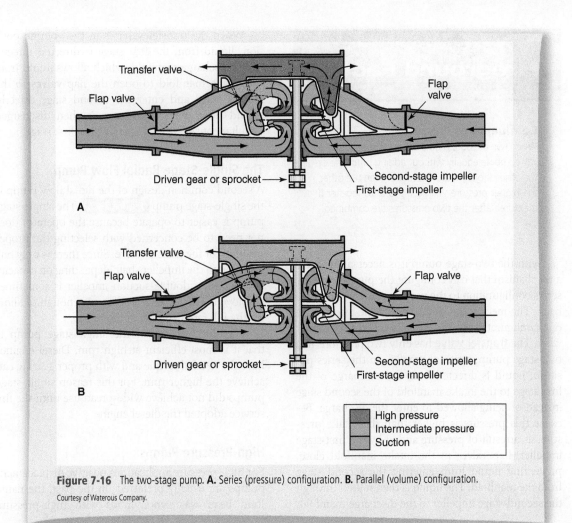

Figure 7-16 The two-stage pump. **A.** Series (pressure) configuration. **B.** Parallel (volume) configuration.
Courtesy of Waterous Company.

Example 7-1

How much pressure will each impeller of a two-stage pump develop in series position while operating from a hydrant, if the residual intake pressure is 60 psi and the discharge pressure is 180 psi?

Answer

To determine how much pressure each impeller will develop, start by subtracting the intake pressure from the discharge pressure. This figure is the total work done by the pump, that is, the net pump pressure:

Discharge pressure − intake pressure = net pump pressure

$$180 \text{ psi} - 60 \text{ psi} = 120 \text{ psi}$$

Now divide the pressure generated by the pump by the number of stages, in this case 2.

$$120 \text{ psi} \div 2 = 60 \text{ psi}$$

Each impeller (stage) will develop 60 psi of pressure.

Fireground Fact

Confluence of Pressures

When two pressures come together, the pressures will combine equally without adding onto one or the other. If one pressure is higher than the other, the higher pressure will be the one to register if measured after the two pressures are combined.

With the two-stage pump it is necessary to have a mechanism that can configure the pump from the series configuration to the parallel configuration and back. The mechanism that transfers the pump from one configuration to the other is called the *transfer valve*. The **transfer valve** has only two positions in a two-stage pump, series or parallel. In the series position, liquid is directed from the discharge of the first stage to the intake manifold of the second stage instead of being allowed to enter the discharge. Because this pressure is higher than the intake pressure, as a result of pressure added by the first-stage impeller, flap valves in the intake manifold close, preventing liquid from entering the second stage from the manifold. The liquid is then routed through the second-stage impeller to the discharge manifold.

When the transfer valve is in the volume position, liquid from the first stage is directed directly to the discharge manifold, which allows liquid from the intake manifold to open the flap valves to the second stage and enter the second stage directly. Liquid from the second stage is then discharged simultaneously with liquid from the first stage.

The Single-Stage Radial Flow Pump

A second common design of the radial flow pump is the single-stage pump **Figure 7-17**. The single-stage pump is easier to operate because the operator does not need to be concerned with selecting the proper position for the transfer valve. Since there is only one impeller on the impeller shaft, depending on capacity and design, a double-suction impeller is sometimes employed. The impeller itself is big enough to allow for the full rated volume of the pump.

The disadvantage of the single-stage pump is that it is most efficient at high rpm. Diesel engines have the needed torque and with proper gearing can achieve the higher rpm. For this reason single-stage pumps did not achieve widespread use until the fire service adopted the diesel engine.

High-Pressure Pumps

For high-pressure applications positive displacement pumps are usually preferred. In the past, the name John Bean was synonymous with high-pressure

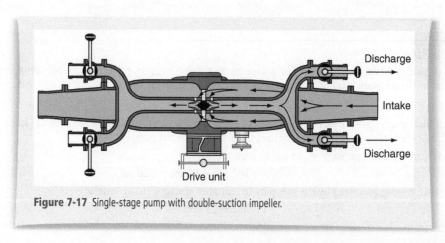

Figure 7-17 Single-stage pump with double-suction impeller.

fire pumps. These were piston pumps designed for low-volume, high-pressure applications. Today there is resurgence in the use of high-pressure pumps for fire suppression purposes. This is a result of European fire service influence and research done by the U.S. Air Force in the use of high-pressure water for fire suppression. Today's high-pressure fire pumps are frequently piston pumps, but one manufacturer makes a four-stage radial flow pump capable of producing more than 1,000 psi. Together these pumps are referred to as *ultra-high-pressure (UHP) pumps*.

End Thrust and Radial Hydraulic Balance

Two conditions associated with radial flow pumps should be mentioned. The first is a condition known as *end thrust*. **End thrust** is produced when the direction of flow of liquid is abruptly changed. As water enters the impeller, the force of the water pushes the impeller against the back wall of the pump housing before the water changes direction by 90°. This end thrust can harm the pump because it causes excessive wear if no compensation is provided. To minimize end thrust, clearance rings are provided at the rear of the impellers. Holes are also provided in the rear of the impeller, between the hub and the vanes, to admit water to the backside of the impeller to partially cushion the force of the water, as shown in Figure 7-12. These two details provide partial hydraulic balance to reduce the forces associated with end thrust. Any remaining force is then easily absorbed by the impeller support bearings.

Another method used to compensate for end thrust is seen in **Figure 7-18**. By placing the two impellers of a two-stage pump on the same shaft, back to back, the end thrust of the first stage is canceled by the end thrust of the second stage. A double-suction impeller is in hydraulic balance as a result of its design.

The second condition associated with the radial flow pump is an issue of radial hydraulic balance. **Radial hydraulic balance** can be defined as the equal discharge of liquid around the circumference of the impeller. Because the volute of the pump accumulates

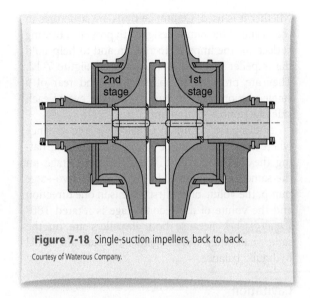

Figure 7-18 Single-suction impellers, back to back.
Courtesy of Waterous Company.

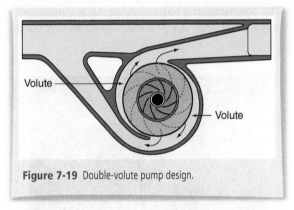

Figure 7-19 Double-volute pump design.

more liquid as the discharge is approached, the load on the impeller is not uniform throughout its circumference. The result is a lack of hydraulic balance around the circumference of the impeller, which can cause excessive and uneven wear on the clearance rings, called **bell mouthing**. To maintain radial hydraulic balance, one of two pump designs is used. First, a double-volute design is sometimes employed **Figure 7-19**. With this design, liquid distribution is mirrored at 180°, which allows the design to be in radial hydraulic balance. The double-volute design is generally limited to single-stage pumps, but even then the capacity and design of the pump determine whether it is used. <u>Clearance rings</u> are structures in the casing of the pump designed to provide a bearing surface for the impeller to ride on and to help hold the impeller in place, as illustrated in Figure 7-12. They are provided at both the eye and rear of a single-suction impeller and around the eye of each side of a double-suction impeller.

A design to counter radial hydraulic balance where there are multiple stages involves alternating the direction of the volute for each stage on the same impeller shaft. In the case of a two-stage pump, the volute of the first stage is in one direction and the volute of the second stage is rotated 180° **Figure 7-20**. Because both impellers are on the same shaft, they work together to maintain radial hydraulic balance.

Cavitation

The operation of a radial flow pump is fundamentally simple. Only a few things can go wrong, other

Figure 7-20 Cutaway of two-stage pump showing orientation of volute.

Courtesy of William F. Crapo.

Fireground Fact

Radial Hydraulic Balance
Figure 7-20 shows the volute rotated 180°. Note that the casing of the impeller on the right is taller than the casing of the impeller on the left. This is so because the larger volume of the volute of the right impeller is on the top, and the larger volume of the volute of the left impeller is on the bottom.

than mechanical breakdowns. The most serious operational mistake is to allow the pump to try to pump more water than it has available. Doing so causes a potentially serious condition known as *cavitation.*

<u>Cavitation</u> is the formation, and collapse, of vapor bubbles in the impeller of the pump. As the pressure on water is reduced, the temperature at which water boils is lowered. Under the right circumstances water can actually boil, creating water vapor, at ambient water temperature inside the impeller of a pump. The circumstances under which this can happen involve trying to make the pump supply more water than is available. This is a condition referred to as letting the pump "run away from the water."

Under normal operation, there is an area of low pressure at the eye of the impeller. Under the right conditions, such as at draft or during operation with a very low intake pressure, this low pressure might even be a vacuum. With sufficient pressure coming into the impeller, the pressure at the eye remains high enough to prevent the formation of vapor bubbles. However, if there is insufficient pressure coming in, such as when the pump is running away from the water, then the pressure can fall below the vapor pressure of the water for its temperature and vapor bubbles will form. These vapor bubbles then travel to the interior of the impeller, where they encounter higher pressures and instantly implode. It's these implosions that can cause significant damage.

Cavitation can occur anytime the pump is operating and the impeller(s) are spinning too fast for the intake pressure. Mild cavitation may hardly be noticeable, but severe cavitation can cause the pumps to shake and make a sound similar to marbles being sucked through the pumps. Any cavitation can cause pitting of the interior impeller surfaces. Severe cavitation, if not stopped when first heard, can cause the impeller to break up, damaging the pumps. Anytime cavitation is heard or even suspected, corrective steps should be taken immediately. The appropriate corrective step is to reduce the rpm of the pump.

Pump Capacity

Fire service pumps are designed to deliver their **rated capacity** at 150 psi discharge pressure. If pressures higher than 150 psi are needed, the amount of water being pumped is reduced. This example is a direct application of Bernoulli's principle of conservation of energy. We can have either a lot of water at a little pressure or a little water at a lot of pressure. The total energy in either evolution will be the same. **Table 7-2** illustrates the breakdown of the pump capacity versus maximum pressure. These figures apply to pumps at draft with maximum lift determined by the pump capacity. **Lift** is the distance between the surface of the water and the centerline of the pump intake. All pumps up to a 1,500 gpm capacity are required to meet their rated capacity at a maximum lift of 10 feet (ft). Pumps with a 1,750 gpm capacity must reach their rated capacity at a maximum lift of 8 ft. All pumps with a capacity of

Table 7-3	Pump Capacity at Maximum Lift
Pump Capacity, gpm	**Maximum Lift, ft**
Up to 1,500	10
1,750	8
2,000 and above	6

2,000 gpm or greater must reach their rated capacity at a maximum lift of 6 ft. **Table 7-3** summarizes pump capacity versus maximum lift.

Although the figures in Table 7-2 apply directly to draft situations, the principle itself applies to hydrant operations too. For hydrant operations, no absolute table of figures can be developed because the actual capacity of the pump varies according to the capacity of the water supply and the residual pressure of the hydrant. For calculating engine pressures and determining maximum lay and maximum flow problems, Table 7-2 can be used as a rule-of-thumb guide.

Note

Pumps are generally designed to deliver a maximum pressure of 300 psi. At this pressure, the maximum capacity is less than 50 percent.

Mixed Flow Pump

Between the axial flow pump and the radial flow pump on our continuum is a third type of nonpositive displacement pump called a *mixed flow pump* **Figure 7-21**. The mixed flow pump is a hybrid between the axial flow pump and the radial flow pump. It employs elements of both pumps to impart energy to water. It pushes the water through the pump to some extent, as its axial flow cousin does. It also throws the water out of the impeller to some extent, as its radial flow cousin does.

Table 7-2	Pump Capacity and Maximum Pressure	
Pressure, psi	**Capacity, Percent**	
0–150	Up to 100	
151–200	Up to 70	
201–250	Up to 50	

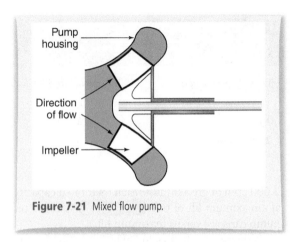

Figure 7-21 Mixed flow pump.

Figure 7-22 Typical pump panel gauges.
Courtesy of William F. Crapo.

Measuring Pressure and Flow

Measuring Pressure

Pressure Gauges

For the operator to know that the correct pressure is being produced by the pump, the pump panel must contain an instrument that reads pressure. To measure pressure, the pump panel will be equipped with not only master pressure gauges, for both intake and discharge, but also pressure gauges for each discharge 1½ in and larger Figure 7-22 . The most fundamental pressure measurement gauge incorporates what is known as a *Bourdon tube* Figure 7-23 .

A Bourdon tube is simply a hollow tube, bent in a partial circle. One end of the tube is closed, and the other end admits pressure into the tube. The tube itself is attached, by gears and levers, to a needle. As pressure is increased, the tube tries to straighten out, which causes the needle on the face of the gauge to rotate, thereby registering the amount of pressure being applied. Although the Bourdon-tube gauge is not a high-tech gadget, it is accurate, reliable, and 100 percent mechanical, and it requires little attention. Bourdon-tube gauges are attached to each discharge, on the discharge side of the gate valve, by thin tubing that allows water to run back to the gauge on the pump panel. The biggest drawback to this arrangement is that in the event of

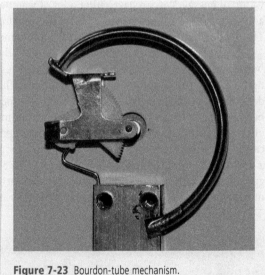

Figure 7-23 Bourdon-tube mechanism.
Courtesy of William F. Crapo.

freezing weather, these lines must be drained to prevent them from freezing and damaging the tubing. Bourdon-tube gauges often outlast the apparatus they were delivered on, without special attention or regular recalibration.

Modern fire apparatus are often equipped with electronic gauges. These gauges have pressure-sensing modules attached on the discharge side at each gate valve. The sensor sends a pressure reading back to the appropriate gauge on the pump panel via wires. This is an advantage over the Bourdon-tube gauge since there is no chance of freezing; however, if you lose electric power, you will lose all your gauges. This author once saw an entire pump panel, with the exception of the intake and master discharge gauges (which were standard Bourdon-tube gauges), go dead after a fuse blew. Electronic gauges may require periodic recalibration to keep them accurate. The good news is that they are easy to recalibrate.

All pump panels have two master gauges to measure pressure. One is the intake gauge, which measures not only positive pressure but also vacuum. Each pump panel also has a master discharge gauge, which easily measures the pressure generated by the pump. NFPA 1901 requires these master gauges to be at least 1 in in diameter larger than the individual gauges, and that the master gauges read a minimum of 300 psi. The master intake gauge must also be positioned either to the left of or beneath the master discharge gauge. The master gauges are the larger gauges at the top of Figure 7-22. We discuss the intake gauge further in Chapter 9.

In addition to master gauges, modern fire apparatus have individual pressure gauges for each discharge. Individual gauges enable the pump operator to regulate pressures to individual lines. Lines that require less pressure can be "gated back" (partially closed) while lines that require full pressure can be left fully open. Gating back may sound tricky, but with a little practice it is a skill that is easily mastered. Individual gauges must be located near the appropriate discharge. Note in Figure 7-22 that each discharge's relevant gauge is positioned either right next to or just above the discharge handle. Also in Figure 7-22, the left top discharge handle is all the way over to the left, indicating that it is fully open. The bottom left and bottom right handles are only about one-quarter of the way open; they have been gated back to prevent the allowance of too high a pressure. Note the pressures on the respective gauges.

Measuring Flow
Flow Meters

Another way to measure the work being done by the pump is to measure the amount of water flowing from each discharge, which can be done with a flow meter. A flow meter utilizes either a paddle wheel or a spring that intrudes into the waterway. The paddle wheel spins and the spring bends as water passes by. If properly calibrated and maintained, flow meters can be reasonably accurate.

Flow meters were first installed on fire apparatus to allow the operator to deliver the correct flow without needing to calculate the correct pressure. As long as the operator knew the amount of water that needed to flow, he or she simply advanced the throttle until the meter registered the specific flow. The problem with this method is that any given flow is dependent on pressure, and if you are flowing multiple lines, it may be necessary to gate back on one or more lines. Since flow and pressure are not directly related to each other, you cannot assume the line requiring the greatest flow will require the highest pressure; instead you must calculate the pressure to determine which line to gate back. Today, NFPA 1901 requires discharges that have flow meters to also have pressure gauges.

A second disadvantage of flow meters is that they demand a great deal of attention to maintain accurate readings. The author's experience with flow meters indicates they should be re-calibrated at least monthly, and with 6, 8, 10, or more gauges, calibrating can be time-consuming. Additionally, flow meters are electric. Because they require a power supply to function, if you lose electric power, you also lose all your gauges.

Combination Pressure Flow Gauge

To comply with NFPA 1901 and provide both pressure reading and flow metering for individual discharges, combination flow meter and pressure gauges were developed Figure 7-24 . This concept has some merit: the operator can pump the calculated pressure and verify it with the flow meter. The challenge with this method is that the flow meter

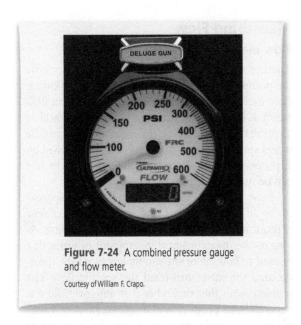

Figure 7-24 A combined pressure gauge and flow meter.

Courtesy of William F. Crapo.

still requires a great deal of attention to maintain calibration. These dual-purpose gauges are also electric and are subject to electrical failure whereby both flow and pressure readings are lost.

Pump Operations

Basic pump operations are very simple. They involve getting water into the pump, from either a positive pressure source or a static water source, and then discharging it at the needed pressure.

Pump operations can be broken down into four main steps: step 1, making pumps ready; step 2, getting water; step 3, pumping; and step 4, shutting down. By following these steps in this order, you will be assured of success. Because this is not a text on pump operations, not every conceivable scenario will be covered. This section is meant to be only a general operational guideline.

Step 1: Making Pumps Ready

This is the most important step in pump operations. Any pump that has not been made ready in advance will cost you valuable time at the incident and has

a higher probability of failing when needed. However, if you take the time to get the pump ready in advance, you arrive at the scene with confidence that everything is working properly and ready to be put into service without delay. Step 1 involves completing these tasks:

- Fill the water tank. This means filling 100 percent. Anything less is unacceptable and should not be tolerated.
- Open the pump-tank valve. Except for freezing weather, the pumps should always be run wet.
- Close and lock all other valves.
- Close all drain valves.
- Prime the pump. Along with the pumps being run wet, they should always be kept fully primed, ready to pump without delay.
- Set the discharge relief valve. By keeping the discharge relief valve set to the pressure required for your most frequently used preconnect line, you will eliminate a step of operation at the scene. (Computer-controlled pumps may not require this step.)
- Fill primer oil tank, if required. (Some modern pumpers have primers that do not require oil to operate, in which case this step can be omitted.)
- Ensure that all blind caps and plugs are no more than hand-tight.
- Open the pump gauge admission valves. Any pump equipped with gauge admission valves must have the admission valves open, or else the gauge will not register a pressure. They should be barely open—just enough for them to work.
- Keep two-stage pumps in the series (pressure) position. With two-stage pumps, operation in the parallel (volume) position is so infrequent that the series position should be considered standard.
- Ensure that apparatus with the potential for pumping from a hydrant have a 15- to 20-ft soft sleeve (4-in or larger large-diameter hose) preconnected at a gated intake, if possible.

In most instances, if step 4, shutting down, has been completed properly, step 1 will only be a matter of ensuring that the above steps have been completed. With these steps completed in advance, you will know the apparatus is ready and capable of working immediately upon arrival. No delay will be experienced in getting things ready to pump at the incident. Whenever possible, these items should be checked each time the driver changes. At a minimum, these items should be checked daily.

Step 2: Getting Water

When all items in step 1 have been verified beforehand, step 2 becomes a relatively minor task. It begins with the apparatus on the scene, brake set, and wheels chocked.

Tip

Whenever valves are opened or closed, the operation of the valve should be smooth, deliberate, and not sudden. Valves must never be jerked open or slammed close. Sudden operation of any valve can cause an undesirable condition known as *water hammer*. Water hammer is a sudden buildup of pressure due to an abrupt change in flow. This buildup in pressure can come from a sudden increase in flow when a valve is suddenly opened, or stoppage of flow when a valve is abruptly closed. Either incident can result in a catastrophic failure of any part of the water supply system. Even water mains have been known to rupture as a result of water hammer.

Operating from Hydrant or Supply Line

- Put the pump in gear. If it is obvious upon arrival that the pump will be needed, put it in gear before leaving the cab.
- Charge the initial line from tank. If the initial line charged is not the one for which the relief valve was preset, it may be necessary to change the setting of the relief valve before getting the full pressure on the initial line.
- Check the relief valve setting. If you charged the line for which the relief valve was preset, this step can be eliminated. Otherwise, reset the relief valve for the required pressure.

- Connect your supply line and have it charged.
- Open the supply line and close the pump-to-tank valve. While you are making the changeover, have one hand on the throttle to throttle down as pressure is admitted from the supply line.
- Fill the tank. Once the supply line is charged, the pressure is stabilized, and the relief valve or pressure-regulating device is set, refill the tank without delay. This will give you a minimum of a full tank of water in case you lose your water supply.

Tip

When you are driving the first-in pumper at a reported structure fire, even if there is no fire showing, it is a good idea to put the pumps in gear before leaving the cab. In the short time it will take to confirm either the presence or the absence of a fire, the pump will not overheat as long as it remains idle. If there is any concern about overheating, open the pump-to-tank valves and let the water circulate through the tank. If fire is discovered, you will not have to take time to enter the cab to put the pump in gear. This will result in a more efficient operation.

Operating from Draft

The steps of operating a pumper from draft are very similar to those of operating from a positive source. The main difference lies in getting water into the pumps. It begins with the apparatus positioned at the water source, brake set and wheels chocked.

Tip

When you are positioning the apparatus to draft, try to make the connection from a side steamer for a midship-mounted pump. Front and rear steamer connections usually have lots of friction loss, because of the added piping, and lots of bends to get from the front or rear to the pump. This will reduce how much water you can supply.

- Close the pump-to-tank valve. (Apparatus used exclusively for drafting may already have this step done.)
- Remove the blind cap from the steamer connection and attach hard sleeves with strainer. The steamer connection is the large, main intake of the pump, usually 4½ in or larger.
- Lower the sleeves and strainer into the water source.
- Disengage relief valve or pressure-regulating device.
- Prime the pump.
- Charge the hose line. Once the pumps are primed and water is flowing, it should be possible to shut down for a short time without losing prime. If you will not be supplying water for more than a few minutes, open an unused discharge to keep water flowing.
- Set the relief valve or pressure-regulating device.
- Fill the water tank if needed.

At this point, you have water flowing at the correct pressure. The next step is to make sure nothing changes unexpectedly.

Step 3: Pumping

Pumping does not require a subset of steps, as seen in step 1 and step 2. The primary goal of step 3 is to make sure things continue as already set. During this step, the driver/operator must remain alert to any changes in the pump or engine that will adversely affect operations. Unless pressure has to be adjusted because lines have been added or shut down, or the flow has changed, this step is one of observation.

Of critical importance during step 3 is an awareness of the condition of the engine driving the pump.

First, the oil pressure and engine temperature must be monitored. The exact acceptable oil pressure and temperature will vary with each pumper. These are things the good driver/operator knows from past experience. New driver/operators should be asking what the acceptable oil pressure and temperature are for each piece they might drive. Also monitor the other engine function gauges: ammeter (or amp meter), fuel level, and transmission temperature gauge, if so equipped. A final, but critical, number the diligent operator must know is the rpm of the engine.

During the course of operation, as long as the discharge of the pump does not change, the engine rpm should not change. If the engine rpm does change without a known change in discharge, begin looking for the problem. Sometimes the answer might be as simple, and harmless, as a line shut down. Other times it can be as critical as a broken line, air leak on the intake side, or even worse, pump damage.

If additional lines are needed, make the necessary changes to the relief valve or pressure-regulating device setting and discharge pressure. Also observe the intake gauge to verify that you have adequate residual pressure.

Fireground Fact

What Is Adequate Residual Pressure?
When you are operating in a relay or two-pump operation receiving water from another pumper, 20 psi is generally considered the minimum acceptable pressure. However, if you are connected directly to a hydrant, a pressure of 5 psi can be considered the minimum. The difference here is due to the source of water. In the relay or two-pump operation, the source just before you is another pumper, and it can calculate 20 psi residual into its discharge pressure. But when you are operating from a hydrant, the goal is to get every last drop of water the hydrant can supply.

If you are pumping to multiple lines, take care to calculate the required pressure for each line. Gate back as needed to prevent overpressurizing lines. If you are pumping supply lines to other apparatus, pump all lines at the same pressure. We discuss this further in Chapter 10.

Finally, be alert for any strange noises, vibrations, or other nonspecific problems. This means asking yourself, What is my general impression about the condition of the pumper? For example, has the sound of the engine changed, indicating a possible problem? Is water leaking from places it should not? Do I have to keep adjusting the discharge pressure? The bottom line is to stay alert and pay attention to what is happening.

Tip

Going down as far as 5 psi residual when operating from a hydrant has an inherent danger. If another pumper begins taking water from the same main you are on, your safe 5 psi might easily become a dangerous 0 psi. Stay alert!

Finally, while you are pumping, always be aware of your water supply. If you lose even a little water, you may experience cavitation, which requires immediate corrective action. Also guard against letting the pump run dry. If pumps run dry, even for a moment, it is not a matter of whether damage has occurred but how much.

Step 4: Shutting Down

Finally, the fire has been extinguished, and you are ready to pick up and go home. Again, a few substeps make it possible to accomplish the shutdown in an efficient manner.

To shut down:

- Throttle down. This will immediately take pressure off all lines.
- Close all discharge gates and open drain valves.
- Take pumps out of gear.
- Close the intake. Or, if at draft, slowly open a discharge gate to allow air into pump to "drop" water.
- Make pumps ready per step 1.

Design credits: RedFlames: Drx/Dreamstime.com; Steel texture: © Sharpshot/Dreamstime.com; Chapter opener photo: Mike Brake/Shutterstock; Orange flames: © Jag_cz/ShutterStock, Inc.; Stacked photo background: © Vitaly Korovin/ShutterStock, Inc.; Paper: © silver-john/ShutterStock, Inc.; Case Study: © Reicaden/iStock; Tip texture: Eky Studio/ShutterStock, Inc.

WRAP-UP

Chapter Summary

- Pumps are divided into two categories: positive displacement and nonpositive displacement.
- Positive displacement pumps operate on the principle of discharging a fixed quantity of fluid with each cycle of the pump.
- Slippage is the tendency of water on the discharge side of the pump to slip back to the intake side.
- Positive displacement pumps include piston pumps, rotary gear pumps, and rotary vane pumps.
- Piston pumps are the only positive displacement pumps that are able to take advantage of incoming pressure.
- Nonpositive displacement pumps can be broadly classified as kinetic energy pumps.
- Kinetic energy pumps come in three types: radial flow, mixed flow, and axial flow.
- The volume of water discharged from a nonpositive displacement pump is dependent on the resistance of movement of the liquid.
- The moving part of the nonpositive displacement pump is called an impeller.
- When water is moving, energy inside a pump simultaneously exists as pressure and velocity.
- End thrust is caused when the direction of flow of water is abruptly changed.
- Radial hydraulic balance is the equal discharge of liquid around the circumference of the impeller(s).
- Cavitation is the formation and collapse of vapor bubbles in the impeller of the pump.
- The amount of work being done by the pump can be measured via pressure gauges and flow meters.
- A pump panel contains master pressure gauges for intake and discharge as well as pressure gauges for each discharge.
- Flow meters measure the amount of water flowing from each discharge.
- Pump operation can be broken down into four major steps of operation: making pumps ready, getting water, pumping, and shutting down.

Key Terms

backflush To pump water into the discharge of a pump and let it come out the intake.

bell mouthing Excessive and uneven wear on the clearance rings due to lack of hydraulic balance.

cavitation The formation of vapor (steam) bubbles in the impeller of the pump.

clearance rings Structures in the casing of the pump designed to provide a bearing surface for the impeller to ride on and help hold the impeller in place.

eccentric Off-center.

eddying Water running contrary to the direction of the main flow.

end thrust Thrust produced when the direction of flow of liquid is abruptly changed.

impeller The part of the nonpositive displacement pump that actually imparts energy to the liquid.

lift The distance between the surface of the water and the centerline of the pump intake.

net pump pressure The pressure generated by the pump.

Newton's first law of motion Every body continues in its state of rest or uniform speed in a straight line unless acted upon by a nonzero force.

nonpositive displacement pump A pump that does not pump a fixed volume of liquid with each cycle of operation.

positive displacement pump A pump that operates on the principle of discharging a fixed quantity of fluid with each cycle of the pump.

radial flow pump A pump in which the liquid is discharged from the center (eye), across the radius of the impeller to its circumference (outer edge).

radial hydraulic balance The equal discharge of liquid around the circumference of the impeller.

rated capacity The maximum amount of water a pump can deliver at 150 psi discharge pressure lift.

slippage The tendency of water on the discharge side of the pump to slip back to the intake side.

tangent A line that touches but does not intersect a circle.

total pump pressure The pressure on the master discharge gauge of the pump.

transfer valve The mechanism that transfers the pump from parallel configuration to series configuration.

water hammer A sudden buildup of pressure due to an abrupt change in flow.

work done by the pump A combination of net pump pressure and how much liquid is being discharged.

Case Study

Training Mishap

One afternoon you have your crew out training on master stream devices. After you have discussed how to set up a monitor nozzle you instruct the crew to set it up with a 1½-in tip. Once the monitor is set up properly, the rookie takes the 1½-in tip to place on the monitor. It takes him a few moments to do so, as he appears to be having trouble getting the tip on, but finally he returns to the pumper. Mission accomplished, or so you think.

Since the monitor is solidly anchored down, you have your crew at the pumper review pump calculations and procedures for pumping to the monitor. After a couple of minutes discussing flow, friction loss, nozzle pressure, allowance for the device, and elevation, you ask the rookie to charge the monitor nozzle. Once the lines are charged using hydrant pressure, the rookie begins to slowly advance the throttle until the correct pressure is obtained.

Less than a minute after the rookie has reached the correct engine pressure, water begins spraying from the base of the 1½-in tip, and then suddenly the tip flies off of the monitor, out into the middle of the lake into which you are pumping. Your only comment is, "Probie, what have you done this time?"

Further investigation reveals that the threads of the tube of the stream straightener were damaged and failed after the monitor was charged, allowing the tip to fly off owing to the force of the water. You use this opportunity to examine the consequences of the loss of the tip. Since the pump's automatic pressure-regulating device had been turned off for the training exercise, it allows you to ask your entire crew the following questions.

1. When the tip flew off, how did it affect the gallons per minute being discharged?

   A. The gpm stayed the same since the pump had already been set.
   B. The gpm increased since there was no longer the constraint of the tip size.
   C. The gpm decreased because without the tip, the friction loss will increase.
   D. It cannot be gauged because of variables concerning the lay, size of hose, and friction loss within the monitor.

2. When the tip flew off, how did it affect revolutions per minute of the engine?

   A. The rpm stayed the same since the pump had already been set.
   B. The rpm increased since, without the tip, the pump did not need to overcome nozzle pressure, and the load on the pump actually decreased.
   C. The rpm decreased since, without the tip, more water could flow, thereby increasing the load on the pump.
   D. The rpm stays the same. With the tip gone, even if more water flows, increased friction loss caused the pump to maintain the same workload.

3. When the tip flew off, how did it affect pressure on the discharge gauge?

   A. Pressure on the discharge gauge stayed the same since the pump had already been set.
   B. Pressure on the discharge gauge increased since friction loss increases with an increased flow. This resulted in more pressure work, which the discharge gauge measures.
   C. Pressure on the discharge gauge decreased. The loss of pressure work to overcome for nozzle pressure was more than the increased friction loss from an increased flow.
   D. It cannot be gauged since it depends on the exact nozzle pressure, how much the flow increased, and how much hose was in the evolution.

4. If, during this mishap, the pumps began to cavitate, what would be your first action?

   A. Immediately close any discharge feeding the monitor.
   B. Slowly close off any and all discharges to the monitor so as to prevent water hammer.
   C. Immediately reduce the throttle.
   D. Immediately shut off water coming into the pump.

Review Questions

1. When did the nonpositive displacement pump first find a place in the U.S. fire service?

2. What general category of pump does not take advantage of incoming pressure?

3. In discussing the characteristics of pumps, reference is made to positive displacement pumps pumping fluid and nonpositive displacement pumps pumping liquid. Why is this distinction made?

4. What is the name of the moving part of a nonpositive displacement pump?

5. What is the name of the inlet of the item referred to in Question 4?

6. What is the purpose of the volute?

7. If a two-stage pump is discharging liquid at 200 psi, how much pressure is each impeller providing if the intake pressure is 60 psi and the pump is in the parallel configuration?

8. During the operation of a pump, if it begins to cavitate, what action should be taken?

9. What is cavitation?

10. What is the name given to the general category of pump installed on all radial flow pumps as "priming pumps"?

11. Give an example of Bernoulli's principle at work in the radial flow pump.

12. Define eddying.

13. What is the name of the part of the discharge pressure generated by the pump?

14. What is the path of water through the two-stage radial flow pump when it is in the series configuration?

15. What condition results when the direction of flow of liquid is abruptly changed in a radial flow pump?

16. Give the name of the device used to measure

A. pressure
B. flow

17. What are the four steps of pump operation?

18. Why is it important to fill the tank as soon as possible?

19. Why is a side steamer connection a preferred connection when drafting?

20. What operational step includes setting the relief valve?

21. How often should pumps be inspected for readiness?

Activities

Put your knowledge of pumps to work to solve the following problems.

1. A pumper connected to a hydrant has a residual intake pressure of 65 psi. If the pumper is discharging water at 235 psi, how much pressure is being added by each impeller of a two-stage pump in the series configuration?

2. At the same fire as the pumper in Activity 1, another pumper is supplying water to a large master stream device. If the pumper has a residual intake pressure of 45 psi and is discharging at 150 psi, how much pressure is each impeller generating if the pump is in the parallel configuration?

3. Prove that the force that accelerates water in a radial flow pump is tangential to the rotation of the impeller.

4. In what step of operations is the tank filled and why?

5. What is considered adequate residual pressure?

6. How do Bourdon-tube gauges and electronic gauges differ?

Challenging Questions

1. A pumper operating from a hydrant has a residual pressure of 30 psi. If it is discharging water at a pressure of 150 psi, how much pressure is being generated by the pump?

2. If the pumper in Question 1 has a two-stage pump, how much pressure does each stage generate in the series position?

3. If the pumper in Question 2 is flowing a large volume of water in the parallel position, how much pressure is each impeller generating?

4. Give an expression for the work done by a 1,500 gpm pump, if it is pumping 70 percent of its capacity at 200 psi.

Formula

To find net pump pressure:

Discharge pressure – intake pressure = net pump pressure

Reference

NFPA 1901, *Standard for Automotive Fire Apparatus*. Quincy, MA: National Fire Protection Association, 2009.

Theory of Drafting and Pump Testing

LEARNING OBJECTIVES

Upon completion of this chapter, you should be able to:

- Understand the laws of physics that permit pumps to draft water.
- Be able to calculate the amount of work a pump has to perform to get water from a draft source.
- Convert inches of mercury to either pressure or feet of lift.
- Interpret the static and dynamic reading on the intake gauge in terms of feet of lift or friction loss.
- Understand pump test procedures.
- Comprehend pump capacity limitations as the discharge pressure increases.

Your company officer has Engine 11 assigned to the training academy today for pump training. He likes to give the younger personnel an opportunity to draft since it is rarely done in your jurisdiction. As he has done before, your company officer asks you to explain the process to the probie so that he can evaluate your knowledge of drafting while passing on much needed information to the probie.

You begin by showing the probie how to position the apparatus at the drafting site. Next you instruct her to make sure the connections for the suction sleeves are tight, and then you have her make the remaining hose connections as needed. Now you are ready to explain the principles of drafting.

1. How is it that water, a substance without tensile strength of its own, is capable of flowing into a pump?
2. What does a compound gauge tell you that a regular gauge does not?
3. Why does the "pump test" actually consist of several tests?

Introduction

This chapter goes hand in glove with material on pump theory and operations. You cannot fully comprehend pump theory without also understanding how water is drafted into a pump. Pump theory (see Chapter 7) and the theory of drafting are divided into two separate chapters to make understanding them easier.

Earlier we discussed the concepts of force and pressure. A thorough understanding of pressure is a prerequisite to the study of this chapter. When we are talking about drafting, it is necessary to understand how the pressure of the atmosphere is responsible for getting water into the pump. Unless you understand force and pressure, this is not possible. It is also necessary to be able to calculate the maximum theoretical lift at atmospheric pressure and the amount of energy loss realized on the intake side of a pump while drafting. From a positive water supply, a fire hydrant, this is not a factor because the pressure of the incoming water is independent of the pump. However, when you are drafting, the pump must not only perform enough work to discharge water, but also perform work to get water into the pump. See Chapter 2 for a discussion of force and pressure.

Previously we have discussed the conservation of energy. In this chapter, you will see it applied to a real-world situation. As the pump discharge pressure is increased, the capacity of the pump must go down. After all, the amount of energy the pump imparts to the liquid being pumped is finite, and it can be in the form of a lot of water at low pressure, or a little water at a lot of pressure. This concept is explained below in the section "Testing Pumps." (See Chapter 3 for a discussion of the conservation of energy.)

Drafting

Drafting is the process of acquiring a water supply from a static source, such as a river, pond, or basin. However, since water has no tensile strength, it cannot be "sucked" into the pump; for water to enter a pump, it must be "pushed" in. From a municipal water system it is pushed into the pump by the pressure of the system. When drafting, we need to rely on another source of pressure to push the water into the pump: the atmosphere.

Recall that there is a difference between relative pressure and absolute pressure. The pressure you read on your apparatus pump panel is relative pressure. As a refresher, recall that this means that when the gauge reads 0 pounds per square inch (psi) relative pressure, it is 14.7 psi absolute. In drafting that

14.7 psi is of critical importance and is what will push the water into the pump. Refer to Chapter 2 for further discussion of pressure.

To draft, we must attach a hard suction sleeve to the pump and place the other end, with a strainer, into the water source. This action will not cause water to automatically flow into the pump, however, because the pressure inside the pump will be the same as that outside the pump and the system is in equilibrium **Figure 8-1**. Since we are relying on the pressure of the atmosphere to push the water into the pump, it is first necessary to reduce the pressure inside the pump. This process of reducing the pressure inside the pump, which allows atmospheric pressure to fill the pump with water, is referred to as **priming the pump**.

Once the pump is primed, the higher pressure outside the pump will push water up the suction sleeve to a point where equilibrium is again achieved **Figure 8-2**. Ideally the pressure reduction inside the pump will be adequate to allow water to be pushed up the entire length of the suction sleeve into the pump. Figure 8-2 illustrates the point that for every 1 psi of pressure we evacuate from the pump, the

atmosphere will push water vertically 2.31 feet (ft). This is a practical application of information learned in Chapter 2.

The pressure indicated on the intake gauge after the pump has been primed is referred to as a *vacuum* because it is below atmospheric pressure. But reading the vacuum on the intake gauge after priming the pump creates a small problem. Since the gauges on the pump panel are designed to read relative pressure, we need at least one gauge to read vacuum on the relative scale. As such, the gauge used to read intake pressure will have the ability to read negative pressure as well positive pressure. A gauge that can read both is referred to as a *compound gauge*. The vacuum, however, is not indicated in pounds of pressure, but rather in inches of mercury, abbreviated as inHg, where Hg is the chemical symbol of mercury **Figure 8-3**.

Recall that mercury is 13.6 times heavier than water. This allows for a more precise measurement of negative pressure than if we tried to calibrate the negative pressure in pounds per square inch. It does, however, mean that we need a way to convert inches of mercury to either pressure or feet of lift. **Table 8-1** is a pressure-mercury-water conversion

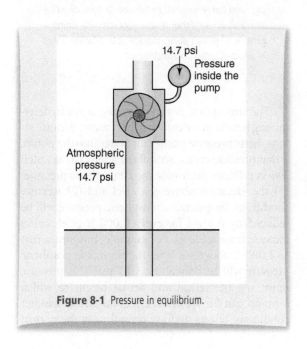

Figure 8-1 Pressure in equilibrium.

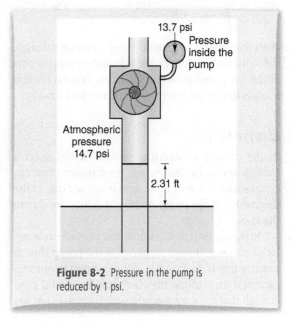

Figure 8-2 Pressure in the pump is reduced by 1 psi.

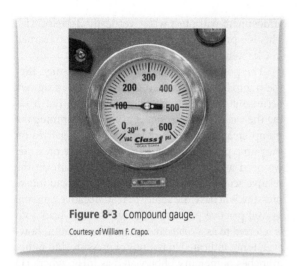

Figure 8-3 Compound gauge.

Courtesy of William F. Crapo.

Table 8-1	Pressure-Mercury-Water Conversion Chart	
Pressure, psi	Mercury, in	Water, ft
0.433	0.882	1.0
0.489	1.0	1.13
1.0	2.037	2.31
14.7	29.94	33.96

chart that will come in handy later when we calculate the work done on the intake side of the pump as part of the net pump pressure, including vertical lift and friction loss in the hard suction sleeve and strainer.

Limitations

In the process of drafting, several factors affect a pump's ability to draft water, even under ideal circumstances. Some of these limitations are due to the mechanics of the pump itself, and others are due to the laws of physics.

In theory, pumps should be able to evacuate every bit of pressure from the interior of the pump, but in practice this is not possible. For the pump to operate, clearance around the impeller shaft has to be great enough that the shaft can turn without restriction, yet

tight enough to prevent air and water leaks. To meet this dual requirement, the pumps cannot be tight enough to prevent all air leakage when drafting. The seals are tight enough to allow the priming pump to evacuate the majority of the air from the pumps, but a perfect vacuum is not possible. Also leakage at gates and valves is a factor in allowing unwanted air into pumps. The older the pump, the greater the air leakage, and so it is necessary to maintain pumps in top condition and be alert for deteriorating performance.

Fireground Fact

Record Keeping

Each piece of pumping apparatus should have a record of ongoing maintenance. This record should include the performance of routine preventive maintenance, problems that have arisen and corrective steps, and a detailed record of annual pump tests as required by NFPA 1911, *Standard for the Inspection, Maintenance, Testing, and Retirement of In-Service Automotive Fire Apparatus*. Together these records will paint a picture of pump performance and indicate areas in need of attention and repair.

The atmosphere is also a limiting factor in determining how high we can lift (draft) water. Because atmospheric pressure pushes the water into the pump, if that pressure varies, so will the ability to lift water. Two conditions determine the atmospheric pressure: (1) the elevation above sea level and (2) weather conditions. In general, atmospheric pressure will be reduced by 0.5 psi for every 1,000 ft of elevation increase from sea level. For example, this means that at 2,000 ft above sea level the normal atmospheric pressure will be only about 13.7 psi. This pressure limits the theoretical and actual height to which a pump can lift water. The second factor, weather conditions, affects drafting regardless of the normal atmospheric pressure. This is so because weather pattern changes cause the atmospheric pressure to rise and fall. When a high-pressure front moves in, the pressure increases and the amount of possible lift increases. A low-pressure front, however, causes the pressure to fall and the amount of possible lift decreases. You can find out how the weather is affecting drafting conditions by checking the weather online or on a mobile device, or by watching the daily news. Weather reports routinely report the barometric pressure as weather patterns change. A word of caution is necessary here. The atmospheric pressure to which local weather reports refer is adjusted for sea level. This is done to maintain uniformity in the interpretation of weather regardless of elevation. This may not be your actual atmospheric pressure. However, as long as you know the change, you will know how much the local atmospheric pressure has changed. Weather reports will usually refer to the atmospheric pressure as **barometric pressure**.

Fireground Fact

Atmospheric Pressure versus Barometric Pressure

Atmospheric pressure and barometric pressure are the same thing. The device used to measure atmospheric pressure is called a *barometer*, thus the name *barometric pressure*.

In Table 8-1, note that 29.94 is the maximum barometric pressure, in inches of mercury, at sea level, assuming an atmospheric pressure of 14.7 psi. As already mentioned, this pressure can increase or decrease to some extent according to weather conditions.

There are two final factors to consider with pumps and drafting. The first is the friction loss on the intake side of the pump. Here, we are generally talking about friction loss in the hard suction

hose and strainer. Just as hose has friction loss, hard sleeves and strainers have a loss associated with them. This loss in the hard suction hose and strainer has to be overcome by atmospheric pressure, which reduces the amount of atmospheric pressure available to push water into the pump. This in turn reduces how high water can be lifted as well as how much water the pump can supply.

Finally, the water temperature is a factor in drafting water. When water is hot, it wants to vaporize. So when you are attempting to draft, as the pressure inside the sleeve and pump is reduced, hot water would rather vaporize than stay in liquid form. This does not mean that hot water cannot be drafted, but it is a factor and may require that the pumper be placed closer vertically to the source of water.

Calculating Lift

One of the most important calculations we need to perform in relation to drafting is that of lift, which includes two different calculations: (1) calculation of maximum theoretical lift and (2) calculation of actual lift.

Regardless of the calculated theoretical lift or calculated actual lift, there are practical limitations to drafting. As a rule of thumb, to obtain the maximum capacity of the pump, a lift of 10 ft should not be exceeded. A lift of 15 ft reduces the capacity of the pump to 70 percent, and a lift of 20 ft reduces the capacity of the pump to 60 percent.

Calculating Theoretical Lift

To calculate the maximum theoretical lift, we need to know two things: what the atmospheric pressure is and how much lift each 1 psi of pressure provides. The second factor has already been provided to us in Table 8-1. Each 1 psi of pressure is capable of lifting water 2.31 ft. Multiply this amount by the atmospheric pressure, and you will find the maximum theoretical lift. The formula $H = 2.31 \times P$ was introduced earlier, and can be used here. Insert the atmospheric pressure in place of P. The maximum theoretical lift will be H and will be given in feet. (See Chapter 2 for discussion of the formula $H = 2.31 \times P$.)

Example 8-1

What is the maximum theoretical lift at sea level?

Answer

At sea level, the atmospheric pressure is 14.7 psi.

$$H = 2.31 \times P$$
$$= 2.31 \times 14.7$$
$$= 33.96 \text{ ft}$$

Please note that the numbers in this chapter are not being routinely rounded off. The reason is that so many numbers are small and rounding would actually introduce, in many instances, a significant error, which is not a good situation when you are trying to teach precision. In real life, round off as the individual situation can tolerate. Also note that the constant 2.31 is assigned units of feet per pound (ft/lb). It should be understood, without including units in the equation, that H will be given in feet—more specifically, in this use of the equation, feet of lift. See Chapter 2 for more information about the units assigned to the constant 2.31.

Example 8-2

What is the maximum theoretical lift at Mile High Stadium in Denver, Colorado?

Answer

Mile High Stadium is approximately 5,000 ft above sea level; therefore the atmospheric pressure is 2.5 psi less than that at sea level.

$$H = 2.31 \times P$$
$$= 2.31 \times 12.2$$
$$= 28.2 \text{ ft}$$

As already mentioned, a good source for up-to-date information on the atmospheric pressure due to weather conditions is the weather service. However, weather forecasters usually give the pressure in terms of inches of mercury, so it is necessary for us to convert the inches of mercury to pressure. To convert inches of mercury to pounds per square inch, we need only multiply the inches of mercury by 0.489 psi. (In Table 8-1, 1 inHg is equal to 0.489 psi.) This creates a new formula, $P = 0.489 \times Hg$. In this formula 0.489 is a constant that represents the pressure equivalent of 1 inHg, Hg represents the barometric pressure or reading in inches of mercury, and P is the pressure equivalent of the barometric reading.

$$P = 0.489 \times Hg$$

Example 8-3

What is the maximum theoretical lift if a storm is moving through and the weather service reports a barometric pressure of 28.5 inHg?

Answer

First find the pressure equivalent of 28.5 inHg.

$$P = 0.489 \times Hg$$
$$= 0.489 \times 28.5$$
$$= 13.94 \text{ psi}$$

The maximum theoretical lift is

$$H = 2.31 \times P$$
$$= 2.31 \times 13.94$$
$$= 32.2 \text{ ft}$$

It is also possible to make a direct calculation from inches of mercury to feet of lift by multiplying the barometric pressure by 1.13 ft, the equivalent lift of water to each inch of Hg. Use the formula $H = 1.13 \times Hg$, where Hg is the current barometric pressure.

$$H = 1.13 \times Hg$$
$$= 1.13 \times 28.5$$
$$= 32.2 \text{ ft}$$

$$H = 1.13 \times Hg$$

Calculating Actual Lift

Because the atmospheric pressure is responsible for pushing water up the sleeve into the pump, we are able to determine actual lift from the mercury reading on the intake gauge. Note that the reading on the intake gauge must be a static reading. At draft, the **static intake reading** is the reading on the intake gauge, in inches of mercury, after the pump has been primed, but before water is flowing. This is the actual lift of the pump. The simple calculation involves multiplying the inches of mercury indicated on the intake gauge by 1.13 ft, using the formula $H = 1.13 \times Hg$.

Fireground Fact

Reading Negative Pressure

Reading negative pressure on a compound gauge of fire apparatus is, by the nature of the gauge, not very precise. The scale below 0 psi, representing 30 inHg, is extremely small. Remember that it only represents 14.7 psi. With a little care and attention to detail you can be precise enough for fireground purposes. In this text, figures given as readings in inches of mercury from a compound gauge are for educational purposes and are therefore given as precise numbers.

Example 8-4

How high is the actual lift if the static intake reading at draft is 6 inHg (**Figure 8-4**)?

Answer

One inch of mercury (inHg) is equal to 1.13 ft of lift.

$$H = 1.13 \times Hg$$
$$= 1.13 \times 6$$
$$= 6.78 \text{ ft}$$

Reverse Lift Calculations

Since we now know how to find the lift when given the mercury reading, it can be useful at times to find the mercury reading when we know the lift. It is just a matter of multiplying the feet of lift by 0.882 inHg. This creates the formula Hg = 0.882 × H, where 0.882 is a constant that represents the mercury equivalent to 1 ft of water from Table 8-1, H is the height of the lift of water, and Hg is the equivalent mercury reading.

$$Hg = 0.882 \times H$$

Example 8-5

If a specific drafting situation requires that the pump be located 12 ft above the surface of the water, how much mercury should be reading on the intake gauge after water has entered the pump (**Figure 8-5**)?

Answer

Essentially you are just calculating the equivalent inches of mercury of 12 ft of lift. From Table 8-1, 1 ft of water is equivalent to 0.882 inHg.

$$Hg = 0.882 \times H$$
$$= 0.882 \times 12$$
$$= 10.58 \text{ in}$$

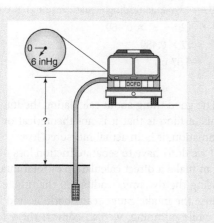

Figure 8-4 Static reading = feet of lift.

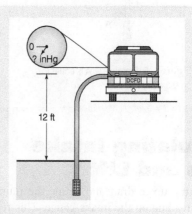

Figure 8-5 What will be the reading on the intake gauge?

When we know the intake reading in inches of mercury, we can calculate exactly how far water has been drawn up the sleeve. This calculation is done by using the formula $H = 1.13 \times Hg$.

Example 8-6

In Example 8-5, where the pumper is 12 ft above the surface of the water, how far up the hard suction hose will water be if the intake gauge only reads 8 inHg **Figure 8-6**? Will water enter the pump?

Answer

Find the feet of lift equivalent to 8 inHg.

$$H = 1.13 \times Hg$$
$$= 1.13 \times 8$$
$$= 9.04 \text{ ft}$$

In this evolution, if the pumper was only able to obtain 8 inHg, it would not be able to get water into the pump.

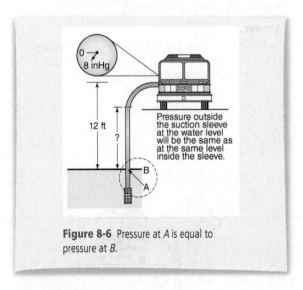

Figure 8-6 Pressure at *A* is equal to pressure at *B*.

Calculating Intake Loss and Lift

After setting up to draft and prime the pump, we obtain a reading on the intake gauge. This reading, assuming you are not yet flowing water, is the static reading used in the previous section to calculate lift.

Once water is flowing, the intake gauge will read an even higher vacuum, representing both the lift overcome and friction loss on the intake side of the pump. This is the <u>**dynamic intake reading**</u>.

To determine the amount of friction loss on the intake side of the pump, we need two pieces of information: (1) the static reading and (2) the dynamic reading on the intake gauge. We have already used the static reading to calculate the lift overcome by the pump. To find the friction loss overcome on the intake side, we simply subtract the static reading from the dynamic reading. This gives us our answer in inches of mercury, so we then need to convert it to pounds per square inch by multiplying by 0.489. These two steps combine to give us the formula FL = (dynamic reading − static reading) × 0.489. The terms of this formula are self-explanatory, with 0.489 used to convert inches of mercury to pounds per square inch. This formula tells us how much work the pump has to do on the intake side to overcome friction loss.

$$FL = (\text{dynamic reading} - \text{static reading}) \times 0.489$$

Example 8-7

How much friction loss is there on the intake side of a pump that has a static reading of 8 inHg and a dynamic reading of 20 inHg?

Answer

$$FL = (\text{dynamic reading} - \text{static reading}) \times 0.489$$
$$= (20 - 8) \times 0.489$$
$$= 12 \times 0.489$$
$$= 5.87 \text{ psi}$$

The good thing about calculating the intake loss in this fashion is that it is not theoretical or an approximation; it is an actual measured loss.

If we don't have to separate friction loss from lift, we can make a direct calculation of total intake loss by using the dynamic reading on the intake gauge. Because the intake gauge reads both lift and friction loss while pumping, we can convert these to pressure to calculate the amount of work done on the intake side of the pump. This is part of the net pressure and an important concept to know when you are conducting pump tests.

Example 8-8

How much work is being done on the intake side of the pump if the dynamic reading is 15.5 inHg?

Answer

Convert inches of Hg to pressure.
$$P = 0.489 \times Hg$$
$$= 0.489 \times 15.5$$
$$= 7.58 \text{ psi}$$

Calculating Total Work

We defined the work done by the pump as a combination of the net pump pressure and how much fluid is being discharged. When calculating the work done by a pump at draft, we define work exactly the same with one addition: when drafting, we need to account for the amount of work (inches of mercury/pressure we have to overcome) on the intake side of the pump in addition to work done on the discharge side. Refer to Chapter 7 for a discussion of the work done by the pump.

On the intake side of the pump, we have to overcome both lift and friction loss in the suction sleeve and strainer. We have just explored how to use static and dynamic readings to find lift and friction loss. This is the total work done on the intake side of the pump and represents the intake side contribution to net pump pressure. The calculation for determining intake loss was performed in Example 8-8.

To find the total work done by the pump, simply add the work done on the intake side of the pump (converted to pounds per square inch) and the work done on the discharge side of the pump (the discharge pressure). This is the net pump pressure from draft. The only thing left is to include the volume of water moving at this pressure. The expression for total work should look something like this: n gpm at y psi. In this expression n is the number of gallons flowing and y is the net pump pressure. For example,

1,250 gpm at 150 psi expresses the total work done by a pump with a rated capacity of 1,250 gpm at 150 psi net pump pressure. This expression is useful for indicating the total work done from both draft and a positive pressure source (remember to use net pump pressure). It is simply the amount of water flowing at a specified net pump pressure and does not have to be used only when meeting pump test criteria.

Example 8-9

What is the total work being done by a pump at draft, if it is flowing n gpm, the discharge pressure is 145 psi, and the intake is reading 18 inHg? The amount of water the pump is flowing is represented by n.

Answer

Add the pressure equivalent of 18 inHg to the discharge pressure to find the net pressure.
$$P = 0.489 \times Hg$$
$$= 0.489 \times 18$$
$$= 8.8 \text{ psi}$$

Now add the intake loss in pounds per square inch to the discharge:
$$\text{work done} = (\text{intake loss} + \text{discharge pressure})$$
$$@ \, n \text{ gpm}$$
$$= (8.8 + 145) @ \, n \text{ gpm}$$
$$= 153.8 \text{ psi}$$

In this problem 153.8 psi represents the *net* pump pressure. As an expression of total work it should be written: n gpm @ 154 psi.

Testing Pumps

Pump tests are not an everyday occurrence, but knowledge of the principles and procedures of pump tests can lend additional insight into how pumps operate. These pump tests should be conducted annually and after any major repair or alteration.

What we think of as "the pump test" is actually eight different tests. These tests are designed to test all aspects of pump performance. An abbreviated description of the complete pump test process is presented here. However, before conducting pump tests, consult NFPA 1911, *Standard for the Inspection, Maintenance,*

Testing, and Retirement of In-Service Automotive Fire Apparatus. Chapter 18 of NFPA 1911 contains all the details concerning testing of apparatus-mounted fire pumps, and it should be read and understood.

While it is possible to test pumps from a hydrant, it is preferred that pump tests be done from draft because the true performance of the pump is easier to evaluate. The water source should be 4 to 8 ft below the level of the pump, and no more than 20 ft of hard suction hose and a strainer is to be used per intake. (Larger-capacity pumps may require as many as four hard suction intakes to flow the capacity of the pump.) The point where the sleeve is placed in the water should have at least 4 ft of water, and the strainer needs to be covered by at least 2 ft of water. In fact, when you are conducting the test, or whenever you are drafting, it is a good practice to have 2-ft clearance in all directions around the strainer.

The maximum amount of lift for a pump being tested is determined by the pump capacity. All pumps up through 1,500 gpm have a maximum lift of 10 ft. Pumps with a capacity of 1,750 gpm are allowed a maximum lift of 8 ft, and pumps of 2,000 gpm and above are allowed a maximum lift of just 6 ft **Table 8-2**. In addition, note that higher-capacity pumps are allowed to use multiple suction sleeves. If you do not know for certain the number or size of sleeves needed for testing your pumps, consult NFPA 1911 prior to testing.

It is important that the water supply not be aerated and be at least 35°F, but not over 90°F. Recall from earlier discussions that the water temperature can affect a pump's ability to draft. Here is a practical application of that principle. The air temperature should also be between 0°F and 110°F, and the barometric pressure should be a minimum of 29 inHg adjusted to sea level.

The exact configuration of any hose lay depends on the capacity of the pump and which test is being conducted. Care should be taken when you are calculating flows from nozzles to know and use the correct nozzle coefficient so the exact flow will be calculated. If the correct coefficients are not used, erroneous results will be found and will invalidate the test.

NFPA 1911 does not specify hose lays for testing pumpers. Larger-capacity pumps may require more than a single nozzle to flow the rated capacity. However, NFPA 1911 does impose one restriction that is important in determining hose lays: It limits the velocity allowed in the hose, regardless of size, to 35 feet per second (fps). (By now, you should be able to calculate velocity without too much difficulty.) When you are setting up the hose lay, use multiple lines or larger lines to keep the velocity below 35 fps.

When it is delivered, each pumper is required to have a test label attached to the pump panel **Figure 8-7**. The test label gives the engine revolutions per minute (rpm) for each of the three tests conducted: 150 psi at full capacity, 200 psi at 70 percent capacity, and 250 psi at 50 percent capacity **Figure 8-8**. The annual service test should be able to duplicate these numbers almost exactly.

When conducting pump tests, we are testing the power source (engine), transmission, and pumps. Therefore, it is necessary to have all running lights,

Table 8-2 Pump Capacity at Maximum Lift	
Pump Capacity, gpm	**Maximum Lift, ft**
Up to 1,500	10
1,750	8
2,000 and above	6

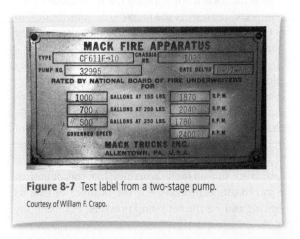

Figure 8-7 Test label from a two-stage pump.

Courtesy of William F. Crapo.

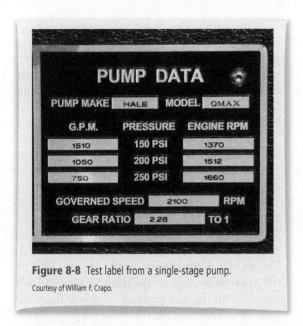

Figure 8-8 Test label from a single-stage pump.
Courtesy of William F. Crapo.

warning lights, and headlights on during the test. Also if the pumper is equipped with an air conditioner, the air conditioner must also be on during the tests. In addition, when you are conducting these tests, any pressure relief valve or equivalent device should be disengaged or otherwise prevented from functioning unless it is being tested.

With one exception, noted below, there is no magic to the order of the various tests; it is necessary only to complete all the tests. The order given here is basically the same order as that of the tests described in NFPA 1911. However, through experience, it is convenient to do certain tests in a given sequence. This sequence will be presented here as appropriate, but it should not be interpreted as the only way to conduct the tests.

Prior to beginning the tests, the following information should be recorded:

- Air temperature
- Water temperature
- Vertical lift
- Elevation of test site
- Atmospheric pressure adjusted to sea level (This is typically done by any credible weather reporting service; you just need to verify they are reporting an adjusted pressure.)

Engine Speed Check

The first test is a no-load governed speed check. If the speed of the engine does not meet the governed speed of the engine when it was new, corrections must be made before continuing with any additional testing. To conduct the engine speed check:

1. Allow the engine to warm up.
2. With the parking brake set, wheel chocked, and engine in neutral, increase the engine rpm until the governor prevents further increase in rpm.
3. The governed speed should be within ±50 rpm as specified on the engine test label.

As mentioned above, if the governed speed is off by more than ±50 rpm, the cause must be found and corrected before any further testing can continue.

Priming Device Test

This is no more than noting the exact time it takes to prime the pump. If a pump is 1,250 gpm or less, it must be fully primed in no more than 30 s, and a pump rated at 1,500 gpm or more must be fully primed in no more than 45 s. If the pump being tested has an auxiliary 4-in or larger intake, with a volume of 1 cubic foot (ft^3) or larger, an additional 15 s is permitted.

This test is most conveniently conducted when priming the pump for the pumping test mentioned below.

Vacuum Test

To conduct this test, it is necessary for all intakes to be capped with all intake valves open. All discharges are to be left uncapped and closed. Operate the primer until at least 22 inHg is obtained. Then let the pump sit for 5 minutes (min). The vacuum cannot fall by more than 10 inHg.

It is convenient to conduct this test just before doing the pumping test. As you prepare the pumper for draft, but before attaching the suction hose, cap off all intakes, close all discharges, and then conduct the test.

Pumping Test

This is the test normally thought of when we talk about pump testing. During this test, the pumps will be required to pump water at 150, 200, and 250 psi. The test capacity, pressure, and duration are listed in **Table 8-3**.

Table 8-3 applies to all radial flow fire service pumps. As stated in earlier chapters, as the needed pressure increases, the capacity of the pump decreases. Conversely, as the volume of water decreases, it is possible to pump at higher pressures.

Table 8-3 Pump Test Criteria

Test	Pressure, psi	Duration, min	Transfer Valve
Full capacity	150	20	Parallel
70%	200	10	Series or parallel
50%	250	10	Series

Fireground Fact

Bernoulli's Principle and the Pump Test

The difference in performance of pumps between the full-capacity test and 50 percent capacity test is a real-life example of Bernoulli's principle. As the discharge pressure increases from one test to the next, the flow experiences a corresponding decrease. At the same time, the energy input to the system in terms of engine rpm changes very little. In Figure 8-7, the change in rpm from the full-capacity test to 50 percent capacity test is only 290 rpm for the single-stage pump. In Figure 8-8, the overall change in rpm is only 260 rpm for the two-stage pump. The amount of energy put into the system in each test is nearly the same. (Don't expect to see the same rpm for each of the three tests, because Bernoulli's principle does not account for friction in the system, and as the rpm of the pump changes, so does its efficiency.)

If the pump being tested is a two-stage pump, it is important that the transfer valve be in the correct position. For the full-capacity test, the pump will be in the parallel position. For the 70 percent capacity test, the pump can be in either series or parallel. During the 50 percent capacity test, the pump must be in the series position.

When you are doing the pumping test, the reading on the discharge gauge will be part of the net pressure. Remember that when you are drafting, part of the work is done on the intake side of the pump. The pressures indicated in the testing criteria—150, 200, and 250 psi—are all *net* pressures. The actual pressures on the discharge gauge will be a few pounds per square inch lower to account for intake loss and lift. Calculate the intake loss and lift from the dynamic reading on the intake gauge, and subtract it from the test pressure to obtain the actual pressure required on the discharge gauge.

Note

Since the pressures are all net pressures, remember to allow for intake loss when you set the discharge pressure. For example, if you conduct the full-capacity test and convert a dynamic intake reading of 9.5 inHg to 4.65 psi, then the actual pressure required on the discharge gauge will be 150 − 4.65 = 145.35 or 145 psi.

Overload Test

This test can be thought of as part of the pumping test and is required to be conducted immediately after the full-capacity portion. After completing the full-capacity test at 150 psi net pressure, adjust your discharge pressure to 165 psi net pressure but gate back your discharges so the flow remains the rated capacity of the pump. After the pressure and capacity have been obtained, maintain the pressure and flow for 5 min.

Pressure Control Test

The pressure control test requires that the discharge relief valve, or equivalent, be tested at three different pressures. First, it will be tested at 150 psi discharge at rated capacity. Once the 20 min of the full-capacity test is completed, set the discharge relief valve to 150 psi and slowly close any open discharges. The pressure control device must not allow the pressure to rise more than 30 psi. This is most conveniently done immediately at the end of the full-capacity test, but before the overload test.

Second, you once again set the pumper to full capacity at 150 psi, but then, by use of the throttle alone, reduce the discharge pressure to 90 psi. Once again, with the pressure control device set, all discharges are slowly closed until there is no water flowing. And again the pressure control device cannot allow an increase in pressure of more than 30 psi.

A final phase of the pressure control test involves testing the device while pumping at 50 percent of capacity at 250 psi discharge pressure. This test is easily completed immediately after completing the 50 percent capacity test. After completing the required 10 min of pumping at 50 percent capacity, set the pressure relief device and close all discharges. As before, the pressure cannot rise more than 30 psi.

Intake Relief Valve System Test

Any apparatus equipped with an intake pressure relief mechanism must have that system tested. The easiest way to do this test is to have a second pumper pump into the test pumper and verify that the intake relief system works at the set pressure.

Gauge and Flow Meter Tests

All apparatus gauges and flow meters need to be tested annually along with the pumps. Pressure gauges themselves should be tested at 150, 200, and 250 psi. Cap each discharge and open the valve to each discharge. Raise the pump pressure to 150 psi on the test gauge. Now check for the accuracy of all other gauges. Pressure gauges cannot be off by more than 10 psi. Repeat this for 200 and 250 psi. At each pressure, the gauges cannot be off by more than 10 psi.

Testing flow meters is a little different. Each flow meter must be tested individually and cannot be off by

more than 10 percent. The test flow for flow meters depends on the size of pipe to the discharge. A table with pipe size and test flows is given in NFPA 1911.

Tank-to-Pump Flow Rate

The final test is the flow rate from the onboard apparatus tank to the pump. NFPA 1911 has a nine-step procedure for conducting this test. However, because it is necessary to determine only how much water will pass from the tank to the pump in 1 min, it is easily accomplished by the following method:

1. With the tank full and a 1½-in tip (or something close) on the deck gun, charge the deck gun.
2. Quickly bring the pump pressure up to the highest pressure the pump can maintain before "running away" from the water.
3. At that point, take a pitot reading of the pressure at the tip.
4. Maintain the flow rate and pressure until the tank is empty, or nearly empty, to ensure the integrity of the tank to pump piping.
5. Use the nozzle pressure obtained from the pitot reading to calculate the flow rate by utilizing Freeman's formula. That will be the flow rate from tank to pump.

The flow rate should be the same as when the pump was new. NFPA 1901, *Standard for Automotive Fire Apparatus,* specifies the required flow rate from tank to pump, depending on the tank size. If the flow rate when the pump was new is not known, then obtain a copy of NFPA 1901 for the year in which the pumper was manufactured and look up the required flow rate. Finally, if a copy of NFPA 1901 for the year of manufacture cannot be located, keep records of all subsequent flow rates and compare after each test for any change.

During each of the tests listed above, all pertinent information and test results must be documented. By maintaining accurate records, year-to-year comparisons can be made that will point to problem areas, so those areas can be corrected before the pump fails on the fireground.

WRAP-UP

Chapter Summary

- Drafting is the process of removing water from a static source by atmospheric pressure.
- For every pound per square inch that we reduce the internal pressure of the pump, the atmosphere will push water up a suction sleeve 2.31 ft.
- Vacuum is given in inches of mercury, abbreviated as inHg.
- The static intake reading represents the actual lift.
- The dynamic intake reading represents the lift, friction loss in the suction sleeve, and friction loss due to the strainer.
- When the work done by a pump at draft is calculated, the work done on the intake side must be added to the discharge pressure to obtain the net pump pressure. Then include the volume of water flowing.
- What is traditionally called the pump test actually consists of eight different tests.
- Pumps are tested at rated capacity at 150 psi, 70 percent of rated capacity at 200 psi, and 50 percent of rated capacity at 250 psi.
- NFPA 1911 specifies the conditions and procedures for testing pumps.

Key Terms

barometric pressure Atmospheric pressure.

drafting The process of acquiring a water supply from a static source, such as a river, pond, or basin.

dynamic intake reading The reading on the intake side of the pump, in inches of mercury; it includes lift and friction loss.

priming the pump The process of reducing the pressure inside the pump, allowing atmospheric pressure to fill the pump with water.

static intake reading The reading on the intake gauge, in inches of mercury, after the pumps have been primed, but before water is flowing.

Case Study

Nature Abhors a Vacuum

The statement "Nature abhors a vacuum" is credited to none other than Aristotle. While we recognize him today as an early philosopher and scholar, he did not have access to the scientific knowledge we have today. The above statement shows a lack of understanding of what we know today to be basic concepts of physics.

To examine the reasoning behind Aristotle's statement, let us consider the example of a bellows. Bellows easily draw in air when opened and then expel the air, with noticeable velocity, when they are closed **Figure 8-9**. But what happens if we close off the opening? The bellows will not open. To do so would require a vacuum to be created in the bellows. Aristotle reasoned that the vacuum could not be achieved—that is, the bellows would not open—because nature would not allow a vacuum to exist.

A second example of this thinking is demonstrated by a syringe. If it is placed in water and the piston is withdrawn, water will rise in the syringe, preventing a vacuum from forming **Figure 8-10**. Again the thought was that nature simply did not like vacuums and therefore would not allow them to exist.

Figure 8-9 Bellows.

Courtesy of William F. Crapo.

Figure 8-10   A syringe.

Courtesy of William F. Crapo.

What really is at play here? Answer the following questions so that Aristotle can understand how his reasoning is flawed. The topics covered in the first eight chapters of this text have provided you with the information you need.

1. Begin by explaining a vacuum to Aristotle.

    A. A vacuum is simply the absence of pressure.

    B. A vacuum is pressure below atmospheric on an absolute scale.

    C. A vacuum is pressure below atmospheric on a relative scale.

    D. A vacuum is simply the presence of any pressure below a given target pressure.

2. Why won't the bellows open up when the opening is closed off?

    A. When the opening of the bellows is closed off, air pressure cannot get into the bellows to push them open.

    B. When the opening of the bellows is closed off, the pressure of the air on the outside is stronger than what is needed to open the bellows and create the vacuum.

    C. When the opening of the bellows is closed off, air cannot be admitted to the bellows to prevent a vacuum from being formed. Thus, nature abhors a vacuum.

    D. When the opening of the bellows is closed off, air cannot be admitted to the bellows to equalize the atmospheric pressure on the outside. Thus, the stronger exterior pressure holds the bellows closed.

3. Why, when it is submerged in water, does the syringe fill with water instead of allowing a vacuum to form?

   A. Water will naturally fill any cavity it finds.
   B. As the piston of the syringe is withdrawn, the size of the syringe cylinder changes. Since there is no air pressure present in the syringe, atmospheric pressure pushing down on the surface of the water pushes water into the syringe.
   C. By withdrawing the syringe, piston water is sucked into the ever-increasing volume of the cylinder. This would even happen if no external pressure were present.
   D. As the piston of the syringe is withdrawn, creating an ever-increasing volume in the cylinder, a vacuum pulls the water into the syringe.

4. Where would you expect an area of low pressure, or a vacuum, to exist in a pump even if operating from a hydrant?

   A. A vacuum will occur in the eye of the impeller.
   B. A vacuum will occur in the entrance to the volute of the impeller.
   C. A vacuum will occur only if cavitation is present.
   D. A vacuum is not possible under any circumstances.

Review Questions

1. How does water enter the pumps when drafting if there is no "intake" pressure?

2. Define *drafting*.

3. What factors limit a pump's ability to draft?

4. Why is it impossible to obtain a perfect vacuum in a pump?

5. One foot (ft) of lift is equivalent to what reading in Hg?

6. What is the maximum amount of hard suction hose allowed, for each intake, during a pump test?

7. Why is the water temperature limited to a maximum of 90°F during the pump test?

8. Why is the pumping portion of the pump test divided into three separate tests?

9. When you are conducting the vacuum test part of the pump test, what is the minimum draw of mercury the pump must be able to achieve?

10. When you are conducting the pressure control test, what is the maximum allowable pressure rise?

11. During the vacuum test, after reaching the required inches of mercury,

    A. How long must the pump sit?
    B. What is the allowable drop?

Activities

It is now time to put your knowledge of drafting to work.

1. What is the maximum theoretical lift at an elevation of 3,000 ft?

2. A barometric pressure of 30.36 inHg is equivalent to how much atmospheric pressure?

3. A pumper at draft shows a static intake reading of 10 inHg. How much lift is the pump overcoming Figure 8-11 ?

4. A sealed glass tube, 20 ft tall and with a 6-in diameter, has water 6 ft up the inside of the tube. What is the pressure in the tube Figure 8-12 ?

5. A pumper operating at draft has a static intake reading of 11 inHg and a dynamic reading of 18 inHg. What is the intake loss?

6. In Figure 8-7, there is a small but steady increase in rpm from the full-capacity test to the 70 percent capacity test and then on to the 50 percent capacity test of the single-stage pump. However, in Figure 8-8, although there is an increase from the full-capacity test to the 70 percent capacity test, there is a remarkable drop in rpm from the 70 percent capacity test to the 50 percent capacity test. Why is this?

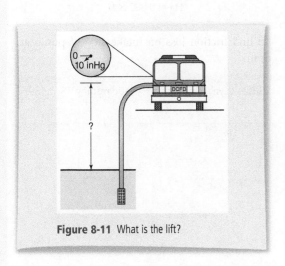

Figure 8-11 What is the lift?

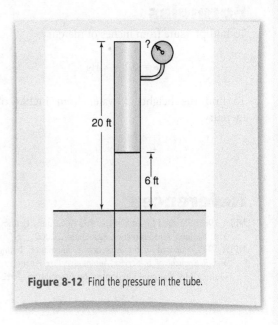

Figure 8-12 Find the pressure in the tube.

WRAP-UP

Challenging Questions

1. How many inches of Hg is equivalent to exactly 0.5 psi?

2. Calculate the amount of lift indicated by a Hg reading of 7.5 in.

3. What is the maximum theoretical lift at 4,000-ft elevation?

4. If the static intake reading is 5.5 inHg and the dynamic reading is 23 inHg, what is the friction loss on the intake side of the pump?

5. How much Hg must the intake gauge read in order to get water into the pump if the lift is 15.5 ft? Atmospheric pressure is 1 atmosphere (atm). Refer to Chapter 2 for a refresher on this unit of pressure.

6. Go back to Challenging Question 5 and calculate the amount of pressure left in the pump.

7. NFPA 1911 limits the flow of water in a hose to 35 fps during pump testing. How many gallons per minute does that equate to in 2½-in hose? Assume $C = 0.997$. If you studied Chapter 4 well, the answer to this question will be easy.

8. During a pump test, you note that the intake is reading 15 inHg. What is the equivalent in pounds per square inch?

9. During the 70 percent capacity test, you have an intake reading of 13.5 inHg. What should your actual discharge pressure be?

Formulas

To find pressure from inches of mercury:

$$P = 0.489 \times Hg$$

To find the height of water from inches of mercury:

$$H = 1.13 \times Hg$$

Reverse lift calculation:

$$Hg = 0.882 \times H$$

To find friction loss on intake side of pump at draft:

$$FL = (\text{dynamic reading} - \text{static reading}) \times 0.489$$

References

NFPA 1901, *Standard for Automotive Fire Apparatus*. Quincy, MA: National Fire Protection Association, 2009.

NFPA 1911, *Standard for the Inspection, Maintenance, Testing and Retirement of In-Service Automotive Fire Apparatus*. Quincy, MA: National Fire Protection Association, 2012.

Fire Streams

LEARNING OBJECTIVES

Upon completion of this chapter, you should be able to:

- Determine what properties are essential for an effective fire stream.
- Calculate the vertical range and horizontal range of streams.
- Calculate the nozzle reaction for smooth-bore and fog nozzles.
- Calculate back pressure and distinguish it from friction loss.

Case Study

As you and the captain crawl down the hot, smoky hallway with the first attack line, you think about the task ahead. When you open the nozzle, will you have sufficient water, will you place the water at the ideal spot for maximum effect, and what pattern will be best?

Finally, you arrive at the door to the room on fire. You position off to the side of the door so you can hit the fire and use the wall for excess steam protection. With the captain cheering you on, you hit the fire and darken it down. The truck has just taken out the windows from the outside and the room is already beginning to vent. All in all, it is a textbook operation and the fire is put out quickly.

1. How would you define an effective fire stream?
2. Explain how Bernoulli's principle applies to a nozzle.
3. Will the nozzle reaction affect how you apply water to the fire?

NFPA 1002 Fire Apparatus Driver/Operator Professional Qualifications, 2014 Edition

This chapter addresses the following requisite knowledge elements within section
5.2.3: proportioning rates and concentration, and foam system limitations.

Introduction

The development of an effective fire stream is arguably the most neglected aspect of firefighting today. In many instances, firefighters think that aiming water at the building is sufficient. This attitude is far from acceptable. Even under the best circumstances, friction of the water against the air will degrade an otherwise ideal fire stream.

Just what makes a stream of water into an effective fire stream? For any fire stream to be considered effective, it must meet the following four goals:

1. *It must flow sufficient gallons per minute (gpm) to absorb the heat or British thermal units (Btu) being generated.* If the water cannot absorb the heat being given off, the fire simply will not go out until all the fuel is consumed. Note that it is not necessary for the water to absorb every last British thermal unit given off; absorption of only 30 to 60 percent of the heat generated will stop the combustion chain process. In addition, although it is generally believed that the cooling effect of water is the primary extinguishing mechanism of water, steam generated and retained will make the atmosphere inert and deprive the fire of oxygen. To get the maximum cooling effect, the majority of the water must be converted to steam.

2. *Water must be applied at the correct point(s) in order to absorb the maximum heat in Btu.* Even if an adequate volume of water is used, if it is not put on the fire or on the superheated atmosphere of the area involved, the fire will not go out. This is often best accomplished by putting multiple smaller streams in service rather than fewer larger ones.

3. *Water must be applied in the correct form.* Regardless of the fire situation, the most effective conversion of water to steam comes when water is applied in the form of small droplets. From the information given earlier, we can calculate that when water is allowed to vaporize,

it will absorb 6.47 times more heat than water that is only allowed to get to 212 degrees Fahrenheit (212°F) without vaporizing. Only where extreme reach or penetration is critical do solid-bore nozzles possibly have a distinct advantage. When you are applying water in the form of fog during an interior attack, you must use caution to avoid generating excessive steam. (See Chapter 1 for a refresher on the latent heat of vaporization of water.)

4. *Water should be applied for the shortest possible time.* If the first three goals are met, fire streams that are operated for too long can cause excessive water damage. In larger firefighting operations, this goal is often difficult to achieve, but on smaller fires it is well within our means to eliminate damage from excess water. During research at Iowa State University in the 1950s, it was determined that when water was applied in the form of fog under the right circumstances, excellent results were obtained when the fire streams were only allowed to operate at the critical flow for only 30 seconds (s). Operating the streams for longer complicated extinguishment efforts.

Effective Fire Streams

In addition to the four goals that help to define an effective fire stream, there are some very specific criteria that aid in defining the effectiveness of a fire stream. These criteria are not new—John Freeman first defined them in 1888—but they are as meaningful today as they were then. As you read and study the following characteristics of a good fire stream, keep in mind that the variables in each criterion can change with each fire. These criteria were written specifically for solid stream nozzles, but the broader principles apply to all fire streams.

An effective solid stream is one that displays the following criteria:

1. *At the limit named, the stream has not lost continuity by breaking into showers of spray.* The reason

is that a broken stream will not reach the fire with the desired mass. If too much water is lost because of breakup, the effectiveness of the stream is compromised.

2. *Up to the limit named, the stream appears to discharge nine-tenths of its volume of water inside a 15-inch (15-in) circle and three-quarters of its volume inside a 10-in circle.* This further defines criterion 1. Because it is impossible for the stream not to experience some breakup due to friction with air, this criterion defines the amount of breakup allowed.

3. *The stream is stiff enough to attain, in a fair condition, the height or distance named even though a fresh breeze is blowing.* Wind affects the quality of the stream. Directing a stream into a breeze can decrease its range, whereas directing it with the wind can increase its range. A wind from the side has an undesirable effect on the stream, but criteria 1 and 2 still apply.

4. *At a limit named, the stream will, with no wind, enter a room through a window opening and just barely strike the ceiling with force enough to spatter well.* As previously stated, we want the stream to stay together as long as possible. When the stream gets to the fire, we want it to break up so it will more efficiently absorb the heat. By specifying that the stream must break up at the fire, we are attempting to break the water down into smaller particles so it will more readily absorb heat.

Range of Streams

The range of a fire stream is dependent on the following five factors, four of which can be controlled by the firefighter:

1. *Nozzle pressure.* All else being equal, the nozzle pressure determines the reach of the stream. The higher the nozzle pressure, the greater is the reach of the stream, to the point where the stream is overpumped and breaks up.

2. *Nozzle diameter.* All else being equal, the nozzle diameter determines the reach of the

stream. The larger-diameter nozzles have a greater flow and a greater reach.

3. *Angle of stream.* All else being equal, the angle of the stream determines the reach of the stream. An angle of 32 degrees (32°) above horizontal provides the maximum horizontal reach.

4. *Wind.* As mentioned in criterion 3, the wind can increase or reduce the range of a stream.

5. *Pattern.* When you are using a fog nozzle, as the angle of the pattern increases, the reach decreases.

Of the five factors, the nozzle pressure is the most critical. It is easy to overpump or underpump a nozzle. If we provide too much pressure on a nozzle, we can cause the stream to break up; and if we underpump it, we can deliver too little water and reduce the range of the stream. Most nozzles can tolerate some degree of excess pressure before the stream begins to break up, but no nozzle will deliver the required flow if it is underpumped, regardless of the condition of the stream.

Another critical factor in determining the range of a stream is the angle. As previously stated, an angle of 32° will provide the maximum horizontal range. However, an angle of 32° is not generally an effective angle for firefighting. To be an effective firefighting stream, the water must enter the building and then be deflected off the ceiling in order to break up the stream to achieve maximum heat absorption. The most effective angle for firefighting is 45°, as the stream gets an acceptable horizontal range as well as a good angle for deflecting water off the ceiling of the room it enters. The best penetration and deflection off the ceiling is achieved when the stream is directed just above the lowest edge of the opening **Figure 9-1** . An advantage of the 45° angle is that it is easy to set up. To get the 45° angle, the nozzle is simply placed as far from the building as the stream will reach up the building. The third floor is generally considered to be the highest story into which a stream can be effectively directed from the ground.

The maximum horizontal range (HR) can be approximately calculated by the formula HR = ½NP + 26* where HR stands for horizontal range

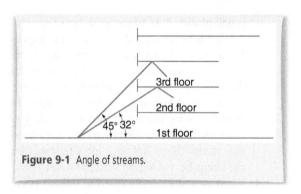

Figure 9-1 Angle of streams.

and NP is the nozzle pressure. The asterisk (star) is there to indicate that for each ⅛ in that the nozzle diameter is more than ¾ in, another 5 is added to the 26. The answer is given in feet.

$$HR = ½NP + 26*$$

Example 9-1

What is the horizontal range of a nozzle with a 1¼-in tip and operating at 50 pounds per square in (psi) nozzle pressure?

Answer

Add 5 for each ⅛ in over ¾ in = 20.

$$HR = ½NP + 26*$$
$$= ½(50) + 46$$
$$= 25 + 46$$
$$= 71 \text{ feet (ft)}$$

The vertical range is also affected by the factors in the list; however, we can do a couple of things to increase our vertical range. First, a stream operating from on top of a pumper has the advantage of being about an additional 8 ft off the ground. This change in vertical reach translates into one additional floor—the fourth floor. Where vertical range

---

*Add 5 for each ⅛ in of nozzle diameter larger than ¾ in.

is critical, it is a definite advantage to leave master stream devices on the apparatus. Another way to gain vertical range is to use an elevated stream, such as a ladder pipe. Just how much additional range is possible depends on the type of elevated stream and its design and length. When you are operating into the upper floors or area of a building, it is still necessary to place the aerial device so that the stream enters the building at a 45° angle. An example of placement for a 100-ft aerial ladder using a ladder pipe assembly would be to place the inside edge of the turntable a distance from the building equal to one-half of the desired reach of the stream. The ladder should then be elevated to an angle of 70° and extended until the nozzle is about 10 to 15 ft below the level of the opening. The stream is then directed into the building so it just clears the windowsill, giving an angle of about 45° **Figure 9-2**. Tower ladders and ladder towers make excellent elevated stream platforms because they do not have the design limitations associated with aerial ladders, such as extension limits and specific angle when operating a stream.

The vertical range (VR) can be calculated by the formula VR = ⅜NP + 26* where VR stands for vertical range and NP is the nozzle pressure. Again the asterisk tells us that 5 must be added for each ⅛ in that the nozzle diameter is more than ¾ in.

$$VR = ⅜NP + 26*$$

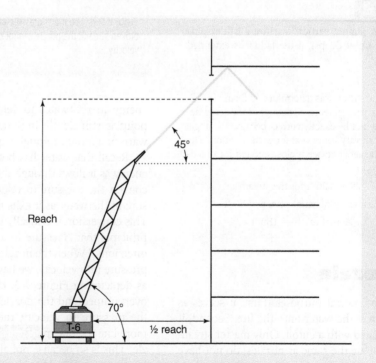

Figure 9-2 Proper placement of elevated stream from aerial ladder.

Example 9-2

What is the maximum vertical reach of a stream placed on the ground if the nozzle has a 1½-in diameter and has a nozzle pressure of 80 psi?

Answer

Add 5 for each ⅛ in over ¾ in = 30.

$$VR = \frac{5}{8}NP + 26^*$$
$$= \frac{5}{8}(80) + 56$$
$$= 50 + 56$$
$$= 106 \text{ ft}$$

If the stream is elevated in any way, simply add the height of the elevation or the extension of the ladder to the answer obtained from the formula.

Example 9-3

What is the maximum vertical reach of a ladder pipe using a 1½-in tip at 80 psi, if the ladder is extended 80 ft?

Answer

The reach of the stream was determined in Example 9-2. Use 80 ft as the vertical reach advantage of the aerial device. (This may not be exactly correct because the angle of the ladder can vary somewhat for each situation. The extension of the aerial device is close enough for practical purposes.)

$$VR = 106 \text{ ft for the stream}$$
$$\text{Extension of aerial device} = 80 \text{ ft}$$
$$\text{Actual reach} = 186 \text{ ft}$$

The Nozzle

The nozzle has several purposes. First, it serves as a means to direct the water onto the fire. Second, it often is associated with a cutoff. Only master stream devices, which are the same as nozzles, do not have a cutoff. Third, the nozzle converts the pressure

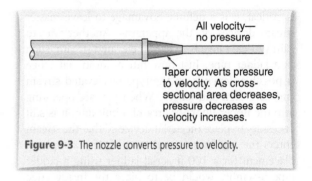

All velocity—no pressure

Taper converts pressure to velocity. As cross-sectional area decreases, pressure decreases as velocity increases.

Figure 9-3 The nozzle converts pressure to velocity.

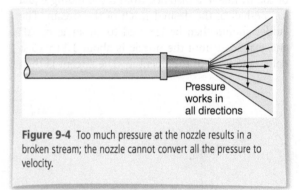

Pressure works in all directions

Figure 9-4 Too much pressure at the nozzle results in a broken stream; the nozzle cannot convert all the pressure to velocity.

energy in the water to velocity energy. This third point is sufficiently important and complicated to warrant a more thorough explanation.

Recall that water has both pressure and velocity energy as it flows through the hose. The nozzle must convert the pressure to velocity so there is no pressure in the water as it exits the nozzle **Figure 9-3**. This conversion is critically important because of the principle that "Pressure in a fluid acts equally in all directions." When the nozzle is able to convert all the pressure into velocity, we have a nice smooth stream, as depicted in Figure 9-3. However, if the nozzle is overpumped and the nozzle is unable to convert all the pressure to velocity, then the stream will look more like the one in **Figure 9-4** as it exits the nozzle. This is so because not all the pressure inside the stream is converted to velocity, and it is trying to push out in all directions after the stream has left the nozzle. When there is no pressure in the stream as the water exits the nozzle, this cannot happen. See Chapter 7

*Add 5 for each ⅛ in of nozzle diameter larger than ¾ in.

for a discussion of pressure and velocity energy as water flows through the hose. Also see Chapter 2 for a discussion of the principle cited.

Types of Nozzles

There are two basic types of nozzles: (1) smooth-bore, or solid stream, and (2) fog nozzles. The **smooth-bore nozzle** is capable of producing only a solid stream of water. Because the stream is solid, it is generally considered to have better penetration than a fog nozzle on straight stream. It is also less affected by wind than the straight stream of a combination fog nozzle.

A **fog nozzle** is one that is capable of producing a spray pattern. The primary advantage of the fog nozzle is that, through design, the nozzle breaks down the water into small droplets that absorb heat much more readily than a solid mass of water. Because the water is broken down into small droplets, a much higher surface area is exposed to the hot atmosphere of the fire. The greater surface area of the droplets is the most advantageous format for water to absorb the heat of the fire and vaporize. For water to be of maximum value, it must vaporize. (Recall from Chapter 1 the calculations of specific heat and latent heat.)

Some fog nozzles today still employ a fixed pattern of spray. These fog nozzles have names such as Navy all-purpose nozzles or low-velocity fog applicators. Nozzles such as cellar nozzles and attic nozzles are often designed with fixed fog discharge set at angles designed to cause the head of the nozzle to spin. This spinning distributes water over a large area, 360° around where the nozzle is inserted.

Another type of cellar or attic nozzle works on the same principle as the one just mentioned, but it does not have fog discharges attached. Instead it spins from the action of several solid stream discharges angled to cause the spin. As the nozzle spins, it causes the water to break up, mimicking the effect of a fog stream. More correctly, it is a broken stream, intended to break down water into droplets, mimicking a fog nozzle intended to absorb heat better. The most famous of this type of nozzle is the *Bresnan distributor* **Figure 9-5**. Since it is designed to create a broken stream, it is included in the section on fog nozzles.

Figure 9-5 Bresnan distributor nozzle.
Courtesy of Elkhart Brass Manufacturing Company, Inc.

In addition to fixed-pattern fog nozzles, there are fog nozzles that have a variable pattern spanning from straight stream to wide fog. These nozzles are referred to in many places as **combination variable fog and straight stream (CVFSS) nozzles**. CVFSS nozzles are capable of producing a pattern from straight stream to 100° fog **Figure 9-6**.

There are two types of CVFSS nozzles: fixed and adjustable gallonage. The fixed-gallonage nozzle is exactly that: it will deliver a set amount of water or gallons per minute (gpm) at the designed nozzle pressure. It is possible to get more or less water out of the fixed-gallonage nozzle by increasing or decreasing the nozzle pressure. This practice is not recommended. While the flow will increase or decrease with a change in nozzle pressure, the resulting flow will not be known. Finally, as noted below, an incorrect nozzle pressure will result in an unacceptable discharge pattern.

Some CVFSS nozzles can vary the flow by turning a ring that changes an aperture within the nozzle, allowing more or less water to flow. There is also a type of CVFSS nozzle, called the *automatic nozzle*, that allows a variable flow of water without manual adjustment. A spring in the nozzle adjusts the pressure to provide an adequate nozzle pressure even as the flow changes.

The **automatic nozzle** is designed to deliver water at any gpm value within its design range and to maintain its designed nozzle pressure within acceptable

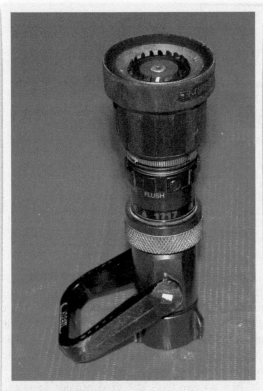

Figure 9-6 The CVFSS (fog) nozzle.
© Jones & Bartlett Learning.

We have just defined smooth-bore and fog nozzles. In practice, the terms *nozzle* and *tip* are often used interchangeably. In general, a cutoff with a tip is called a *nozzle* and is defined by the type of tip it has. For example, a cutoff with a smooth-bore tip is called a smooth-bore nozzle.

Nozzle Pressure

<u>Nozzle pressure</u> is the pressure required at the nozzle/tip to allow it to deliver the designed gpm and produce an effective fire stream. For the nozzle to operate properly and deliver the correct amount of water, it is critical that the nozzle receive the correct pressure. Nozzle pressures do not vary much and are usually specific to a particular type. For example, hand line smooth-bore nozzles usually have a 50 psi nozzle pressure while master stream smooth-bore nozzles have an 80 psi nozzle pressure. These pressures are not absolutes and can be varied to some extent. If the nozzle pressures for smooth-bore tips are varied, use caution not to overpump the nozzle, or else a poor, ineffective stream will result. If too low a pressure is used, insufficient water will flow and the extinguishing capacity will be minimized.

Fog nozzles require a pressure of 100, 75, or 50 psi, depending on the manufacturer's design. In general, 100 psi is the standard pressure for a fog nozzle, regardless of whether it is a hand line nozzle or master stream tip. The 75 and 50 psi nozzles are hand line nozzles specifically designed to provide required flows at lower nozzle reaction. The primary difference between a high-pressure fog nozzle (100 psi) and a low-pressure fog nozzle (75 or 50 psi) is the water droplet size. As the nozzle pressure is reduced, the water droplet size increases.

Nozzle pressures for fog nozzles are specific and cannot be varied from the manufacturer's design pressure. A 100 psi nozzle cannot be pumped at 75 psi and get an acceptable pattern and flow. Similarly, a 50 psi nozzle pumped at 100 psi will be overpumped and produce a poor fog pattern. In short, give a fog nozzle the pressure for which it is designed. Even automatic nozzles need adequate nozzle pressure to perform properly. With automatic nozzles it is the flow that varies, not the nozzle pressure.

limits. When pumping to an automatic nozzle, you calculate the pump discharge pressure based on friction loss for the desired flow and a nozzle pressure between 50 and 100 psi, depending on the nozzle design. The advantage of the automatic nozzle is that to change the flow, you only need to recalculate the pump discharge pressure and adjust the discharge to the new pressure. The nozzle automatically adjusts the size of the opening to accommodate either more or less water.

A word of caution is in order about automatic nozzles. With a standard nozzle, if you have a kink in the line, a poor pattern will indicate inadequate flow. With the automatic nozzle, the nozzle will adjust to the lower flow and maintain the nozzle pressure, disguising the less than expected flow. To avoid this situation, you must be extra alert for kinks in hose or any other situation that may result in inadequate pressure.

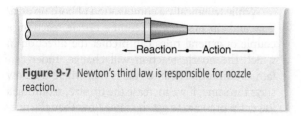

Figure 9-7 Newton's third law is responsible for nozzle reaction.

Nozzle Reaction

As the water exits the nozzle/tip, regardless of whether it is a straight stream or fog stream, there is an opposing reaction called the *nozzle reaction* **Figure 9-7** . This condition is an example of **Newton's third law of motion**: "Whenever one object exerts a force on a second object, the second exerts an equal and opposite force on the first." In a hose line or fire stream frame of reference, the first object would be the hose line. Technically the hose line is not creating any force; however, it is directing the force of the water created by velocity and the weight of the water, much as a rocket engine directs exhaust to create thrust. The second object would be the water as it exits the nozzle. Because the hose line and nozzle direct the force created by the velocity and weight of the water, the water exiting the nozzle creates an equal and opposite reaction against the nozzle. In practical terms, **nozzle reaction** is the opposing force created by water exiting the nozzle, as predicted by Newton's third law.

Technically, what we refer to as nozzle reaction is a function of both pressure at the point of discharge (nozzle pressure) and the amount of water flowing. Since the amount of water that flows is a function of both the discharge pressure and the area of the nozzle, a formula should come to mind: $F = P \times A$. Force is a static measurement and assumes that nothing is moving. A reaction force, however, is a dynamic measurement, in which water or other fluid must be flowing. To calculate the dynamic nozzle reaction from the static force we multiply force by a factor of 2. The result is the dynamic nozzle reaction. See Chapter 2 for more about the formula $F = P \times A$.

For example, the area of a 1-in tip is found by calculating $A = 0.7854 \times D^2$, or $A = 0.7854 \times 1^2 = 0.7854 \times 1$. The area of a 1-in tip, then, is 0.7854

square inch (in^2). The resulting force, for a nozzle operating at 50 psi nozzle pressure, if the tip were capped off would be $F = P \times A$, or $F = 50 \times 0.7854 = 39.27$ psi. To find the nozzle reaction, a dynamic force, we multiply 39.27×2 to get a nozzle reaction of 78.54 pounds (lb). Compare this number with the answer in Example 9-4 below.

From our work above we could write a formula for nozzle reaction that would look something like this: $NR = 2 \times P \times A$. By inserting the area of a 1-in tip for A, we can multiply 2×0.7854 and develop a constant of 1.57, or $NR = 1.57 \times P$. But this is useful for only a 1-in tip. We can make the formula universal for any size tip by inserting the term D^2, giving us the formula $NR = 1.57 \times D^2 \times P$. This formula can be used when the tip size and pressure are known. When you are calculating the nozzle reaction, remember that it is a force and is therefore measured in pounds.

$$NR = 1.57 \times D^2 \times P$$

Example 9-4

What is the nozzle reaction for a 1-in tip operating at 50 psi nozzle pressure?

$$\begin{aligned} NR &= 1.57 \times D^2 \times P \\ &= 1.57 \times 1^2 \times 50 \\ &= 1.57 \times 1 \times 50 \\ &= 78.5 \text{ lb} \end{aligned}$$

The formula $NR = 1.57 \times D^2 \times P$ is a commonly used formula to calculate the nozzle reaction from a smooth-bore nozzle where the pressure and tip size are known. Calculating the nozzle reaction for fog nozzles requires another formula.

The separate formula for calculating nozzle reaction from a fog nozzle begins with the formula $NR = 1.57 \times D^2 \times P$. In this formula we need to insert an expression for finding an equivalent nozzle diameter when the flow and nozzle pressure are known. Fortunately we developed just such a formula earlier for finding the diameter when pressure and flow are known. That formula is $D = \sqrt{gpm/(29.84 \times C \times \sqrt{P})}$. See Chapter 5 for information about the development of this formula.

Since we need an expression to replace D^2 in the formula for nozzle reaction, we simply square the equation $D = \sqrt{gpm/(29.84 \times C \times \sqrt{P})}$ to obtain $D^2 = gpm/(29.84 \times C \times \sqrt{P})$. We substitute this for D^2 in the above formula, and we end up with NR = 1.57 × $[gpm/(29.84 \times C \times \sqrt{P})] \times P$. By inserting an average C factor of 0.97, the formula becomes NR = 1.57 × $[gpm/(29.84 \times 0.97 \times \sqrt{P})] \times P$. Multiply 29.84 by 0.97 and the formula becomes NR = 1.57 × $[gpm/(28.94 \times \sqrt{P})] \times P$. Now divide 1.57 by 28.94 to get a constant of 0.0543. The final manipulation is to divide P by $\sqrt{P}$ to get $\sqrt{P}$. Put them together and the formula for nozzle reaction becomes NR = 0.0543 × gpm × $\sqrt{P}$.

There is no need to learn two formulas for nozzle reaction. The constant 0.0543 will give you extremely accurate results for either straight stream or fog nozzles. The only criticism is that for master stream devices there can be an error of a few pounds due to a C factor for master stream devices of 0.997. However, given that the error is only a few pounds when the total nozzle reaction can be several hundred pounds, the error is too small to be considered significant. This formula is also extremely accurate for smooth-bore nozzles when flow and pressure are known. In short, the formula is valid whenever you know the flow and pressure, regardless of the type of nozzle.

$$NR = 0.0543 \times gpm \times \sqrt{P}$$

Example 9-5

What is the nozzle reaction for a 1-in tip flowing 205 gpm at 50 psi nozzle pressure?

Answer

$$\begin{aligned} NR &= 0.0543 \times gpm \times \sqrt{P} \\ &= 0.0543 \times 205 \times \sqrt{50} \\ &= 0.0543 \times 205 \times 7.07 \\ &= 78.7 \text{ lb} \end{aligned}$$

Compare this answer to the answer in Example 9-4 (78.5 lb); the difference is due to rounding.

While technically a combination of both tip area/flow and tip pressure determine nozzle reaction, a couple of general rules can predict the direction in which the nozzle reaction will change under certain conditions. The first rule is that where the flow stays the same, if we increase the tip size, requiring a lower tip pressure to get the same flow, then the nozzle reaction will go down. And if we decrease the tip size, requiring a higher tip pressure to get the same flow, the nozzle reaction will go up.

The second rule concerns events when the tip size stays the same but the tip pressure is changed to get a different flow. When the nozzle pressure is increased, the flow increases along with nozzle reaction. And when the nozzle pressure is decreased, the flow will decrease along with the nozzle reaction.

Under either condition described above, it will be necessary to calculate the nozzle reaction to find out the amount of change, such as in Example 9-6.

Example 9-6

What would be the nozzle reaction for 205 gpm if the tip size were 1⅛ in?

Answer

The 1-in tip in Example 9-5 flows 205 gpm at 50 psi nozzle pressure. Here we are going to find the nozzle reaction for the same flow through a larger tip, thus requiring a lower nozzle pressure. Will the nozzle reaction be higher or lower? We need to start by determining the nozzle pressure for the 1⅛-in tip to deliver 205 gpm. Assume a C factor of 0.97.

$$\begin{aligned} P &= [gpm/(29.84) \times D^2 \times C)]^2 \\ &= \{205/[29.84 \times (1\tfrac{1}{8})^2 \times 0.97]\}^2 \\ &= [205/(29.84 \times 1.27 \times 0.97)]^2 \\ &= (205/36.76)^2 \\ &= (5.58)^2 \\ &= 31.14 \text{ psi nozzle pressure} \end{aligned}$$

Now find the nozzle reaction for 205 gpm from a 1⅛-in tip at 31.14 psi nozzle pressure.

$$\begin{aligned} NR &= 1.57 \times D^2 \times P \\ &= 1.57 \times (1\tfrac{1}{8})^2 \times 31.14 \\ &= 1.57 \times 1.27 \times 31.14 \\ &= 62.1 \text{ lb} \end{aligned}$$

The difference in nozzle reaction between Example 9-5 and Example 9-6 is 16.6 lb. This illustrates that if flow is maintained, reducing the nozzle pressure will cause the nozzle reaction to decrease.

A note of caution is in order here. As the nozzle pressure is reduced, the reach of the stream will also be reduced. However, where reach is not a primary factor, this trade-off may be acceptable.

Nozzle Reaction and Fog Pattern

The nozzle reaction calculated for fog nozzles by the above formula gives the nozzle reaction for a straight stream pattern. However, fog nozzles can be adjusted to any pattern from straight stream to 100° fog. As the pattern shifts away from straight stream, the angle of the water leaving the nozzle changes, causing the nozzle reaction to change.

Because the nozzle reaction is an opposite reaction, we need to find a way to determine it for patterns other than straight stream. To do so, we need to analyze the vector forces exerted by the water as it leaves the nozzle. **Vectors** are the measure of strength and direction of forces. Figure 9-8 is an example of how the forces for a 60° fog pattern would look if analyzed by their component vectors. Vector **B** is perpendicular to the nozzle reaction and is in equilibrium because it exists equally all around the nozzle. Because this force is in equilibrium, it produces no net force. Vector **C** of the stream is at an angle, and because it is produced equally around the periphery of the nozzle, it reacts against itself and is also in equilibrium. This is easy to understand if you visualize vector **C** extending down into the nozzle, as illustrated by the dotted lines. The two vectors will intersect at a point, effectively canceling each other out. Vector **C** is the force calculated by the nozzle reaction formula. Vector **A** of the vector forces is the only force not in equilibrium and is responsible for the nozzle reaction for the 60° fog pattern.

The strength of vector **A** is a function of the strength of vector **C** and angle *BC*. More specifically, the nozzle reaction will be the nozzle reaction calculated by the formula above, multiplied by the sine of angle *BC*. **Sine** is a trigonometric function of an angle that when multiplied by the length of the hypotenuse of a right triangle will find the length of the

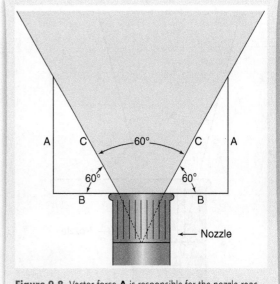

Figure 9-8 Vector force **A** is responsible for the nozzle reaction you feel.

side opposite the angle. By breaking the forces of a fog nozzle into vectors, a right triangle is formed. In Figure 9-8, the calculated nozzle reaction is used in place of the length of the hypotenuse, in this case vector **C**, and the opposing force (recall Newton's third law), the nozzle reaction, is represented by vector **A**. The sine of an angle can be looked up in any good math book that has trigonometric functions, and most scientific calculators can calculate sine.

For practical purposes, it is only necessary to understand how the nozzle reaction can change as the angle of the pattern changes. At a 30° pattern, the nozzle reaction is calculated by finding the sine of angle *BC*, which is 75° Figure 9-9 . The sine of 75° is 0.97. At a 60° pattern, the sine of angle *BC*, which is 60°, is 0.87. At a 90° pattern the sine of angle *BC*, which is 30°, is 0.7. Finally, at 100°, which is the widest fog pattern, the sine of angle *BC*, which is 20°, is 0.64. In each example you can think of the sine as a percentage of the nozzle reaction that a fog pattern of a given angle will have. For example, a fog pattern of 30° will have a nozzle reaction that is 97 percent that of a straight stream pattern.

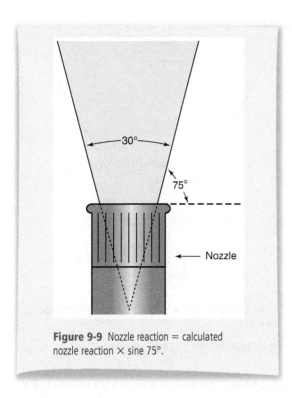

Figure 9-9 Nozzle reaction = calculated nozzle reaction × sine 75°.

Example 9-7

What is the nozzle reaction of a fog nozzle flowing 225 gpm at 100 psi nozzle pressure if it is on a 60° pattern?

Answer

First, calculate the nozzle reaction.

$$NR = 0.0543 \times gpm \times \sqrt{P}$$
$$= 0.0543 \times 225 \times \sqrt{100}$$
$$= 0.0543 \times 225 \times 10$$
$$= 122.18 \text{ lb}$$

Now multiply the answer by 0.87 to get the actual nozzle reaction on the 60° pattern.

$$\text{Actual nozzle reaction} = 122.18 \times 0.87$$
$$= 106.3 \text{ lb}$$

or a reduction in nozzle reaction of 15.88 lb from straight stream.

Stream Straighteners

To get the best possible stream from a smooth-bore nozzle, it is important that the water flow be as straight as possible as it leaves the nozzle. The problem is that fluids have a natural tendency to rotate in a counterclockwise direction in the Northern Hemisphere. This natural tendency of fluids to rotate as a result of the rotation of Earth is called the *Coriolis effect*. In the Southern Hemisphere, the **Coriolis effect** causes water to rotate clockwise. (There is no rotation at the equator.)

To prevent the Coriolis effect from trying to degrade stream integrity and to reduce the effects of turbulence in the water, stream straighteners are installed in master stream devices. A stream straightener is simply a pattern, such as the one in **Figure 9-10A**, that is made of thin metal or plastic and extends several inches up the barrel of a master stream device **Figure 9-10B**. The barrel is then attached to a master stream device, and the tip is attached to the barrel. The stream straightener will negate the Coriolis effect and allow for a smoother flow of water. Some hand line nozzles used with smooth-bore tips also have stream straighteners.

Back Pressure and Forward Pressure

Back pressure is defined as "the pressure exerted by a column of water against the discharge of a pump." This concept is essentially the same as the one introduced earlier that allowed us to calculate the pressure at the base of a column of water. The pressure exerted by the column of water is useful because it can create hydrant pressure, but when a pump has to pump water to an elevated position, this pressure works against us. See Chapter 2 for a refresher of that discussion.

In **Figure 9-11**, the pumper at ground level is pumping to a line on the third floor of a building. Assuming each floor is 8 ft and 1 ft for the thickness of each floor, the nozzle on the third floor would be about 18 ft above the pumper. A column of water this tall would exert $P = 0.433 \times H$, where $H = 18$ ft,

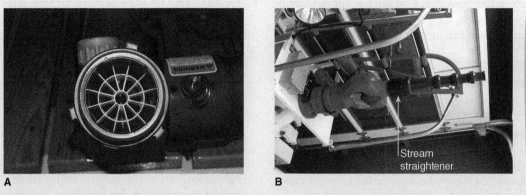

Figure 9-10 A. End view of stream straightener. Exact internal design may vary. **B.** Stream straightener for monitor nozzle.

Courtesy of William F. Crapo.

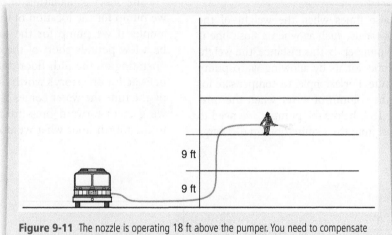

9 ft

9 ft

Figure 9-11 The nozzle is operating 18 ft above the pumper. You need to compensate for the back pressure.

or 7.79 psi pressure. The problem now is that when we begin to calculate the amount of pressure needed to deliver the specified gpm, we must also overcome the back pressure. In the example where the nozzle is operating on the third floor, in addition to pumping the correct nozzle pressure and friction loss, we need to add 7.79 lb of pressure to overcome the back pressure.

The back pressure must be accounted for in all instances where the nozzle is operating higher than the level of the pumper. It does not matter that the nozzle is attached to a preconnect attach line, on a 100-ft line hooked to a standpipe system, or at the tip of an aerial ladder; the back pressure will be the same in all instances as long as the vertical distance from the pump to the nozzle is the same. Even in instances where a fire department pumper is supplementing the water supply for a sprinkler system on an upper floor, the back pressure must be taken into account.

Example 9-8

How much back pressure must be compensated for on a ladder pipe if the stream is at 80-ft elevation?

Answer

Even though the angle of the ladder can actually change the actual height that the stream is off the ground, use the extension of the ladder to calculate the elevation.

$$P = 0.433 \times H$$
$$= 0.433 \times 80$$
$$= 34.64 \text{ psi}$$

There are also times when the weight of the water can work for us, such as when a hose line is 18 ft below the pumper. In this instance, the weight of the water works for us by allowing us to pump at a lower pressure. For example, to compensate for the weight of the column of water when the nozzle is operating 18 ft below the pumper, we need to subtract 7.79 psi from the required pump discharge

pressure. This pressure working for us is sometimes called the *forward pressure*. In fact, we can define **forward pressure** as the pressure created by a column of water that reduces the work required of the pump.

The amount of pressure to add to compensate for back pressure or to subtract to compensate for forward pressure depends on where the nozzle is operating. In **Figure 9-12** we have a hose line operating off a standpipe on the fifth floor while the nozzle is on the fourth floor. How much back pressure is added, and how much forward pressure is subtracted? Since the hose is being taken off the standpipe system on the fifth floor, do we need to overcome the back pressure to the fifth floor? If we do calculate the back pressure to the fifth floor, how do we account for the forward pressure gained by having the line descend one story?

The short explanation of this problem is that we pump for the location of the nozzle. In the example, if we pump for the fourth floor, we will be a few pounds short of the correct pressure for operating on the fifth floor due to failure to compensate for one story's worth of back pressure. But by the time the water comes back down one story, we gain in forward pressure from the fifth floor to the fourth floor what we originally lost in back

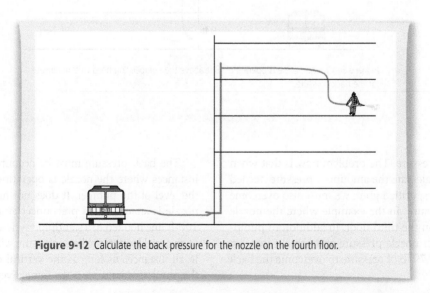

Figure 9-12 Calculate the back pressure for the nozzle on the fourth floor.

pressure from the fourth floor to the fifth floor. This will result in the correct pressure at the nozzle on the fourth floor.

Elevation

The need to compensate for pressure, either added as in back pressure or subtracted as in forward pressure, is called *elevation*. **Elevation** can be defined as the pressure compensation due to the vertical position of the nozzle in reference to the pump.

When you are calculating elevation, simply add pressure when the nozzle is above the pumper and subtract pressure when the nozzle is below the pumper. Keep in mind that the elevation is calculated for the position of the nozzle in relation to the pump, regardless of how it got there. For example, if a line goes into a building, up to the third floor, then down to the second, you calculate elevation for the second floor. [The proper use of elevation when calculating pump discharge pressure will be illustrated in Chapter 10.]

> **Note**
>
> When you are calculating elevation, simply add pressure when the nozzle is above the pumper and subtract pressure when the nozzle is below the pumper.

Special Appliances

When calculating the required pressure to provide the best stream, we often have to include allowances for special appliances. Special appliances are devices such as wagon pipes, monitor nozzles, and ladder pipes. These appliances require special consideration because of friction loss considerations in the device. For instance, friction loss in a monitor nozzle may be as high as 20 psi for the device itself. This 20 psi must be included in any calculations to obtain the correct pump discharge pressure to deliver the correct gpm.

Sprinkler systems and standpipe systems are also included in the category of special appliances. It is

critical that sprinkler systems not be overpumped, so particular attention must be paid to sprinkler systems when the water supply is supplemented. If the pressure to a sprinkler system is too high, the water flowing from each head will exit with such force that it will be atomized. The water droplets will then be too small and light to penetrate the fire plume to the seat of the fire and control it. In short, if a sprinkler system is overpumped, the water spray will become so fine as to be carried away by the heat generated by the fire, and the fire will claim the building.

Standpipe systems are less critical. The standpipe system in a building can be thought of as an extension of the pump discharge. It does have friction loss that must be accounted for, but that is easy to do. We usually account for the friction loss in a standpipe system by assigning a fixed pressure for the system. Then when we calculate the pressure for the line in use, we simply add the fixed system pressure as an appliance loss requirement.

Today some sprinkler and standpipe systems in the same building use a common riser from the fire department siamese. Any time a combination system is encountered, it is critical that the pressure to the system be limited. Because it is possible that hose lines taken off the riser may easily require a pressure higher than the system maximum, the sprinkler system must be given priority. Once the fire is declared under control and hose lines are sent in to overhaul, the system can be left to serve solely as a standpipe. To comply with this, it is important to know the maximum pressure for the system being used because each system may be different.

> **Note**
>
> If the combination system maximum pressure is not known, a pressure of 175 psi should not be exceeded.

Foam

A unique category of special appliances is foam. At one time when the word *foam* was mentioned around

Figure 9-13 Foam eductor.
© Jones & Bartlett Learning. Photograph by Glen E. Ellman.

firefighters, the natural inclination was to think in terms of Class B fires or high-expansion foam. Today that has changed. In addition to Class B and high-expansion foam, there are now Class A foam systems and compressed air foam systems (CAFS) for Class A fires.

If any kind of built-in foam system is used, the manufacturer's specifications should be followed. This caveat includes Class A foam, CAFS, or foam proportion systems for Class B foam. The most common foam-producing device is the simple foam eductor similar to the one illustrated in **Figure 9-13**. Using a foam eductor successfully requires attention to a few simple rules:

1. *Limit the amount of hose that comes off the eductor.*
2. *Have a flow of water that at least meets the minimum required by the eductor.* For example, the eductor may need a flow of 95+ gpm to work. If you try to flow only 80 gpm through this eductor, there will not be enough velocity to create the necessary vacuum in the eductor. Without enough vacuum, you will be unable to draft adequate foam solution out of the container. This will result in a foam solution that does not have the proper mixture of concentrate, producing an unreliable foam solution.

3. *Pump 200 psi to the eductor and do not be concerned about friction loss in the hose coming off the eductor or the nozzle pressure.*
4. *Calculate friction loss only for the hose supplying the eductor.*

Foam eductors usually require 200 psi at the inlet of the eductor. At this pressure, it is possible to overcome friction loss in the device, as well as have adequate pressure left over for the hose and nozzle on the discharge side of the eductor. For the eductor to work properly, the pressure required on the discharge side of the eductor cannot exceed 65 percent of the inlet pressure. This pressure will account for nozzle pressure, friction loss, and elevation. If the hose lay on the discharge side of the eductor requires more than 65 percent of the inlet pressure, insufficient water will flow for the eductor to work properly. Knowing this, you can change the hose size or length, depending on needs, as long as the pressure required on the discharge side does not exceed 65 percent of the pressure at the inlet. Foam eductors can actually work at pressures lower than 200 psi; however, the foam delivery rate and concentration will be adversely affected.

Example 9-9

How much 1¾-in hose can be taken off an eductor if it will be working 15 ft above the level of the pumper and flowing 95 gpm?

Answer

First, find the allowable pressure on the discharge side of the eductor.

$$200 \times 0.65 = 130 \text{ psi}$$

Now calculate the friction loss FL in 1¾-in hose for 95 gpm.

$$FL\ 100 = CF \times 2Q^2$$
$$= 7.76 \times 2 \times (0.95)^2$$
$$= 7.76 \times 2 \times 0.9025$$
$$= 14 \text{ psi}$$

With only 130 psi available on the discharge side of the eductor,

$$130 - 100 \text{ psi nozzle pressure} = 30 \text{ psi left}$$

(continues)

(*Example 9-9 continued*)

Now calculate the elevation loss.

$$P = 0.433 \times H$$
$$= 0.433 \times 15$$
$$= 6.5 \text{ psi}$$

Pressure available for friction loss is

$$30 - 6.5 = 23.5 \text{ psi}$$

To find how much hose to use, divide the pressure available for friction loss by the FL 100.

$$\text{Feet of hose} = 23.5 \div 14$$
$$= 1.67 \text{ ft, rounded down}$$
$$\text{to 150 ft of hose to account for 50-ft sections}$$

Fireground Fact

Foam Eductors and Bernoulli's Principle

It has already been mentioned how the foam eductor is a practical application of Bernoulli's principle. When you studied Bernoulli's principle, you learned that it did not take into account friction loss. But we also understand that, in real life, friction loss is a certainty. In the eductor, the cost of Bernoulli's principle in terms of friction loss is 35 percent of the inlet pressure.

Standard CVFSS nozzles can be used with some foams, although other foams require special foam nozzles. Even those foams that will work without special foam nozzles will produce better foam with dedicated foam nozzles or foam tubes attached to CVFSS nozzles.

Dedicated foam nozzles can be either handheld or master stream devices. **Figure 9-14** is a master stream foam nozzle capable of delivering 340 gpm. It is also a self-educting nozzle, meaning an eductor is built into the nozzle. This nozzle requires 100 psi nozzle pressure to operate, including picking up the

Figure 9-14 Self-educting, master stream foam nozzle.
Courtesy of William F. Crapo.

foam. The male quick-connect coupling at the lower right of the nozzle is where the pickup tube attaches.

Foam Application Rate

For foam application to be effective, it must be applied at a rate that ensures complete coverage. This application rate can vary across different kinds of foam. For example, protein foam and fluoroprotein foam must be applied at an application rate of 0.16 gallon per minute per square foot (gpm/ft^2) to be effective, whereas aqueous film-forming foam (AFFF) and film-forming fluoroprotein foam (FFFF) only need to be applied at an application rate of 0.1 gpm/ft^2. From the application rate, we can calculate how large an area we can cover with a single line/nozzle. This can then be used to calculate the minimum number of lines we need to cover a large Class B fire. A simple formula will help us determine how much area a single point of discharge will cover. The formula for how much area a single line/nozzle will cover is area of cover (AC) = gpm (of the nozzle)/AR, where AC is the area of coverage, gpm is the flow of the nozzle, and AR is the application rate in square feet.

$$AC = \text{gpm (of the nozzle)/AR}$$

Example 9-10

How many square feet of area can be covered by a foam nozzle that is delivering 95 gpm of AFFF?

Answer

$$AC = gpm/AR$$
$$= 95/0.1 \, gpm/ft^2$$
$$= 950 \, ft^2$$

A single line of AFFF at 95 gpm can be expected to cover an area of 950 ft^2.

If a single line of AFFF at 95 gpm can produce enough foam to cover an area of 950 ft^2, you can now calculate how many lines are needed to cover a fire covering a known area. For example, if the area of the fire is only about 500 ft^2, a single line will be adequate. However, if the fire covers 1,200 ft^2, two lines will be needed.

In addition to being able to meet the application rate of foam, you will need sufficient foam to maintain the application rate for an adequate time. Generally speaking, foam should be applied for 15 minutes (min). This brings us to one last point concerning the application of foam: Never begin applying foam until enough foam is on hand to be effective. The question is, how much foam must be on hand? If you have a foam unit with several hundred gallons of foam in a foam tank, this should be sufficient unless the fire is of biblical proportions. However, when you are operating with hand lines and a foam eductor, knowing how much foam to have on hand becomes critical.

Calculating the amount of foam needed requires a two-step approach. First, we must find out how much foam solution we can make from a 5-gallon (5-gal) can of foam at the chosen application rate. For the purpose of this illustration we will be using a 3 percent AFFF solution and applying it at 95 gpm. We must figure out how many gallons of foam concentrate will be used in 1 min. Remember, Example 9-10 told us that application of AFFF at 0.1 gpm/ft^2 will cover 950 ft^2 if applied at 95 gpm. Begin by finding how many gallons are contained in 3 percent of 95 gpm:

$$95 \times 0.03 = 2.85 \, gal$$

The number 2.85 represents how many gallons of concentrate will be used in 1 min. Divide that into 5 gal, the size of the foam can, to get the number of minutes it will take to empty the can:

$$5/2.85 = 1.75 \, min \, (1 \, min \, 45 \, s)$$

To find how many 5-gal cans of foam will be needed to apply foam for 15 min, divide 15 min by the time each can will last:

$$15/1.75 = 8.57, \, or \, 9 \, cans \, containing \, 5 \, gal$$
$$of \, foam \, concentrate \, are \, needed$$

Of course, if the area of the spill is larger than 950 ft^2, you will have to double, triple, and so forth, the quantity, depending on the size of the spill. Also, if the size of the spill is substantially smaller than 950 ft^2, you will not need the entire 9 containers of concentrate. You can prorate the amount of foam concentrate required by the size of the spill relative to 950 ft^2, but keep in mind that you still have to flow the foam solution at 95 gpm. Having a little more foam than you need is better than coming up short.

Example 9-11

How many cans of foam will be needed to apply a protein foam to an oil fire of about 500 ft^2 if we are applying a 6 percent foam at 95 gpm?

Answer

First, find AC for 95 gpm.

$$AC = gpm/AR$$
$$= 95/0.16$$
$$= 593.75 \, ft^2 \, Only \, one \, line \, will \, be \, necessary.$$

Now find how long a can of foam will last.

$$95 \times 0.06 = 5.7 \, gpm \, of \, concentrate$$

Next, find how many minutes the 5-gal can will last.

$$5/5.7 = 0.877 \, min \, (53 \, s)$$

Finally, find how many 5-gal containers of foam will be needed.

$$15/0.877 = 17.10, \, or \, 18, \, containers \, of \, foam$$
will be needed.

Chapter Summary

- The range of fire streams is determined by the nozzle pressure, nozzle diameter, angle of the stream, and pattern of the stream.
- The most effective angle for a fire stream is 45°.
- The nozzle converts pressure energy to velocity energy.
- Nozzle reaction is the equal and opposite reaction to the nozzle.
- The nozzle reaction varies as the pattern changes.
- Stream straighteners correct for the Coriolis effect.
- Foam application rates vary with the type of foam.

Key Terms

automatic nozzle A nozzle designed to deliver water at any gpm rate within its design range and to maintain its designed nozzle pressure.

combination variable fog and straight stream (CVFSS) nozzle A nozzle that is capable of producing any pattern of water, from a straight stream to 100° fog.

Coriolis effect The natural tendency of fluids to rotate according to the rotation of Earth.

elevation The pressure compensation due to the vertical position of the nozzle in reference to the pump.

fog nozzle A nozzle that is capable of producing a spray pattern.

forward pressure Pressure created by a column of water that reduces the work of the pump.

Newton's third law of motion Whenever one object exerts a force on a second object, the second exerts an equal and opposite force on the first.

nozzle pressure The pressure required at the nozzle to allow it to deliver the designed gpm and spray pattern.

nozzle reaction The opposing force created by water exiting the nozzle.

sine A trigonometric function of an angle that when multiplied by the length of the hypotenuse of a right triangle will find the length of the side opposite the angle.

smooth-bore nozzle A nozzle that is capable of producing only a solid stream of water.

vector The measure of strength and direction of forces.

Case Study
Reach of Streams

It is widely believed that to obtain the greatest reach, a solid stream or smooth-bore tip is a necessity. During a training exercise with my department in the late 2000s, that opinion was called into doubt.

The exercise was designed to demonstrate the difference in reach between a smooth-bore nozzle and a CVFSS nozzle flowing the same gpm. To duplicate as accurately as possible the conditions for testing both nozzles, the following setup was used. Even though the tips being tested were hand line tips, they were attached to a monitor nozzle. This setup allowed us to duplicate the angle of the stream and provided a stationary position for the nozzle.

The gpm flow was determined by the selection of smooth-bore tip (1¼ in). The exact flow was determined by pumping the line from a discharge with a flowmeter and was verified by a pitot gauge. After the reach of the smooth-bore tip was marked, the CVFSS tip was placed on the nozzle. Using an automatic CVFSS tip, the pump operator then pumped the tip until the same gpm flow was obtained. To everyone's amazement, the reach of the two streams was essentially the same.

The one element in question was that the CVFSS tip seemed to be losing more water in the form of spray than the smooth-bore tip. However, it was impossible for us to verify just how much water had actually reached the extreme reach of either tip.

1. What factor does NOT affect reach of a fire stream?

   A. Nozzle pressure
   B. Angle of stream
   C. Age of nozzle
   D. Wind

2. Why was the hand line nozzle put on a monitor nozzle base?

   A. It provided ease in operating the nozzle.
   B. It provided a consistent angle and a stationery platform for the nozzle.
   C. Fewer personnel were required to accomplish the evolution.
   D. This is a local training evolution.

3. What was the purpose of taking the pitot reading when using a flowmeter?

   A. It was used to verify the flow.
   B. Flowmeters are not to be trusted under any circumstances.
   C. It verified the correct nozzle pressure.
   D. It is always good practice to let personnel get experience using various types of pressure-measuring devices whenever possible.

4. What was the one drawback the fog nozzle had in achieving the same reach as the smooth-bore nozzle?

   A. The required angle of the stream was different.
   B. A higher-flow fog nozzle was required.
   C. It appeared that the fog nozzle was losing more water in the form of spray.
   D. There were no differences; virtually everything was the same.

Review Questions

1. Why is it so important that water be applied at precisely the correct area of a fire?

2. Why is it so important that the water be applied in the correct form?

3. Up to the limit named, how much water should be inside a 15-in circle?

4. Why is it considered necessary for a fire stream to penetrate an opening and be deflected off the ceiling?

5. With all else being equal, how does the nozzle diameter affect the range of a stream?

6. What is the correct angle for the maximum horizontal range?

7. What is the best angle for an effective firefighting stream?

8. Which of Newton's three laws explains nozzle reaction?

9. How is the formula NR = 1.57 × D^2 × P converted to find the nozzle reaction for a fog nozzle?

10. Which has the greater nozzle reaction, a 30° pattern or a 60° pattern?

11. How are back pressure and forward pressure different from friction loss?

12. Is compensation for elevation added or subtracted from the required pump discharge pressure?

13. What are special appliances?

Activities

Apply your knowledge of fire streams to solve the following problems.

1. The division chief orders you to have your company direct a stream into the third floor of a building. If the window is approximately 20 ft off the ground, how far from the building will you have to position the nozzle?

2. What is the maximum horizontal range for a $^{15}⁄_{16}$-in tip at 50 psi?

3. What is the maximum vertical range for the line in Activity 2?

4. What is the nozzle reaction on a ladder pipe at 80-ft extension if it is using a 1½-in tip at 80 psi?

5. What is the nozzle reaction for a fog nozzle flowing 205 gpm with a nozzle pressure of 31.14 psi? Compare your answer to the answer in Example 9-6.

6. How much back pressure would be created in a standpipe system in a seven-story building if each story were 12 ft and we were pumping to a line on the fifth floor?

7. How many 5-gal containers of 6 percent fluoroprotein foam will be needed for a flammable liquid fire that covers 900 ft²? All you have available are hand lines and 60 gpm foam eductors.

WRAP-UP

Challenging Questions

1. What is the maximum horizontal range for a 1-in tip at 50 psi nozzle pressure?

2. What is the maximum vertical range for a 1-in tip at 50 psi nozzle pressure?

3. Calculate the nozzle reaction for a ⅞-in tip at 40 psi nozzle pressure.

4. How many gallons of foam would the master stream nozzle in Figure 9-13 use in 1 min, using foam at a 3 percent concentration?

5. What size area could the master stream foam nozzle in Figure 9-13 cover with foam using FFFF foam?

6. Go back to Example 9-9. Recalculate the feet of hose needed for a nozzle working 15 ft below the level of the pumper.

Formulas

To find a horizontal range of streams:

$$HR = \tfrac{1}{2}NP + 26*$$

To find a vertical range of streams:

$$VR = \tfrac{5}{8}NP + 26*$$

To calculate the nozzle reaction:

$$NR = 1.57 \times D^2 \times P$$

To find the nozzle reaction when gpm and pressure are known:

$$NR = 0.0543 \times gpm \times \sqrt{P}$$

To calculate the area of coverage AC for foam:

$$AC = gpm \text{ (of the nozzle)}/AR$$

*Add 5 for each ⅛ in of nozzle diameter more than ¾ in.

Reference

Särdqvist, Stefan. *Water and Other Extinguishing Agents*. Husk-varna, Sweden: Swedish Rescue Services Agency, 2002.

Calculating Pump Discharge Pressure

LEARNING OBJECTIVES

Upon completion of this chapter, you should be able to:

- Understand the pump discharge pressure (PDP) formula.
- Select and insert the correct components into the PDP formula.
- Calculate the correct PDP for a variety of basic situations.
- Calculate the correct PDP by using the relay formula.

Case Study

Training time in the fire station is always a mixed bag. It can be informative and a chance to practice infrequently used skills. But at times it can be less than exciting.

The captain begins by presenting a simple pump problem. He asks you to calculate the correct PDP. You walk up to the board and easily calculate the total friction loss and add the correct nozzle pressure to find the PDP. "Correct," says the captain. "However, why didn't you write out the entire PDP formula?" he asks.

1. What are the advantage and the disadvantage of writing out the entire PDP formula when you are working through even the most basic problem?
2. Why does elevation have a plus-or-minus sign (±) in front of it?
3. What are some examples of special appliances?

NFPA 1002 Standard for Fire Apparatus Driver/Operator Professional Qualifications, 2014 Edition

This chapter addresses the following requisite knowledge elements within sections

5.2.1, 5.2.2, 5.2.3, and **5.2.4:** hydraulic calculations for friction loss and flow; foam system limitations; calculation of pump discharge pressure; location of fire department connection; and operating principles of sprinkler systems as defined in NFPA 13, NFPA 13D, and NFPA 13R.

Introduction

Up to this point, this book has been concerned with a specific principle, rule, or theory of hydraulics. Now we put this knowledge to work to find the correct PDP, or engine pressure, for various hose layouts.

The point of learning hydraulics is to be able to accurately calculate the pressure needed in hose lines on the fireground. That process actually began in previous chapters that were dedicated to a single process, such as calculating the gallons per minute (gpm) or friction loss. In this chapter we put the entire process together. The end result will be a PDP that is as accurate as the process allows. Keep in mind, when you are doing these problems, that a high degree of accuracy is expected here because these are not fireground calculations. The problems presented here are for training purposes, and accuracy is part of the training. Thoughts on adapting the formula presented in this chapter and the next to the fireground are addressed as appropriate, but are more thoroughly presented in Chapter 13.

It is the author's hope that the way this text is presented, by first explaining scientific principles and showing the evolution of various formulas, will motivate firefighters to be as accurate as practical when doing hydraulic calculations. In fact, for the purposes of this chapter and the rest of this text, total friction loss will be calculated before it is rounded off and inserted in the PDP formula. Elevation will also be calculated and rounded off in similar fashion. If we practice being exact when we do calculations on paper, then when we transfer the calculations to real-world applications, we will be well within practical and acceptable parameters. If, however, we practice with friction loss charts that have been rounded off or hand methods that only approximate friction loss, then we can never expect a high degree of accuracy. Aristotle expressed this sentiment best when he said, "We are what we repeatedly do. Excellence, then, is not an act, but a habit."

Over the years, several formulas have been developed to calculate PDP. Some require knowledge of special factors or are specific to limited situations. The best formula, however, is one that is accurate, easy to remember without knowledge of special factors, and as useful in the classroom as on the fireground. Just such a formula exists and is used to calculate PDP in this chapter. It is referred to as the *PDP formula*.

The Pump Discharge Pressure Formula

The purpose of the PDP formula is to give us a consistent, easy-to-follow, and reliable formula for calculating the PDP. To correctly calculate the PDP, it is necessary to add or subtract nozzle pressure, friction loss, elevation pressure, and allowances for special appliances (appliance loss). The formula used here is simple, easy to remember, accurate, versatile, and nearly infallible. If you use this formula both in the classroom and on the fireground, it will become second nature and the only formula you will need to learn.

The PDP formula is PDP = NP + FL 1 + FL 2 ± EP + AL, where PDP = pressure the engine will be pumping, NP = nozzle pressure, FL = total friction loss, EP = elevation pressure, and AL = appliance loss. Friction loss appears twice because we often have to deal with situations where there is more than one size hose involved or one size hose may have more than one friction loss. In real life, there could actually be more than two occurrences of friction loss. For example, standpipe evolutions always have two friction losses. However, it is the author's experience that when this formula is used on the fireground, the need for more than the two friction losses will be almost nonexistent. Think of the two occurrences of friction loss in the formula as a reminder to look for the need for multiple friction losses. Elevation pressure is either plus or minus (±) because it can be added or subtracted depending on the location of the nozzle in relation to the pumper. Finally, AL is pressure loss due to friction in various appliances.

PDP = NP + FL 1 + FL 2 ± EP + AL

In this chapter, we solve basic calculations using progressively more elements of the PDP formula. Each element is explained in further detail as it is used.

Solving Problems

When you are solving for PDP, the key to accuracy is to follow a defined process that can be repeated under all circumstances. The following steps are recommended to provide consistency and accuracy to solving problems:

1. Read the entire problem at least twice to determine what is being asked.
2. Draw a schematic of the evolution. This will allow you to have a visual representation of the elements of the problem, putting the problem into context.
3. Label your drawing with known facts such as pressures, hose lengths, elevations, and nozzle pressures. Once your drawing is labeled with the known facts, identify what is missing. The missing information in your drawing should be the same as what you determined was being asked in step 1.
4. Select the formula needed to solve for the unknown(s). Multiple formulas may be needed to solve a single problem.
5. Solve for each unknown. In problems requiring solutions for multiple unknown factors, once they are all determined, insert the relevant information into the PDP formula and calculate the answer.

PDP = NP + FL 1 + FL 2 ± EP + AL

The most fundamental PDP calculation involves simply adding the nozzle pressure and the required friction loss. Earlier we learned how to calculate friction loss. The friction loss we calculated was for every 100 feet (ft) of hose. When calculating friction loss to put into our PDP formula, we need to determine the amount of friction loss for the total hose lay. For this we use the formula FL = FL 100 × L, where FL 100 is the friction loss per 100 ft and L is the length of hose in hundreds. Remember, to determine the length of line in hundreds, simply move the decimal point two places to the left. A line 650 ft long has an L of 6.5. Refer to Chapter 6 for more information on how to calculate friction loss.

$$FL = FL\ 100 \times L$$

Example 10-1

Find the PDP needed for a 350-ft 1½-inch (1½-in) hose line with a CVFSS nozzle, if the nozzle pressure is 100 pounds per square inch (psi) and it is flowing 100 gpm Figure 10-1.

Answer

Calculate FL 100 by the formula FL 100 = CF × $2Q^2$. (See Chapter 6.)

FL 100 for 100 gpm in 1½-in hose is

$$
\begin{aligned}
FL\ 100 &= CF \times 2Q^2 \\
&= 12 \times 2 \times (1)^2 \\
&= 12 \times 2 \times 1 \\
&= 24\ \text{psi} \\
FL &= FL\ 100 \times L \\
&= 24 \times 3.5 \\
&= 84\ \text{psi}
\end{aligned}
$$

Now find the PDP:

$$
\begin{aligned}
PDP &= NP + FL\ 1 + FL\ 2 \pm EP + AL \\
&= 100 + 84 + 0 \pm 0 + 0 \\
&= 184\ \text{psi}
\end{aligned}
$$

The correct PDP for a 1½-in hose line flowing 100 gpm at 100 psi nozzle pressure is 184 psi. This is the most basic PDP calculation, and it is the one used for the majority of all fireground situations.

PDP = NP + FL 1 + FL 2 ± EP + AL

It may be necessary at times to calculate PDP when a hose line is made up of more than one size hose. A practical fireground scenario might involve making a knockdown with a 2½-in hand line and then extending the line with 1¾-in hose. For example, after a fire has been knocked down with the larger line, it is extended with a smaller hose for overhaul. Or the larger hose may be extended with something smaller to give it needed mobility. In either case the flow in each size hose will be the same (that of the smaller hose), but the friction loss in each size hose will be different and must be calculated independently.

Example 10-2

What is the PDP for a line that consists of 150 ft of 2½-in hose that has been extended with 100 ft of 1¾-in hose? The 1¾-in CVFSS nozzle is operating at 100 psi nozzle pressure and 125 gpm Figure 10-2. (This is a good example of how multiple friction losses are handled in the PDP formula. Note how the friction loss for the 2½-in hose is designated as FL 1 and the friction loss for the 1¾-in hose is designated as FL 2.)

Answer

First calculate the FL 100.

FL 100 for 2½-in hose flowing 125 gpm:

$$
\begin{aligned}
FL\ 100 &= CF \times 2Q^2 \\
&= 1 \times 2 \times (1.25)^2 \\
&= 1 \times 2 \times 1.56 \\
&= 3.12\ \text{psi} \\
FL &= FL\ 100 \times L \\
FL\ 1\ (2\tfrac{1}{2}) &= 3.12 \times 1.5 \\
&= 4.68\ \text{or 5 psi total friction} \\
&\quad\ \text{loss in the 2½-in hose}
\end{aligned}
$$

This is FL 1.

(continues)

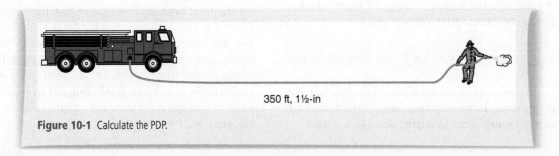

350 ft, 1½-in

Figure 10-1 Calculate the PDP.

(Example 10-2 continued)

Now calculate the friction loss in the 1¾-in hose:

$$FL\ 100 = CF \times 2Q^2$$
$$= 7.76 \times 2 \times (1.25)^2$$
$$= 7.76 \times 2 \times 1.56$$
$$= 24.21\ psi$$
$$FL\ 2\ (1¾) = 24.21 \times 1$$
$$= 24.21\ or\ 24\ psi\ total\ friction$$
$$loss\ in\ the\ 1¾\text{-in hose}$$

This is FL 2.

Now find the PDP:

$$PDP = NP + FL\ 1 + FL\ 2 \pm EP + AL$$
$$= 100 + 5 + 24 \pm 0 + 0$$
$$= 129\ psi$$

In Example 10-2, the larger-diameter hose is laid first and then the smaller-diameter hose is used to extend it. If, for some reason, the hose had been laid in with the smaller-diameter hose first and then the larger-diameter hose, then the problem would be solved in the same way. The order of the different diameter hose is irrelevant. It is not likely that hand lines will be extended with larger-diameter hose, but it is possible that supply lines can have two different diameters of hose in a single line in any order.

PDP = NP + FL 1 + FL 2 ± EP + AL

Earlier we introduced the concept of back pressure. Later back pressure and forward pressure were mentioned, where they evolved into the concept of elevation. Here is where elevation is applied. To arrive at the correct PDP, frequently elevation above or below the pumper must be taken into consideration. (See Chapter 3 for a discussion of back pressure and Chapter 9 for more about both back pressure and forward pressure.)

Where the exact elevation is known, we can use the formula $P = 0.433 \times H$ to determine the exact pressure adjustment required. In most cases we do not know the exact height and we only know how many stories the nozzle is above or below the pumper. If we assume the average story of a building to be 12 ft from floor to floor, using the formula $P = 0.433 \times H$ gives us a standard pressure per floor of $P = 0.433 \times 12$, or 5.2 psi. This amount is usually rounded off to a constant of 5 psi for each floor above the first. The actual distance between the floors of a building can vary. Some texts use a figure of 10 ft, and the author is aware of buildings where the distance between floors is actually 14 ft. If you have only one- and two-family dwellings in your jurisdiction, you may want to use a constant of 4 psi per story above the first story because the floor-to-floor height of the average house is only about 9 ft. For the purposes of this text, when you are calculating elevation, 5 psi per story is used. (The formula $P = 0.443 \times H$ is discussed in detail in Chapter 2.)

There will also be situations where elevation is estimated. Where possible, the formula $P = 0.433 \times H$ should be used to calculate the exact pressure adjustment needed. However, there are situations on the fireground in which quick calculations are necessary. In such cases, it is common practice to use ½ psi per foot (psi/ft) of elevation. For example, where an aerial ladder is extended with the ladder pipe in operation, if the ladder is extended 80 ft, the required compensation for elevation will be 40 psi.

One final reminder on elevation: remember that elevation is calculated for the position of the nozzle in relation to the pump. If the line goes to the fourth floor and then back down to the second floor, you calculate elevation for the second floor.

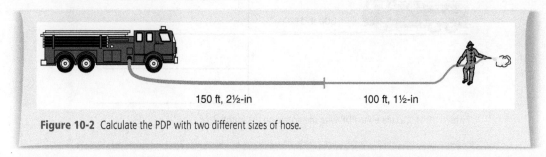

150 ft, 2½-in   100 ft, 1½-in

Figure 10-2 Calculate the PDP with two different sizes of hose.

Example 10-3

What PDP is required at the pumper if it is pumping to a 300-ft 1¾-in line with a CVFSS nozzle, flowing 150 gpm at 100 psi nozzle pressure, operating on the fourth floor **Figure 10-3** ?

Answer

FL 100 for 150 gpm in 1¾-in hose:

$$FL\ 100 = CF \times 2Q^2$$
$$= 7.76 \times 2 \times (1.5)^2$$
$$= 7.76 \times 2 \times 2.25$$
$$= 34.92\ psi$$
$$FL = FL\ 100 \times L$$
$$= 34.92 \times 3$$
$$= 104.76\ or\ 105\ psi\ total\ friction\ loss$$

Now find the PDP. Remember, the nozzle is on the fourth floor, or three floors above the first. Elevation pressure will be 5 psi for each floor above the first, or 15 psi.

$$PDP = NP + FL\ 1 + FL\ 2 \pm EP + AL$$
$$= 100 + 105 + 0 + 15 + 0$$
$$= 220\ psi$$

It is important to remember that elevation pressure can be calculated just as easily for a location below the position of the pumper. Again, do not count the first floor, but count every floor below the first. A basement requires a −5 psi of elevation while a subbasement requires a −10 psi of elevation.

Example 10-4

An engine company has advanced its 250-ft preconnect 1¾-in line with a CVFSS nozzle to a second-level underground parking garage for an automobile fire. What is the required PDP if the line is flowing 150 gpm at 100 psi nozzle pressure **Figure 10-4** ?

Answer

We already know from Example 10-3 that FL 100 for the 1¾-in line is 34.92 psi.

$$FL = FL\ 100 \times L$$
$$= 34.92 \times 2.5$$
$$= 87.3\ or\ 87\ psi\ total\ friction\ loss$$

Now find the PDP. Elevation pressure is −10 psi.

$$PDP = NP + FL\ 1 + FL\ 2 \pm EP + AL$$
$$= 100 + 87 + 0 - 10 + 0$$
$$= 177\ psi$$

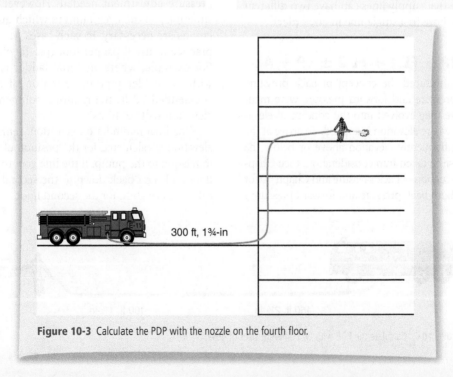

300 ft, 1¾-in

Figure 10-3 Calculate the PDP with the nozzle on the fourth floor.

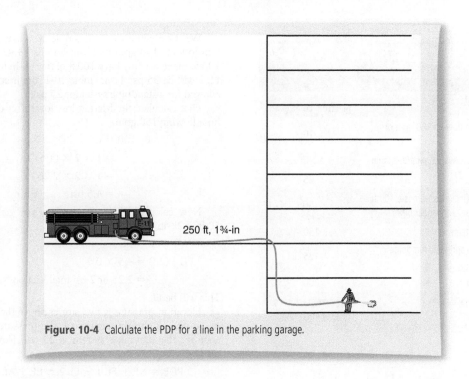

250 ft, 1¾-in

Figure 10-4 Calculate the PDP for a line in the parking garage.

PDP = NP + FL 1 + FL 2 ± EP + AL

In addition to nozzles and hose, we often lose friction in other devices, such as monitor nozzles or ladder pipes. To deliver the correct gpm at a pressure that will produce the desired pattern, friction loss in these special appliances must be accounted for. **Table 10-1** shows friction loss allowances for selected devices. Keep in mind that the figures given in this table are not absolute figures because they can vary from one manufacturer to another. Instead, they are provided for illustrative purposes and are used through the remainder of this text for the purpose of standardization.

In general, the pressures in Table 10-1 were determined through testing. Some of the pressures, such as maximum pressures on sprinkler systems and combination sprinkler and standpipe systems, were calculated by engineers to provide maximum flow without compromising the effectiveness of the systems. The pressure on the foam eductor is a design requirement in order to obtain a sufficient venturi effect to create a vacuum to draft foam concentrate from a container.

For water tower operations on aerial devices that have piping installed to deliver water to the nozzle, the manufacturer usually stipulates a pressure. It is usually a specific pressure that must be maintained at the intake of the waterway, and it includes all friction loss in the system, loss in the monitor/turret, and nozzle pressure. The pump operator needs only to add friction loss for the hose supplying the system and elevation. The required intake pressure must be made available to pump operators supplying the aerial device.

When you are calculating PDP when a special appliance is used, all the elements of the PDP formula are employed. Even some special appliances, such as ladder pipes and monitor nozzles, have a

Table 10-1	Special Appliance Loss Allowances	
Appliance	**Friction Loss, psi**
Wye/Siamese flowing 350 gpm or greater	10
Monitor nozzle/wagon pipe/deck gun	20
Standpipe	25
Ladder pipe (includes Siamese, 100 ft of 3-in hose, and ladder pipe)	50
Sprinkler systems	150
Combination sprinkler/standpipe system	175
Foam eductor	200
Prepiped aerial device	Use manufacturer's recommendation

nozzle pressure. There is also friction loss in the hose to the appliance, and the hose can be located above or below the pumper. Only sprinkler systems and combination systems will have one set pressure that covers everything.

What is the PDP where a pumper is pumping to a 1¾-in hand line with a CVFSS nozzle on the third floor of a building? The line is 100 ft long, flowing 150 gpm at 100 psi nozzle pressure and is connected to a standpipe system. The standpipe system is supplied by a 50-ft section of 2½-in hose **Figure 10-5**.

Answer

A standpipe system should be thought of as a water distribution system. While they are usually vertical, horizontal standpipe systems are also used in special applications. Regardless, the PDP will be calculated in the same way.

We already know from Example 10-3 that the friction loss for 150 gpm in 1¾-in hose is 34.92 psi per 100 ft. Since we only have 100 ft of the 1¾-in hose, the FL 1 will be 35 psi. From Table 10-1, the friction loss allowed for a standpipe system is 25 psi.

First calculate the friction loss for 100 ft of 2½-in hose flowing 150 gpm.

$$FL\ 100 = CF \times 2Q^2$$
$$= 1 \times 2 \times (1.5)^5$$
$$= 1 \times 2 \times 2.25$$
$$= 4.5\ psi$$

Now calculate FL 2. Remember, you only have 50 ft of 2½-in hose.

$$FL = FL\ 100 \times L$$
$$FL\ 2\ (2\frac{1}{2}) = 4.5 \times 0.5$$
$$= 2.25\ or\ 2\ psi\ total\ friction\ loss$$

This will be FL 2.

Elevation pressure is two stories above the pump, or 10 psi. As already mentioned, the pressure allowance for the standpipe system is 25 psi. Now solve for PDP.

$$PDP = NP + FL\ 1 + FL\ 2 \pm EP + AL$$
$$= 100 + 35 + 2 + 10 + 25$$
$$= 172\ psi$$

Sprinkler Systems

When a sprinkler system is used, the one instance in which a special appliance will be encountered is where it will be the only element, with rare exceptions for elevation, of the PDP formula used. The recommended course of action for a sprinkler system is to pump a set pressure: 150 psi (50 psi less than test pressure) is recommended unless specific knowledge dictates a different pressure. Because it is unknown how many heads may have opened, where they are located, and how much water is flowing, the set pressure is used. A note of caution is in order here: If you don't get sufficient water to the seat of the fire, extinguishment or even control is unlikely. If you are pumping to a sprinkler system with a two-stage pump, the transfer valve should be in the parallel position.

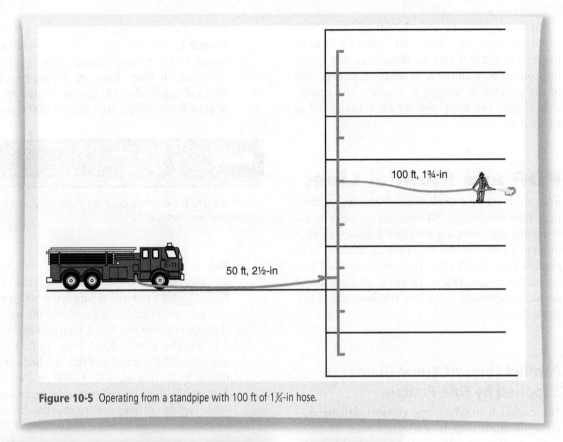

Figure 10-5 Operating from a standpipe with 100 ft of 1¾-in hose.

Make certain the sprinkler system is not overpumped. If too much pressure is put into the system, then water will come out of the sprinkler head at too high a velocity and the water will be atomized. When water is atomized, the droplets are too small and light to penetrate the fire plume and fall to the seat of the fire.

Foam Eductor

One final special case under special appliances is the foam eductor. As mentioned earlier, foam eductors are designed to work at a specified pressure and flow. When calculating the PDP for an eductor, you only need to add friction loss and elevation pressure to the point of the eductor. See Chapter 9 for more on foam eductors.

Example 10-6

Calculate the PDP for a foam line that has 200 ft of 2½-in hose supplying a foam eductor designed to flow a minimum of 95 gpm at 200 psi.

Answer

In this example the line is flowing 95 gpm. FL 100 for 95 gpm through the 2½-in hose is

$$FL\ 100 = CF \times 2Q^2$$
$$= 1 \times 2 \times (0.95)^2$$
$$= 1 \times 2 \times 0.90$$
$$= 1.80 \text{ psi}$$
$$FL = FL\ 100 \times 2$$
$$= 1.80 \times 2$$
$$= 3.6 \text{ or } 4 \text{ psi total friction loss}$$
$$PDP = NP + FL\ 1 + FL\ 2 \pm EP + AL$$
$$= 0 + 4 + 0 + 0 + 200$$
$$= 204 \text{ psi}$$

In Example 10-6, with 200 ft of 2½-in hose supplying the foam eductor, the eductor will work properly at a PDP of 204 psi. Remember that we are not concerned with the hose on the discharge side of the eductor, as long as it is within the acceptable length. The acceptable length is calculated in Chapter 9.

PDP and Parallel Lines

Real-life fireground evolutions often involve multiple hose evolutions. Where one pumper is responsible for pumping multiple lines to another pumper or device, the evolution is referred to as *parallel lines*. Parallel line lays can be divided into two different categories: (1) where all lines are of equal diameter and (2) where the lines are of unequal diameter.

Parallel Lines of Equal Diameter Supplied by One Pumper

The evolution in which one pumper will be supplying multiple lines to a master stream device is common. However, three rules governing these situations must be taken into consideration to arrive at the correct PDP. These rules apply to evolutions where all lines have the same diameter.

- Rule 1. *Each line into the device is assumed to be carrying an equal share of the water.* For instance, if two lines are supplying a monitor

nozzle that is flowing 600 gpm, the gpm is divided by 2 to get the quantity of water assumed to be flowing through each hose, or 300 gpm. If three lines are supplying the monitor nozzle, the 600 gpm is divided by 3 to get a flow through each hose of 200 gpm.

> **Note**
>
> Each line into the device is assumed to be carrying an equal share of water.

- Rule 2. *When the lengths of the parallel lines are unequal, the average length is used.* In Figure 10-6 a pumper is supplying 600 gpm to a monitor nozzle. It is supplying the water through one 3-in line 400 ft long and another 3-in line that is only 300 ft long. To find the average, add the lengths of the two lines and divide by 2, that is, 400 + 300 = 700 ÷ 2 = 350 ft. This procedure works regardless of the number of lines coming into the device, as long as they are from the same pumper.

> **Note**
>
> When the lengths of the parallel lines are unequal, the average length is used.

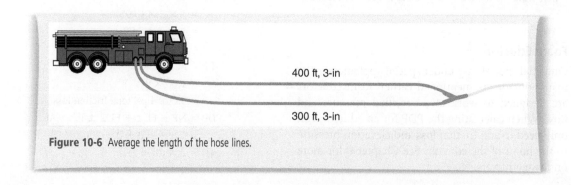

400 ft, 3-in

300 ft, 3-in

Figure 10-6 Average the length of the hose lines.

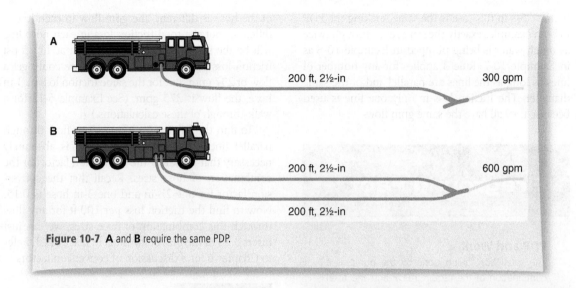

Figure 10-7 A and **B** require the same PDP.

- Rule 3. *When you are calculating PDP where Rule 1 and Rule 2 apply, friction loss through only one line is used.* Where lines are parallel, friction loss is not cumulative. The best way to explain this concept is with an illustration, as shown in .

Example 10-7 and Example 10-8 validate Rule 3 by illustrating how it is possible to have the same PDP in both Figure 10-7A and Figure 10-7B.

Example 10-7

In Figure 10-7A, the pumper is supplying a monitor nozzle with a flow of 300 gpm at 80 psi nozzle pressure through a 200-ft-long, 2½-in hose line. If the FL 100 in 2½-in hose at 300 gpm is 18 psi, what will the PDP be?

Answer

In Figure 10-7A only one line is being used, and it is carrying all the water.

$$FL = FL\ 100 \times L$$
$$= 18 \times 2$$
$$= 36\ \text{psi total friction loss}$$
$$PDP = NP + FL\ 1 + FL\ 2 \pm EP + AL$$
$$= 80 + 36 + 0 \pm 0 + 20$$
$$= 136\ \text{psi}$$

Note

When you are calculating PDP where Rule 1 and Rule 2 apply, friction loss through only one line is used.

Example 10-8

In Figure 10-7B the pumper is supplying a monitor nozzle with a flow of 600 gpm at 80 psi nozzle pressure through two 200-ft-long, 2½-in hose lines. Each line is flowing only 300 gpm, so the friction loss is 18 psi. What is the required PDP?

Answer

$$FL = FL\ 100 \times L$$
$$= 18 \times 2$$
$$= 36\ \text{psi total friction loss}$$
$$PDP = NP + FL\ 1 + FL\ 2 \pm EP + SA$$
$$= 80 + 36 + 0 \pm 0 + 20$$
$$= 136\ \text{psi}$$

Notice that both the nozzle pressure and the appliance loss are the same in Example 10-7 and Example 10-8. The only possible variable is the friction loss, but because each line is flowing the same gpm, even friction loss in this case is the same, making the PDP in both examples exactly the same, even though twice as much water is being pumped in Example 10-8 as in Example 10-7. Rule 3 applies for any number of lines as long as the lines are parallel and of the same diameter. The friction loss in only one line is used because they all have the same gpm flow.

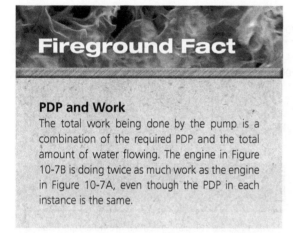

Fireground Fact

PDP and Work

The total work being done by the pump is a combination of the required PDP and the total amount of water flowing. The engine in Figure 10-7B is doing twice as much work as the engine in Figure 10-7A, even though the PDP in each instance is the same.

Parallel Lines of Unequal Diameter Supplied by One Pumper

When a pumper is supplying parallel lines of unequal diameter, only Rule 2 applies. Each of the lines handles a different amount of water because, as mentioned earlier, the pressures in parallel lines will equalize at the point where both lines have the same total friction loss. This point cannot be over emphasized. When the diameter of the hose is the same, it means the gpm in each line is the same. However, when the diameter

of the hose is different, the gpm flow in each line is different, but the total friction loss in each hose line will be the same. Earlier we found that at 11.25 psi friction loss per 100 ft in 2½-in hose we could get a flow of 237 gpm. But for the same friction loss in 3-in hose, the flow is 375 gpm. (See Example 6-12 for a walk-through of these calculations.)

To find the friction loss for a given flow through parallel lines of unequal diameter, it is absolutely necessary that we know the conversion factor for the combination of hose sizes. Recall that the conversion factor for one 2½-in and one 3-in hose is 0.15. Now to find the friction loss per 100 ft for any flow through this combination of hose sizes, we can just insert 0.15 in the FL 100 formula in place of CF. Refer to Chapter 6 for a discussion of conversion factors.

Example 10-9

What is the PDP for a pumper that is supplying 500 gpm to a monitor nozzle at 80 psi nozzle pressure if it is pumping through 300 ft of parallel lines, one 2½-in and one 3-in Figure 10-8 ?

Answer

First find the friction loss for 500 gpm through this combination of hose.

$$FL\ 100 = CF \times 2Q^2$$
$$= 0.15 \times 2 \times (5)^2$$
$$= 0.15 \times 2 \times 25$$
$$= 7.5\ psi$$

Now find the total friction loss.

$$FL = FL\ 100 \times L$$

(continues)

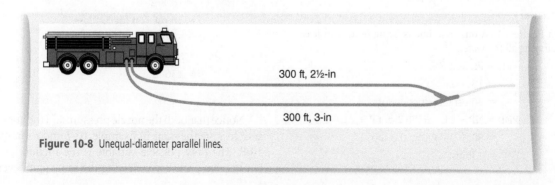

Figure 10-8 Unequal-diameter parallel lines.

(Example 10-9 continued)

$$= 7.5 \times 3$$
$$= 22.5 \text{ or } 23 \text{ psi total}$$
friction loss

Finally, find the PDP.

$$PDP = NP + FL\ 1 + FL\ 2 \pm EP + AL$$
$$= 80 + 23 + 0 \pm 0 + 20$$
$$= 123 \text{ psi}$$

When parallel lines of unequal diameter and unequal length are used, Rule 2 still applies. Add up the lengths of all the lines, then divide by the number of lines to get the average length.

Example 10-10

In Example 10-9, if the 2½-in line had only been 250 ft, what would the PDP be?

Answer

We already know the FL 100 for a 2½-in and 3-in hose, so now the first step is to determine the average length of lines. The 2½-in hose is 250 ft long and the 3-in hose is 300 ft long.

$$250 + 300 = 500/2 = 275 \text{ ft}$$

Now find the total friction loss.

$$FL = FL\ 100 \times L$$
$$= 7.5 \times 2.75$$
$$= 20.6 \text{ or } 21 \text{ psi total friction loss}$$

Now find the PDP.

$$PDP = NP + FL\ 1 + FL\ 2 \pm EP + AL$$
$$= 80 + 21 + 0 \pm 0 + 20$$
$$= 121 \text{ psi}$$

Relay Formula

A **relay** is any situation in which one pumper is supplying water to another pumper. Relays often involve only two pumpers, one at the hydrant or other water source and other one at the fire, such as illustrated in **Figure 10-9**. This situation is often referred to as a *two-pump operation* and is the most common relay evolution. The more common idea of a relay is several pumpers in line, such as shown in **Figure 10-10**. What makes either evolution a relay is that water is being supplied from the pumper at the water source to the pumper at the fire with any number of pumpers in between.

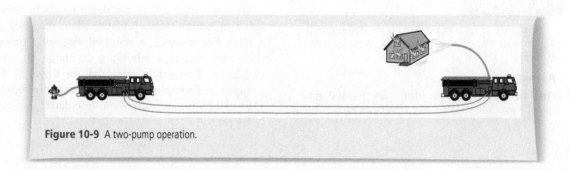

Figure 10-9 A two-pump operation.

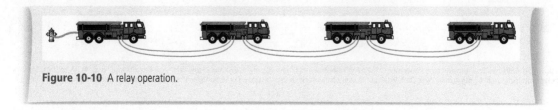

Figure 10-10 A relay operation.

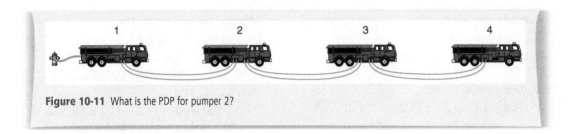

Figure 10-11 What is the PDP for pumper 2?

The relay formula for use in any relay application is PDP = 20 + FL ± EP. Here PDP, FL, and EP have the same definition as in the PDP formula; however, 20 is a new element. The 20 is actually 20 psi and is the minimum amount of pressure we want at the intake of the next pumper in line. (Some jurisdictions may use a pressure higher than 20 psi, but in no case should a pressure less than 20 psi be used.) In the relay formula, friction loss is calculated just as it would be in the PDP formula. The PDP for pumper 2 in **Figure 10-11** is calculated in Example 10-11.

$$PDP = 20 + FL \pm EP$$

Example 10-11

What is the PDP for pumper 2 in Figure 10-11, pumping through two 500-ft, 3-in supply lines delivering 800 gpm?

Answer

First, we need to find the friction loss for 800 gpm through two 3-in supply lines. Remember, each line is only flowing 400 gpm, so we only need to find friction loss for 400 gpm through 3-in hose.

$$
\begin{aligned}
FL\ 100 &= CF \times 2Q^2 \\
&= 0.4 \times 2 \times (4)^2 \\
&= 0.4 \times 2 \times 16 \\
&= 12.8\ psi
\end{aligned}
$$

Now find the total friction loss.

$$
\begin{aligned}
FL &= FL\ 100 \times L \\
&= 12.8 \times 5 \\
&= 64\ psi\ total\ friction\ loss
\end{aligned}
$$

Now solve for PDP using the relay formula.

$$
\begin{aligned}
PDP &= 20 \pm FL \pm EP \\
&= 20 + 64 \pm 0 \\
&= 84\ psi
\end{aligned}
$$

It is not unusual for fire departments to establish a minimum pump pressure when pumping in a relay. For instance, a standard operating procedure may require a minimum discharge pressure of 125 psi. This would mean that the driver of the pumper in Example 10-11 would actually have a PDP of 125 psi instead of the calculated pressure of 84 psi.

WRAP-UP

Chapter Summary

- The PDP formula is PDP = NP + FL 1 + FL 2 ± EP + AL.
- When you are pumping multiple lines to the same device, each line is assumed to be carrying an equal share of the water.
- When parallel lines are of unequal length, the average length is used.
- When parallel lines of the same size are flowing the same gpm, the friction loss in only one line is used.
- When parallel lines of unequal diameter are used, it is necessary to know the conversion factor for the combination of hose sizes.
- The relay formula is PDP = 20 + FL ± EP.

Key Term

relay Any situation in which one pumper is supplying water to another pumper.

Case Study

Getting It Right

Getting the correct PDP on the fireground has always been a challenge. With the confusion and noise that dominates the emergency scene, it is necessary to have a method of finding the correct PDP that ensures accuracy and ease of use. This method should also require the fewest mental calculations.

The PDPs for frequently used evolutions and preconnect lines should be well known. Preconnect lines should be so well prerehearsed that the exact PDP is known without having to calculate it every time the line is used. In fact, some departments actually label the discharge gauge for each preconnect line with the required PDP. The only thing that would need to be added would be the elevation pressure. Even standard evolutions, such as standpipe evolutions, can be precalculated to the point that only the elevation needs to be added.

Of course, even with preconnect hose lines, the operator is not exempt from knowing how to calculate the correct pressure. If a hose line length is changed, or if a tip size is changed, the operator needs to make immediate corrections.

Calculating PDP requires an easy-to-understand fireground formula that also makes sense for training and does not require any special knowledge or correction factors. The PDP formula meets this need.

1. At minimum, how many elements of the PDP formula are necessary to find a PDP?
   A. 5
   B. 4
   C. 3
   D. 2

2. Which of the following is NOT a factor in making the PDP formula easy to use on the fireground?
   A. It is also used during classroom training.
   B. It is easy to remember.
   C. It works in most instances.
   D. It covers everything needed to get the correct PDP.

3. In the PDP formula, what is the symbol for appliance loss?

  A. NP
  B. EP
  C. FL 2
  D. AL

4. What is the difference between FL 1 and FL 2?

  A. FL 1 is for a primary flow and FL 2 is for a secondary flow, from a variable flow nozzle.
  B. FL 1 is friction loss in hose, and FL 2 is friction loss in appliances.
  C. FL 1 and FL 2 are friction losses for two different sizes of hose.
  D. FL 1 and FL 2 are friction losses for two different appliances.

Review Questions

When you are solving the problems below, use the following conversion factors to calculate FL 100 for dual lines:

  1–2½-in and 1–3-in line   CF = 0.15
  1–3-in and 1–3½-in line   CF = 0.062

1. You are operating the second pumper in a three-pumper relay. If pumper 3 is on a hydrant and is pumping the correct pressure, what should your intake pressure be?

2. What is the PDP for a pumper pumping to a sprinkler system that covers a three-story building?

3. When you are factoring in a special appliance, how much pressure is allowed for a monitor nozzle?

4. When you are pumping parallel lines, when will the pressure equalize?

Activities

You are now ready to begin solving problems using the PDP and relay formulas. In these problems, you will be required to solve each element of the problem, including gpm and FL 100, as well as the PDP.

1. What is the PDP for a 250-ft, 2½-in line with a 1-in tip at 45 psi nozzle pressure? ($C = 0.97$)

2. What is the PDP for 150 ft of 1¾-in hose being supplied by 200 ft of 2½-in hose? The nozzle pressure is 75 psi for a fog nozzle that is flowing 180 gpm **Figure 10-12**.

3. Calculate the PDP for both the pumper at the hydrant and the pumper at the fire in the following evolution. Pumper 1 is hooked up to a hydrant and is pumping through two supply lines, one 300 ft of 3-in hose and the other 400 ft of 3½-in hose, to pumper 2. Pumper 2 is supplying a monitor nozzle with a 1¾-in tip at 80 psi nozzle pressure through two 250-ft, 3-in lines. Pumper 2 is also supplying a 150-ft, 2½-in hand line with a 1⅛-in tip at 50 psi tip pressure.

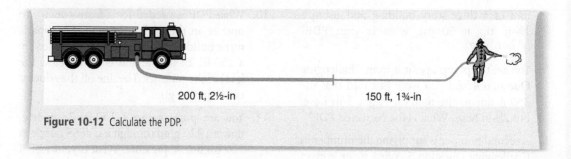

Figure 10-12 Calculate the PDP.

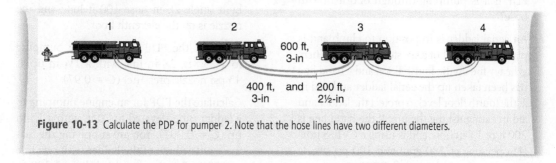

Figure 10-13 Calculate the PDP for pumper 2. Note that the hose lines have two different diameters.

4. What is the PDP for a ladder pipe being operated off an aerial ladder if the ladder is extended to 70 ft at a 70° angle, operating a fog nozzle flowing 500 gpm and being supplied by two 100-ft 2½-in lines?

5. Calculate the PDP for pumper 2 in Figure 10-13. Pumper 2 is supplying 1,000 gpm through parallel lines. One line is 600 ft of 3-in hose, and the second line is 400 ft of 3-in hose and 200 ft of 2½-in hose.

Challenging Questions

1. Calculate the PDP for a preconnect 200-ft, 1¾-in line with a CVFSS nozzle rated at 150 gpm at 75 psi nozzle pressure.

2. You are operating a 300-ft, 2½-in hand line on the fourth floor of a building; if it has a 1-in tip and is operating at 50 psi nozzle pressure, what is the PDP? ($C = 0.97$)

3. You are operating a pumper at a multiple-alarm fire. After the fire has been knocked down and the decision is made to begin overhaul,

your officer orders 100 ft of 1¾-in hose to be used to extend a 2½-in hand line that is 200 ft long. What will the new PDP be if the line is operating on the roof of a warehouse building approximately 25 ft tall? The 1¾-in line is using a CVFSS nozzle designed to flow 125 gpm at 100 psi nozzle pressure.

4. If you are operating an engine pumping through two 3-in lines, one 300 ft and the other 350 ft long, to a monitor nozzle on the

roof of a three-story building and using a 1⅜-in tip at 80 psi, what is your PDP? ($C = 0.997$)

5. You are the pump operator from Challenging Question 4 and you were just told that the 350-ft line to the monitor is not 3-in hose, but 2½-in hose. What is the corrected PDP?

6. A second pumper is supplying the pumper in Challenging Question 5. What is the correct PDP if it is pumping through 600 ft of 4-in hose?

7. An aerial ladder is in position to climb and is placed to the roof of a six-story building. There is heavy fire on the fourth floor, and a hand line has been taken up the aerial ladder and tied in at the fourth floor level to protect the ladder and aid in extinguishing the fire. If the hand line is 200 ft of 1¾-in hose and is using a CVFSS nozzle operating at 180 gpm and 100 psi nozzle pressure, what is the correct PDP?

8. You are pumping to a ¹⁵⁄₁₆-in smooth-bore nozzle ($C = 0.97$) on a 1¾-in line. The line has been taken to the roof of a five-story building and then taken down a set of stairs to the fourth floor. If the line is 400 ft long and the nozzle pressure is 45 psi, what is the correct PDP?

9. Pumper 1 is at a fire supplying a 2½-in hand line operating with a 1-in tip ($C = 0.97$) at 50 psi nozzle pressure, a 1¾-in hand line with a CVFSS nozzle flowing 150 gpm, and a wagon pipe with a 1½-in tip ($C = 0.997$) at 80 psi nozzle pressure. This pumper is being supplied by pumper 2 connected to a hydrant and supplying the first pumper through parallel lines of one 3-in and one 3½-in hose, each 750 ft long. What is the PDP of pumper 2?

10. What PDP is needed for a foam eductor to operate in the parking garage of an apartment building? The eductor is supplied by a 250-ft, 2½-in line and is 15 ft below the level of the pumper. The line off the eductor is flowing 95 gpm.

11. You are pumping to a 150-ft, 2½-in line, flowing 225 gpm through a CVFSS nozzle at 100 psi nozzle pressure. What is your PDP if the line is operating off a standpipe supplied by a single 2½-in hose 100 ft long and the nozzle is on the eleventh floor?

12. Calculate the PDP for an engine pumping to a 250-ft, 2½-in line with a 1¼-in tip at 45 psi nozzle pressure. ($C = 0.97$)

13. Calculate the PDP for an engine pumping to a ladder pipe elevated to 70 ft, with a 1½-in tip ($C = 0.997$). You are supplying the Siamese at the base of the ladder with one line of 150-ft, 3-in hose and one line of 200-ft, 2½-in hose. The tip pressure is 80 psi.

14. What is the PDP for a 300-ft, 1¾-in line that is two stories above the pumper? The nozzle is a CVFSS at 125 gpm at 100 psi.

15. What is the PDP if you are pumping to a standpipe system where the nozzle is on the third floor after being connected to the riser on the second floor? The 150-ft, 2½-in hose has a 250 gpm CVFSS 75 psi nozzle. Connection to the standpipe is made with 100 ft of 2½-in hose.

16. You are pumping to a foam eductor through 350 ft of 2½-in hose. On the discharge side of the eductor, you have 200 ft of 1¾-in hose. The nozzle is located 10 ft below the pumper. What is your PDP if you are flowing 95 gpm?

17. What is the PDP for a 200-ft, 2½-in hose with 1½-in tip operating from the roof of a three-story building? (*C* = 0.97)

18. Calculate the PDP for pumpers 1 and 2 in the following relay. Pumper 1 is operating off the hydrant, supplying water through two 3-in lines, each 300 ft long, to pumper 2. Pumper 2 is pumping to pumper 3 through 1,000 ft of 4-in hose. Pumper 3 is supplying a monitor nozzle with 1¾-in tip at 80 psi nozzle pressure. (*C* = 0.997)

Formulas

Pump discharge pressure formula:

$$PDP = NP + FL\ 1 + FL\ 2 \pm EP + AL$$

To find the total friction loss:

$$FL = FL\ 100 \times L$$

Relay formula:

$$PDP = 20 + FL \pm EP$$

Advanced Problems in Hydraulics

LEARNING OBJECTIVES

Upon completion of this chapter, you should be able to:

- Apply the pump discharge pressure (PDP) formula to evolutions involving Siamese lines.
- Apply the PDP formula to evolutions involving wyed lines.
- Solve maximum lay problems.
- Solve maximum flow problems.
- Understand the value of two-pump operation.
- Calculate actual nozzle pressure, flow, and friction loss when lines are overpumped or underpumped.

Case Study

As the morning of working pump problems progresses, the problems are becoming more and more complicated. As if calculating the PDP for wyed lines isn't confusing enough, the captain has thrown in a maximum lay problem.

As an added challenge, the captain even throws in a problem involving calculation of the correct nozzle pressure when the PDP had been incorrectly calculated in the first place. The captain asks you to explain to everyone how to get the correct PDP for this line.

1. What special considerations need to be accounted for when you are pumping to wyed lines?
2. Maximum lay and maximum flow calculations are not true PDP calculations. Where is their use most valuable?
3. What is the value of being able to find the correct nozzle pressure and friction loss when the PDP had been originally miscalculated?

NFPA 1002 Fire Apparatus Driver/Operator Professional Qualifications, 2014 Edition

This chapter addresses the following requisite knowledge elements within sections
5.2.1, 5.2.2, 5.2.3, and **5.2.4:** hydraulic calculation for friction loss and flow; foam system limitations; calculation of pump discharge pressure; location of fire department connection; and operating principles of sprinkler systems as defined in NFPA 13, NFPA 13D, and NFPA 13R.

Introduction

Previously we learned how to use the PDP formula for basic calculations. Each element of the PDP formula was explained, in turn, as the full formula was developed. Those examples represent the most common evolutions for the PDP formula. (Refer to Chapter 10 for more on the PDP formula.)

In this chapter we explore evolutions of the PDP formula that are a bit more complicated or less common. Some of the evolutions in this chapter have greater practical application for preplanning than for actual fireground operations, but all the evolutions discussed have important application to the study of hydraulics.

As before, when you are solving for the PDP in this chapter, use the entire PDP formula. Some elements may not be needed, but they should never be left out of the formula.

Siamese Lines

Siamese line evolutions are very similar to parallel-line evolutions. The difference is that when we define an evolution as Siamese lines, we have two or more pumpers pumping into the same device. That device can be anything from an actual Siamese to another pumper.

Rule 1 for parallel lines of equal diameter applies to Siamese lines as well: "Each line into the device will be assumed to be carrying an equal share of the water." The size of the line being used is irrelevant. For example, if two pumpers are pumping into a monitor nozzle flowing 600 gallons per minute (gpm) and one is using 2½-inch (2½-in) line and the other is using 3-in line, they will both calculate the PDP for 300 gpm.

In Example 11-1, both pumpers are pumping the same amount of water to the same device, which

Example 11-1

Calculate the PDP for pumper 1 and pumper 2 in **Figure 11-1**. The monitor nozzle is flowing 600 gpm at 80 pounds per square inch (80 psi) (1½-in tip). Pumper 1 is pumping through 200 feet (200 ft) of 3-in hose and pumper 2 is pumping through 250 ft of 2½-in hose.

Answer

First, find the friction loss for 300 gpm in 3-in hose.

$$FL\ 100 = CF \times 2Q^2$$
$$= 0.4 \times 2 \times (3)^2$$
$$= 0.4 \times 2 \times 9$$
$$= 7.2 \text{ psi}$$

Total friction loss is then

$$FL = FL\ 100 \times L$$
$$= 7.2 \times 2$$
$$= 14.4 \text{ or } 14 \text{ psi}$$

Now find the PDP for pumper 1.

$$PDP = NP + FL\ 1 + FL\ 2 \pm EP + AL$$
$$= 80 + 14 + 0 \pm 0 + 20$$
$$= 114 \text{ psi}$$

We already know from an earlier example that a flow of 300 gpm in 2½-in hose has a FL 100 of 18 psi. (See Example 10-7.)

Total friction loss for the 2½-in hose is

$$FL = FL\ 100 \times L$$
$$= 18 \times 2.5$$
$$= 45 \text{ psi}$$

Now find the PDP for pumper 2.

$$PDP = NP + FL\ 1 + FL\ 2 \pm EP + AL$$
$$= 80 + 45 + 0 \pm 0 + 20$$
$$= 145 \text{ psi}$$

is a monitor nozzle. However, because they are each using a different size and length of hose, their PDPs are different.

A situation similar to a Siamese arises when a single pumper is receiving water from more than one other pumper. Even though this situation mimics a Siamese scenario in most respects, it has one major difference. With Siamese lines, each line into a device carries an equal share of the water whereas with multiple pumpers pumping into another pumper, the pumpers can each supply a different volume of water. Generally speaking, in situations where multiple pumpers are pumping into another pumper, each will be supplying as much water as it can or however much is needed **Figure 11-2**.

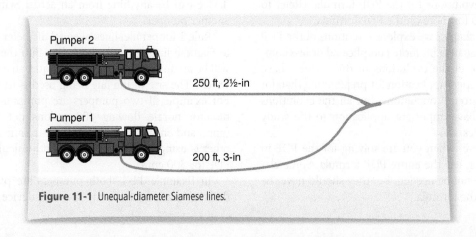

Figure 11-1 Unequal-diameter Siamese lines.

Pumper 2

250 ft, 2½-in

Pumper 1

200 ft, 3-in

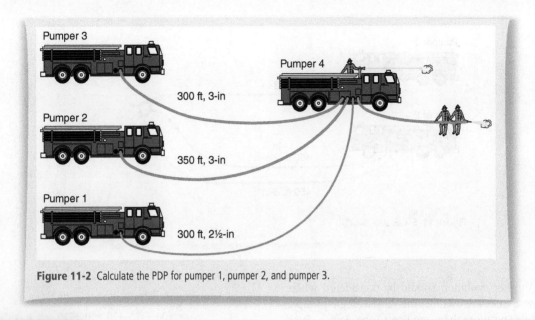

Figure 11-2 Calculate the PDP for pumper 1, pumper 2, and pumper 3.

Example 11-2

In Figure 11-2, pumper 4 is pumping to a combination of devices and lines that are using a total of 900 gpm. Calculate the PDP for pumper 1 if it has a 300-ft, 2½-in line supplying pumper 4 and is supplying 300 gpm. Calculate the PDP for pumper 2 if it has a 350-ft, 3-in line supplying pumper 4 with 200 gpm and is located 20 ft above pumper 4. Finally, calculate the PDP for pumper 3 if it has a 300-ft, 3-in line supplying pumper 4 with 400 gpm.

Answer

First, find the PDP for pumper 1. We already know from our previous example that the FL 100 for 300 gpm in 2½-in hose is 18 psi per 100 ft.

$$FL = FL\ 100 \times L$$
$$= 18 \times 3$$
$$= 54 \text{ psi total friction loss}$$

The PDP for pumper 1 is

$$PDP = 20 + FL \pm EP$$
$$= 20 + 54 \pm 0$$
$$= 74 \text{ psi}$$

Now find the friction loss and then PDP for pumper 2.

$$FL\ 100 = CF \times 2Q^2$$
$$= 0.4 \times 2 \times (2)^2$$
$$= 0.4 \times 2 \times 4$$
$$= 3.2 \text{ psi}$$

$$FL = FL\ 100 \times L$$
$$= 3.2 \times 3.5$$
$$= 11.2 \text{ or } 11 \text{ psi total friction loss}$$

Do not forget to account for the elevation.

$$P = 0.433 \times H$$
$$= 0.433 \times 20$$
$$= 8.66 \text{ or } 9 \text{ psi}$$

Remember, because pumper 2 is higher than pumper 4, you must subtract the gain caused by elevation/forward pressure.

$$PDP = 20 + FL \pm EP$$
$$= 20 + 11 - 9$$
$$= 22 \text{ psi}$$

Finally, calculate the friction loss and PDP for pumper 3.

$$FL\ 100 = CF \times 2Q^2$$
$$= 0.4 \times 2 \times (4)^2$$
$$= 0.4 \times 2 \times 16$$
$$= 12.8$$

$$FL = FL\ 100 \times L$$
$$= 12.8 \times 3$$
$$= 38.4 \text{ psi or } 38 \text{ psi total friction loss}$$

$$PDP = 20 + FL \pm EP$$
$$= 20 + 38 \pm 0$$
$$= 58 \text{ psi}$$

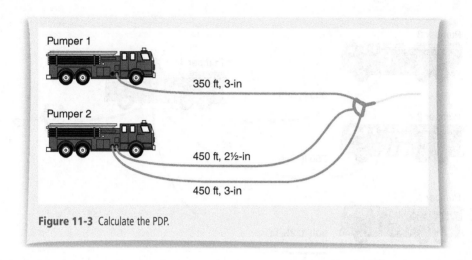

Figure 11-3 Calculate the PDP.

Another evolution should be considered where Siamese lines are used. In this case a single pumper is pumping more than one line to the device. Such an evolution is illustrated in **Figure 11-3**. In this figure, pumper 1 is pumping to a monitor nozzle through 350 ft of 3-in hose, and pumper 2 is pumping through 450-ft parallel lines of one 2½-in hose and one 3-in hose.

Example 11-3

Calculate the PDP for both pumpers in Figure 11-3. Assume the monitor nozzle is flowing 900 gpm with a fog nozzle.

Answer

Recall the rule that says that each line must handle a proportional amount of water. In this instance the rule still applies. Each line is expected to carry 300 gpm, so pumper 1 will be supplying 300 gpm and pumper 2 will be supplying 600 gpm.

We already know FL 100 for 300 gpm in 3-in hose is 7.2 psi.

$$FL = FL\ 100 \times L$$
$$= 7.2 \times 3.5$$
$$= 25.2\ or\ 25\ psi$$

Now calculate the PDP for pumper 1.

$$PDP = NP + FL\ 1 + FL\ 2 \pm EP + AL$$
$$= 100 + 25 + 0 \pm 0 + 20$$
$$= 145\ psi$$

Now solve for the PDP of pumper 2 just as in any parallel lay problem. First we need to find the FL 100 for 2½- and 3-in parallel lines.

$$FL\ 100 = CF \times 2Q^2$$
$$= 0.15 \times 2 \times (6)^2$$
$$= 0.15 \times 2 \times 36$$
$$= 10.8\ psi$$
$$FL = FL\ 100 \times L$$
$$= 10.8 \times 4.5$$
$$= 48.6\ or\ 49\ psi\ total\ friction\ loss$$

The PDP is

$$PDP = NP + FL\ 1 + FL\ 2 \pm EP + AL$$
$$= 100 + 49 + 0 \pm 0 + 20$$
$$= 169\ psi$$

In one last evolution of Siamese lines we consider the case when more than two pumpers are pumping to the same device, such as in **Figure 11-4**. In this scenario, there are three pumpers supplying water to a ladder tower and flowing a total of 1,200 gpm from its two turrets/monitors.

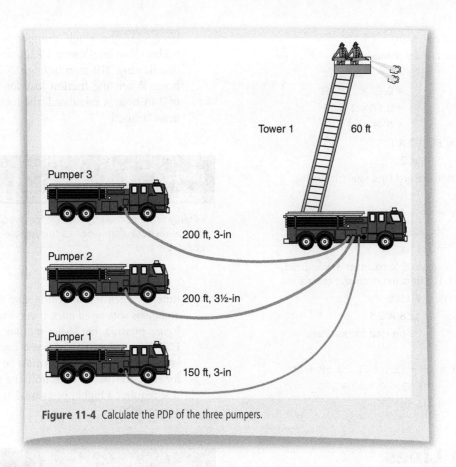

Tower 1   60 ft

Pumper 3

200 ft, 3-in

Pumper 2

200 ft, 3½-in

Pumper 1

150 ft, 3-in

Figure 11-4 Calculate the PDP of the three pumpers.

Example 11-4

In Figure 11-4, tower 1 is flowing water from its two turrets, totaling 1,200 gpm. The ladder tower is extended 60 ft. Pumper 1 is supplying it through 150 ft of 3-in hose, pumper 2 is supplying it through 200 ft of 3½-in hose, and pumper 3 is supplying it through 200 ft of 3-in hose. Find the PDP for each of the three pumpers.

Answer

Since this is an aerial device with a prepiped waterway, a set intake pressure at the connection for the waterway must be maintained. In this scenario, use 170 psi for the appliance loss. The 170 psi will account for all friction loss within the piping on the ladder tower, loss in the monitor/turret, and nozzle pressure. In addition to friction loss in the hose, only elevation pressure will have to be added.

With three pumpers supplying an equal share of water in this evolution, each pumper will be supplying 400 gpm.

First find the PDP for pumper 1. In Example 11-2 we calculated the FL 100 for 400 gpm in 3-in hose to be 12.8 psi.

$$FL = FL\ 100 \times L$$
$$= 12.8 \times 1.5$$
$$= 19.2\ or\ 19\ psi\ total\ friction\ loss$$

When you are finding the PDP, remember to calculate EP at ½ pound per foot (lb/ft) of extension.

$$PDP = NP + FL\ 1 + FL\ 2 \pm EP + AL$$
$$= 0 + 19 + 0 + 30 + 170$$
$$= 219\ psi$$

(*continues*)

(Example 11-4 continued)

Next find the PDP for pumper 2.

$$FL\ 100 = CF \times 2Q^2$$
$$= 0.17 \times 2 \times (4)^2$$
$$= 0.17 \times 2 \times 16$$
$$= 5.44\ \text{psi}$$

$$FL = FL\ 100 \times L$$
$$= 5.44 \times 2$$
$$= 10.88\ \text{or}\ 11\ \text{psi total friction loss}$$

The PDP is

$$PDP = NP + FL\ 1 + FL\ 2 \pm EP + AL$$
$$= 0 + 11 + 0 + 30 + 170$$
$$= 211\ \text{psi}$$

Last, calculate the PDP for pumper 3. We already know the FL 100 from the pumper 1 calculation.

$$FL = FL\ 100 \times L$$
$$= 12.8 \times 2.5$$
$$= 32\ \text{psi total friction loss}$$

The PDP is

$$PDP = NP + FL\ 1 + FL\ 2 \pm EP + AL$$
$$= 0 + 32 + 0 + 30 + 170$$
$$= 232\ \text{psi}$$

Wyed Lines

Calculating the PDP for wyed lines is fairly straight-forward; however, two rules simplify the process.

- Rule 1. *When you are calculating the PDP for wyed lines, include the total gpm when calculating the friction loss to the wye.* For example, in

Figure 11-5 a single pumper is supplying both a 1¾-in line flowing 150 gpm and a 2½-in line flowing 319 gpm through 200 ft of 3-in hose. When the friction loss for the 200 ft of 3-in hose is calculated, the total 469 gpm must be used.

> **Note**
>
> When you are calculating the PDP for wyed lines, include the total gpm when calculating the friction loss to the wye.

- Rule 2. *When you are calculating the PDP for evolutions with wyed lines, if one line requires a higher pressure, the higher pressure will be the PDP.* When the PDP for wyed lines is calculated, separate calculations must be performed for each line that extends from the wye. A line that requires a higher pressure determines the correct PDP.

> **Note**
>
> When you are calculating the PDP for evolutions with wyed lines, if one line requires a higher pressure, the higher pressure will be the PDP.

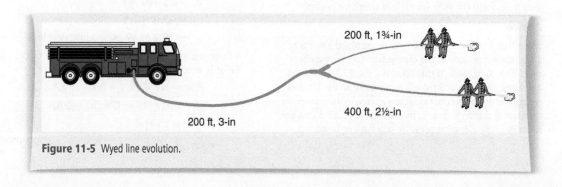

200 ft, 1¾-in

400 ft, 2½-in

200 ft, 3-in

Figure 11-5 Wyed line evolution.

Wyed Lines and Hydraulic Balance

When you are calculating pressures for the separate lines off the wye, it is theoretically possible to gate back any line that requires a significantly lower pressure. However, in practice this is highly impractical. Every reasonable effort should be made to ensure that both lines off a wye require the same, or nearly the same, pressure. The ideal situation with wyed lines is to make sure both lines off the wye are the exact same size, length, and elevation and require the same nozzle pressure. When this is done, or whenever both lines off a wye require the same pressure, the layout will be in hydraulic balance.

Example 11-5

Calculate the correct PDP for the evolution in Figure 11-5 if the 1¾-in line is 200 ft long, using a combination variable fog and straight stream (CVFSS) nozzle at 75 psi nozzle pressure at 150 gpm, and the 2½-in line is 400 ft long with 1¼-in tip at 50 psi nozzle pressure. We know that the flow of a 1¼-in tip at 50 psi is 319 gpm. The total flow is therefore 150 + 319 = 469 gpm. (See Example 5-6.)

Answer

We first need to calculate friction loss in the 3-in hose for the total 469 gpm. The friction loss for the 3-in hose will be FL 1.

$$FL\ 100 = CF \times 2Q^2$$
$$= 0.4 \times 2 \times (4.69)^2$$
$$= 0.4 \times 2 \times 22.00$$
$$= 17.60\ psi$$

Now find the total friction loss.

$$FL = FL\ 100 \times L$$
$$= 17.60 \times 2$$
$$= 35.20\ or\ 35\ psi\ total\ friction\ loss$$

This is FL 1. We have previously calculated the FL 100 for 150 gpm in 1¾-in hose to be 34.92 psi per 100 ft.

$$FL = FL\ 100 \times L$$
$$= 34.92 \times 2$$
$$= 69.84\ or\ 70\ psi\ total\ friction\ loss$$

This is FL 2 for the 1¾-in side of the wye.

Now calculate the PDP for the 1¾-in side. Don't forget AL for the wye.

$$PDP = NP + FL\ 1 + FL\ 2 \pm EP + AL$$
$$= 75 + 35 + 70 \pm 0 + 10$$
$$= 190\ psi$$

This is the correct PDP for the 1¾-in portion of the hose line.

When calculating the needed PDP for the 2½-in hose, we use the same FL 1 as for the 3-in hose. Our next step is to calculate the friction loss for the 2½-in hose.

$$FL\ 100 = 2Q^2$$
$$= 2 \times (3.19)^2$$
$$= 2 \times 10.18$$
$$= 20.36\ psi$$

The total friction loss will be

$$FL = FL\ 100 \times L$$
$$= 20.36 \times 4$$
$$= 81.44\ or\ 82\ psi\ total\ friction\ loss$$

This is FL 2 for the 2½-in side of the wye.

The correct PDP needed for the 2½-in portion of the hose line is then

$$PDP = NP + FL\ 1 + FL\ 2 \pm EP + AL$$
$$= 50 + 35 + 82 \pm 0 + 10$$
$$= 177\ psi$$

Because the 1¾-in line requires the highest pressure, 190 psi is the PDP.

In situations where the pressures vary as illustrated here, the higher pressure determines the actual PDP. Without gauges, gating back is a hit-or-miss proposition at best, and it should be avoided. With the requirements for these two lines having only 13 psi difference, the best course of action is to pump the higher pressure. If the pressures were significantly different, it would be best to adjust one line or the other until equal pressures were obtained. This is why it is best, when using wyed lines, for both lines coming off the wye to require the same pressure. The easiest way to do this is to make sure all lines are of equal length, size, and flow and have the same nozzle pressure. While, generally speaking, any elevation difference should be accounted for, in the case of wyed lines, a small elevation difference, such as a single floor, would not be significant enough to worry about on the fireground.

A practical application of this principle is what some departments refer to as a **skid load**. A skid load is two 1½-in or 1¾-in hand lines, both with the same type of nozzle and same length of hose, attached to a wye with a 2½-in or 3-in supply line. The skid load can be advanced up a long drive to a house or left in front of a building while the pumper goes to the hydrant. When the supply line is charged, one or both of the hand lines can be used.

Example 11-6

The pumper in **Figure 11-6** has deployed a skid load of two 1¾-in lines, each 200 ft long, flowing 125 gpm at 100 psi nozzle pressure. The wye is supplied by 400 ft of 2½-in hose. What is the correct PDP for the evolution?

Begin by finding the FL 100 for 250 gpm in the 2½-in hose.

$$FL\ 100 = 2Q^2$$
$$= 2 \times (2.5)^2$$
$$= 2 \times 6.25$$
$$= 12.5\ psi$$

Total friction loss for the 2½-in hose is

$$FL = FL\ 100 \times L$$
$$= 12.5 \times 4$$
$$= 50\ psi\ total\ friction\ loss\ for\ 2½\text{-in hose}$$

This is FL 1.

Now find the FL 100 for 125 gpm in 1¾-in hose.

$$FL\ 100 = CF \times 2Q^2$$
$$= 7.76 \times 2 \times (1.25)^2$$
$$= 7.76 \times 2 \times 1.56$$
$$= 24.21\ psi$$

Now find the total friction loss for the 1¾-in hose.

$$FL = FL\ 100 \times L$$
$$= 24.21 \times 2$$
$$= 48.42\ or\ 48\ psi\ total\ friction\ loss\ for\ 1¾\text{-in hose}$$

This is FL 2.

Now calculate the PDP for the lay:

$$PDP = NP + FL\ 1 + FL\ 2 \pm EP + AL$$
$$= 100 + 50 + 48 \pm 0 + 10$$
$$= 208\ psi$$

One additional advantage of having hose of equal size and length, and with the same elevation and nozzle pressure, on each outlet of the wye is that you only have to calculate the pressure for one of the lines. Just remember to include the total flow in the hose supplying the wye.

Maximum Lay Calculations

The purpose of maximum lay problems is to determine how long a hose line we can lay that will still give us a specified amount of water. This information can be used for preplanning specific target hazards, water supply training, or calculating the benefits of large-diameter hose (LDH).

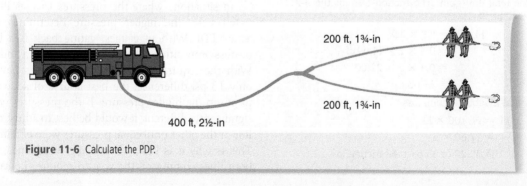

200 ft, 1¾-in

200 ft, 1¾-in

400 ft, 2½-in

Figure 11-6 Calculate the PDP.

Note

The purpose of maximum lay problems is to determine how long a hose line we can lay and still give us a specified amount of water.

To do maximum lay problems, we need to know three pieces of information: (1) the desired quantity of water, (2) the hose size, and (3) the pressure available to supply the water in the first place. The pressure can be either hydrant pressure or pressure from a pumper at the water supply.

To determine the maximum lay, follow this procedure:

1. Determine the gpm needed.
2. Determine the size of hose available to get the water from the source to the scene.
3. Calculate the FL 100 in the available hose for the gpm needed.
4. Subtract 20 psi from the maximum pressure available. This amount is the intake pressure for the pumper at the scene, determined by using the relay formula.
5. Divide the friction loss from step 3 into the pressure remaining after 20 psi has been subtracted. The answer is the maximum lay in hundreds.
6. Finally, convert the maximum lay, in hundreds, into feet by multiplying by 100 and rounding off to the last 50 increment. For example, if the lay works out to be 3.75 in hundreds, it is 375 ft. However, there are no 25-ft sections of hose, so the answer would be rounded down to the next 50-ft increment, or 350 ft. Large-diameter hose usually comes in 100-ft sections, so you would have to round it down to the next 100-ft increment.

Example 11-7

If it is necessary to deliver a minimum flow of 700 gpm, what is the maximum lay of 4-in hose that can do this? The hose is directly connected to a hydrant that has a static pressure of 60 psi.

Answer

We know the gpm and the size hose, so the next step is to determine the friction loss per 100 ft. The conversion factor for 4-in hose is 0.1.

$$FL\ 100 = CF \times 2Q^2$$
$$= 0.1 \times 2 \times (7)^2$$
$$= 0.1 \times 2 \times 49$$
$$= 9.8\ psi$$

Now subtract 20 psi from the hydrant pressure, to determine how much pressure is available for friction loss, and then divide by 9.8 psi.

$$60 - 20 = 40 \div 9.8 = 4.08\ or\ 408\ ft$$

This would be rounded down to a maximum lay of 400 ft.

In Example 11-7, we discovered that even with 4-in hose, if we need to deliver 700 gpm, we can only lay hose a maximum of 400 ft. If we had to go farther than that, we would have to do it in a relay. If a relay were set up from the first pumper, the pumpers forming the relay would have a maximum pressure much higher than the hydrant pressure. Two things would determine their top pressure: (1) the capacity of the pump and (2) the maximum pressure limits of LDH.

The capacity of the pump is important because it determines the maximum discharge pressure. Recall that as the pump capacity increases, the maximum pressure possible decreases. For instance, if we were to use a 1,250 gpm pump in Example 11-7 and it were necessary to relay water to a second pumper, then the 1,250 gpm pump would be limited to a maximum discharge pressure of 200 psi when delivering 700 gpm. We determine this by calculating that 700 gpm is 56 percent of the capacity of a 1,250 gpm pump. If we do not exceed 50 percent of the capacity of the pump, we can pump at a maximum of 250 psi pressure. But because we need to pump at 56 percent of the capacity of the pump, we have to lower our maximum pressure. Because we can pump up to 70 percent of the capacity of the pump at 200 psi, this becomes the maximum discharge pressure. (See Table 7-1 for data on pump capacity and maximum pressure.)

Fireground Fact

Hydrant Supply and Pump Capacity

Earlier, when talking about pump capacity, we went to great lengths to explain how discharge pressure affected pump capacity. The same is true when supplying water from a hydrant, but with some "wiggle room" built in. This is so because from draft, the pump must do all the work to get water into and out of the pump. From a hydrant, however, the pump takes advantage of the incoming pressure, and the discharge pressure and flow are not entirely the result of work done by the pump. In both instances, from draft and from a hydrant, the key is the net pump pressure. From draft, the net pump pressure is easy to calculate. When you are operating from a hydrant, however, calculating net pump pressure is unpredictable because just how much additional help the pump will get depends largely on the water supply capacity and pressure. For example, if the hydrant is on a larger main, the hydrant can supply more water even if the static pressure is the same as that of another hydrant on a smaller main. Keep this in mind when you are calculating PDPs versus capacity when the supply is a positive pressure. When the pressure versus discharge doesn't work out exactly as illustrated earlier but is close, it is possible/probable that it will work from a pressurized source nonetheless.

Another restriction we need to be aware of is that LDH has pressure limitations. It is not uncommon to find that LDH is limited to a maximum working pressure of 185 psi. In fact, unless otherwise known, it should always be assumed that a limitation exists. Before you do maximum lay problems in real-life situations, determine what maximum pressure exists.

Note

It is not uncommon to find that LDH has a maximum working pressure of 185 psi.

Example 11-8

If we were to establish a relay off the hydrant in Example 11-7 and the first pumper had a maximum capacity of 1,250 gpm, how far could the second pumper be from the first pumper **Figure 11-7** ?

Answer

All the important facts are already known. The maximum pressure, as previously explained, is 185 psi because the hose is LDH, and the friction loss for the 700 gpm in 4-in hose is 9.8 psi.

Because this hose is limited to 185 psi, pressure available for friction loss is

$$185 - 20 = 165$$

The maximum lay is

$$165 \div 9.8 = 16.84 \text{ or } 1,684 \text{ ft}$$

This would be rounded down to a maximum of 1,600 ft. Remember LDH comes in 100-ft sections. (You may carry a single 50-ft section of LDH on each pumper to better complete evolutions such as this one.)

400 ft, 4-in   ? ft, 4-in

Figure 11-7 Find the length of hose between the two pumpers.

Maximum Flow Calculations

Maximum flow calculations allow us to find the maximum amount of water we can deliver under specified hose layouts. By using maximum flow calculations, we can calculate the advantage of using larger-diameter supply line, or how to get the most out of a water supply by placing a pumper on the hydrant.

Note

Maximum flow calculations allow us to calculate the maximum amount of water we can deliver under specified hose layouts.

To do the maximum flow calculations, we need to know the following:

- What size of supply line is being used?
- How long is the hose lay?
- What is the maximum pressure available at the source?
- If there is a pumper at the source, what is its capacity?
- If the source is a hydrant, what is its capacity?

With these five pieces of information, we can calculate the maximum amount of water we can deliver with any specified hose lay. Just as with the maximum lay calculations, maximum flow calculations can be useful for preplanning of target hazards and can be used to calculate maximum flows at fire scenes.

To calculate the maximum flow, follow this procedure:

1. Determine the length and size of hose.
2. Determine the maximum discharge pressure at the source. With a line directly connected to a hydrant, this is the hydrant static pressure. If a pumper is at the source, either at draft or connected to a hydrant, it may require some trial-and-error calculations to determine the maximum pressure. (This principle

is illustrated in Example 11-9.) With practice, you will be able to get a feel for the correct pressure without so much trial and error.

3. Subtract 20 psi from the maximum pressure to find the amount of pressure available for friction loss.
4. Divide the amount of pressure available for friction loss by the amount of hose in hundreds, which gives the friction loss available per 100 ft.
5. From the available friction loss, calculate the gpm.

Example 11-9

In , pumper 1 (rated at 1,000 gpm) is attached to a hydrant pumping through two 3-in lines to pumper 2. If the 3-in lines are 600 ft long, how much water can pumper 1 supply to pumper 2?

Answer

We were given both the diameter and length of the lines. The first step is to determine the maximum discharge pressure of pumper 1. If pumper 1 is rated at 1,000 gpm at 150 psi, then at 200 psi it can only pump 700 gpm, and at 250 psi it can only pump 500 gpm. It may be necessary to start at 150 psi and calculate the flow at each pressure/flow range until you find one that works. In this particular problem, unless you have a conversion factor for dual 3-in hose lines, you will actually calculate the gpm for one line. To determine the total gpm, you must double the quantity.

If we assume a maximum pressure of 150 psi, maximum pressure available for friction loss is

$$150 - 20 = 130 \text{ psi}$$

The FL 100 is

$$130 \div 6 = 21.66 \text{ psi}$$

To find the gpm from the FL 100:

$$Q = \sqrt{FL\ 100\ /(CF \times 2)}$$
$$= \sqrt{21.66/(0.4 \times 2)}$$
$$= \sqrt{21.66/0.8}$$
$$= \sqrt{27.08}$$
$$= 5.20 \text{ or } 520 \text{ gpm}$$

For both lines the total capacity will be 1,040 gpm.

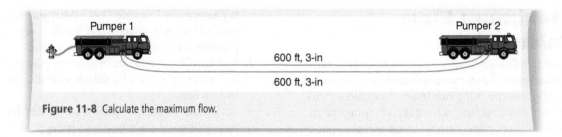

Figure 11-8 Calculate the maximum flow.

In Example 11-9 even though the pumper is rated at only 1,000 gpm, it is so close that we will accept this answer. If we were to go to a maximum pressure of 200 psi, the friction loss available per 100 ft would be much higher. This would be associated with a flow much higher than the 700 gpm possible at 200 psi.

In Example 11-9, the correct answer is 1,040 gpm at a pressure of 150 psi. At 150 psi the pumper can pump its maximum capacity. Even with a maximum rated capacity of 1,000 gpm, the extra 40 gpm is not a concern for two reasons. First, it is possible to simply calculate friction loss for a maximum of only 1,000 gpm and adjust the discharge pressure accordingly. Second, from a hydrant with a residual pressure above 20 psi it is probable that the 1,000-gpm-rated pumper can actually pump 1,040 gpm. Remember, the rating of a pumper indicates its performance at draft with a negative intake, not a positive intake as with a hydrant.

Example 11-10

Calculate the maximum gpm possible if pumper 1 is attached to a hydrant and is pumping through 1,200 ft of 4-in LDH to pumper 2. Pumper 1 is rated at 1,250 gpm. Remember, unless otherwise stated, LDH is limited to a maximum pressure of 185 psi.

Answer

We begin by calculating the maximum flow at a discharge pressure of 150 psi.

$$150 - 20 = 130 \text{ psi}$$
$$130 \div 12 = 10.83 \text{ psi} = FL \ 100$$
$$Q = \sqrt{FL \ 100/(CF \times 2)}$$
$$= \sqrt{10.83/(0.1 \times 2)}$$

$$= \sqrt{10.83/0.2}$$
$$= \sqrt{54.15}$$
$$= 7.36 \text{ or } 736 \text{ gpm}$$

This amount is only 59 percent of the capacity of the pump. We can obviously pump more water than this. We need to recalculate, using a maximum pressure of 185 psi.

$$185 - 20 = 165 \text{ psi}$$
$$165 \div 12 = 13.75 \text{ psi} = FL \ 100$$
$$Q = \sqrt{FL \ 100/(CF \times 2)}$$
$$= \sqrt{13.75/(0.1 \times 2)}$$
$$= \sqrt{13.75/0.2}$$
$$= \sqrt{68.75}$$
$$= 8.29 \text{ or } 829 \text{ gpm}$$

Because 829 gpm is not higher than 70 percent of the rated capacity of a 1,250 gpm pumper (875 gpm), this is the maximum flow we can get.

In Example 11-10, we needed to calculate the maximum flow twice to find the correct answer. The first time the gpm was too low. By recalculating at the higher discharge pressure of 185 psi, we were able to use a higher FL 100, which had a flow that is almost exactly 70 percent of the pump capacity. This amount fits perfectly with the pump capacity guidelines introduced previously. If we were now to try the problem again, using 250 psi, assuming no pressure restriction, we would find the FL 100 is 19.17 psi, which works out to a gpm of 992 gpm. Because at 250 psi the pump can flow only 625 gpm, 992 gpm is beyond the capacity of the pump at that pressure. Refer to Table 7-2 for the pump capacity guidelines.

Two-Pump Operation

Earlier we introduced the concept of a two-pump operation, the most common relay evolution. Examine the difference between the maximum lay obtained in Example 11-7 and that achieved in Example 11-8. The maximum lay increased nearly fourfold by putting a pumper on the hydrant. In Example 11-7, the maximum lay will increase to some extent with a higher hydrant static pressure, and there are some variables that could alter things slightly, but not enough to make up for the fourfold increase in lay in Example 11-8. (See Chapter 12 for further discussion of a two-pump operation.)

The primary variables that could affect the evolution somewhat would be a higher hydrant static pressure and the hydrant being on a larger main. A larger main would supply the water to the hydrant with less friction loss in the system, giving us more water at the same initial static pressure.

We can demonstrate the advantage of putting a pumper on a hydrant by comparing maximum flow calculations. In Example 11-9, we calculated a flow of 1,040 gpm with a pumper on the hydrant. In Example 11-11 we will recalculate the same problem with a 60 psi static pressure at the hydrant instead of a pumper.

Example 11-11

It is necessary to deliver water to the incident scene through two 600-ft lines of 3-in hose. If the lines are directly connected to the hydrant and it has a static pressure of 60 psi, how much water will this evolution provide?

Answer

First, we calculate the amount of pressure from the hydrant available for friction loss.

$$60 - 20 = 40 \text{ psi}$$

With 40 psi available to overcome friction loss, we next determine the FL 100.

$$40 \div 6 = 6.67 \text{ psi}$$

Then we calculate the flow in 3-in hose that has an FL 100 of 6.67 psi.

$$Q = \sqrt{FL\,100/(CF \times 2)}$$
$$= \sqrt{6.67/(0.4 \times 2)}$$

$$= \sqrt{6.67/0.8}$$
$$= \sqrt{8.34}$$
$$= 2.89 \text{ or } 289 \text{ gpm}$$

Since there are two 3-in lines, the total flow will be 578 gpm. This is only 56 percent as much water as was supplied in Example 11-9, which had a pumper on the hydrant.

There are situations in which it is not necessary to put a pumper on the hydrant. Variables that will need to be taken into consideration are the amount of water needed, the size of hose being used, the static pressure at the hydrant, and most importantly, with or without the pumper, whether the hydrant can supply the needed water. One other point to keep in mind, if you are pumping directly from the hydrant, is that as other pumpers hook up to the water supply and start pumping, the pressure will drop. As long as there is enough water in the system, a pumper on a hydrant can compensate for this drop in pressure.

Finding Correct Nozzle Pressure, Gallons per Minute, and Friction Loss

There are times when we discover we have been pumping an incorrect PDP to a hand line for any number of reasons. Because we already know how important it is to use the correct gpm to calculate friction loss, it is easy to understand how an incorrect PDP can change the gpm. As part of a thorough understanding of hydraulics, we should be able to determine just how much water is flowing when an incorrect PDP is used. From the gpm we can then calculate what the friction loss should have been.

To calculate the actual flow, we first need to subtract for both elevation and special appliances. The excess or shortage of pressure is distributed proportionally between nozzle pressure and friction loss. Because the allowances for both elevation and special appliances are constants, they cannot change and should not be a part of the calculation.

The first step in calculating the correct flow is to find out how much nozzle pressure we actually had. Begin by establishing a ratio of the correct nozzle pressure to the correct PDP, which is also the ratio of the incorrect nozzle pressure to the incorrect PDP. Mathematically, this ratio is written CNP/CPDP = INP/IPDP, where CNP is the correct nozzle pressure, CPDP is the correct PDP, INP is the incorrect nozzle pressure, and IPDP is the incorrect discharge pressure. Because this formula represents a proportion, the values CPDP and INP are referred to

as *means* and the values CNP and IPDP are referred to as *extremes*.

$$\frac{CNP}{CPDP} = \frac{INP}{IPDP}$$

To work this problem, we need to know *any* three of the values. Generally, we will know the correct nozzle pressure and can then determine the friction loss for the calculated flow, from which we can calculate the correct PDP. The next step is to determine

Example 11-12

Earlier we determined that a 250-ft, 2½-in line with a 1-in tip at 45 psi will flow 194 gpm, giving us a FL 100 of 7.6 psi, or 19 psi for 250 ft. The PDP was 64 psi. In the heat of the moment, the pump operator allowed the PDP to rise to 100 psi. What was the actual nozzle pressure at this PDP? (See Chapter 10, Activity 1.)

Answer

$$CNP = 45 \quad CPDP = 64 \quad INP = ? \quad IPDP = 100$$
$$CNP/CPDP = INP/IPDP$$
$$45/64 = INP/100$$

Multiply the extremes and then divide by the known means.

$$(45 \times 100)/64 = 70.31 \text{ or } 70 \text{ psi}$$

This is the actual nozzle pressure, or INP, when the line has the incorrect PDP of 100 psi.

Now let us prove that 70 psi is the actual nozzle pressure. Assume a C of 0.97. The gpm for a 1-in tip at 70 psi is

$$
\begin{aligned}
gpm &= 29.84 \times D^2 \times C \times \sqrt{P} \\
&= 29.84 \times (1)^2 \times 0.97 \times \sqrt{70} \\
&= 29.84 \times 1 \times 0.97 \times 8.37 \\
&= 242.27 \text{ or } 242 \text{ gpm}
\end{aligned}
$$

Now calculate the friction loss for 250 ft of 2½-in hose flowing 242 gpm.

$$
\begin{aligned}
FL\ 100 &= CF \times 2Q^2 \\
&= 1 \times 2 \times (2.42)^2 \\
&= 1 \times 2 \times 5.86 \\
&= 11.72 \text{ psi} \\
FL &= FL\ 100 \times L \\
&= 11.72 \times 2.5 \\
&= 29.3 \text{ or } 29 \text{ psi}
\end{aligned}
$$

Finally, we need to verify the PDP.

$$
\begin{aligned}
PDP &= NP + FL\ 1 + FL\ 2 \pm EP + AL \\
&= 70 + 29 + 0 \pm 0 + 0 \\
&= 99 \text{ psi}
\end{aligned}
$$

the incorrect PDP. This is easy; it will be what the discharge gauge indicated. Now fill in the formula CNP/CPDP = INP/IPDP with the appropriate values for CNP, CPDP, and IPDP and solve for INP. We solve for INP by multiplying the extremes and then dividing by the known means.

Example 11-13

The PDP is calculated to be 163 psi. Find out what the new nozzle pressure will be and verify the final PDP if the line were to be underpumped at 150 psi. See Chapter 10 Activity 2 for calculation of the PDP.

Answer

Calculate the incorrect nozzle pressure.

$$\text{CNP/CPDP} = \text{INP/IPDP}$$
$$75/163 = \text{INP}/150$$
$$\text{INP} = \frac{75 \times 150}{163}$$
$$= 69.02 \text{ or } 69 \text{ psi}$$

Knowing that the incorrect nozzle pressure is 69 psi, we can now verify the answer. However, this problem presents a twist because it has a fog nozzle. We have to start by finding an equivalent diameter for the nozzle. We do this by using the following formula with a C factor of 0.97. (See Chapter 5 for a discussion of the formula.)

$$D = \sqrt{\text{gpm}/(29.84 \times C \times \sqrt{P})}$$
$$= \sqrt{180/(29.84 \times 0.97 \times \sqrt{75})}$$
$$= \sqrt{180/(29.84 \times 0.97 \times 8.66)}$$
$$= \sqrt{180/250.66}$$
$$= 0.72$$
$$= 0.85\text{-in equivalent diameter}$$

The 0.85 fits directly into the gpm formula without conversion, and we do not need to know what size tip it is since it is a fog nozzle, not a smooth-bore tip.

In Example 11-12 we can accept that the answer is off by 1 psi due to rounding. Considering that the incorrect PDP of 100 psi is 56 percent higher than the correct PDP of 64 psi, this is remarkably accurate. We have just verified the process.

Using the equivalent diameter, find the flow of the nozzle with a nozzle pressure of 69 psi.

$$\text{gpm} = 29.77 \times D^2 \times C \times \sqrt{P}$$
$$= 29.84 \times (0.85)^2 \times 0.97 \times \sqrt{69}$$
$$= 29.84 \times 0.72 \times 0.97 \times 8.31$$
$$= 173.18 \text{ or } 173 \text{ gpm}$$

Calculate the friction loss for the 1¾-in hose.

$$\text{FL } 100 = \text{CF} \times 2Q^2$$
$$= 7.76 \times 2 \times (1.73)^2$$
$$= 7.76 \times 2 \times 2.99$$
$$= 46.40 \text{ psi}$$
$$\text{FL} = \text{FL } 100 \times L$$
$$= 46.40 \times 1.5$$
$$= 69.60 \text{ or } 70 \text{ psi}$$

Now calculate the friction loss for the 2½-in hose.

$$\text{FL } 100 = \text{CF} \times 2Q^2$$
$$= 1 \times 2 \times (1.73)^2$$
$$= 1 \times 2 \times 2.99$$
$$= 5.98 \text{ psi}$$
$$\text{FL} = \text{FL } 100 \times L$$
$$= 5.98 \times 2$$
$$= 11.96 \text{ or } 12 \text{ psi}$$

Finally, verify the PDP.

$$\text{PDP} = \text{NP} + \text{FL } 1 + \text{FL } 2 \pm \text{EA} + \text{AL}$$
$$= 69 + 70 + 12 \pm 0 + 0$$
$$= 151 \text{ psi}$$

The difference can be attributed to rounding. Once again the process works.

Chapter Summary

- When two or more pumpers are pumping to a Siamese, each line is assumed to be carrying an equal share of water regardless of the hose size.
- When lines are wyed, regardless of the hose size, all lines off the wye require the same pressure.
- Maximum lay calculations allow us to calculate how long a hose lay can be and still flow a specified amount of water.
- Maximum flow calculations allow us to calculate the maximum amount of water we can deliver under specified hose lays.
- A two-pump operation allows us to maximize the capacity of a hydrant.
- When an incorrect PDP is used, before we can determine the actual amount of water flowing, we need to find the actual (incorrect) nozzle pressure.

Key Term

skid load Two 1½-in or 1¾-in hand lines, both with the same type nozzle and same length of hose, attached to a wye with a 2½-in or 3-in supply line.

Case Study

Putting It All Together

You are the driver on a pumper that responds to a three-alarm fire involving three 150-year-old row houses. As the fire has progressed to a fourth unit, you have already taken on a second hand line in addition to a master stream device. Your division commander is now calling on you to pump yet another hand line that may tax your ability to adequately supply each of your current commitments.

The additional line will be a 2½-in hand line approximately 200 ft long with a 1¼-in tip. You are currently supplying a 1½-in tip from you deck gun, a 1¾-in hand line flowing 150 gpm, and a 2½-in line with a 1⅛-in tip. Your pump is rated at 1,250 gpm, and your current PDP is 205 psi. Can you supply an additional 2½-in hand line with a 1¼-in tip? You are being supplied from a hydrant through dual 3-in supply lines 400 ft long, which are being pumped by a 1,500 gpm pumper. The pumper on the hydrant is down to 10 psi intake.

1. Why is the capacity of the pump given above important in determining whether you can pump the additional line?

   A. As additional lines are added, the PDP will have to be increased to accompany the additional load.
   B. Any additional lines cannot require a pressure higher than is already being pumped.
   C. The rated capacity of the pump will limit how much water it can pump at a given pressure.
   D. From a hydrant, the rated capacity of the pump is not really relevant.

2. Besides knowing how much water you will be flowing, what additional piece of information is needed to determine if the pump can handle the additional line?

   A. Amount of water available at the source
   B. Highest required discharge pressure
   C. Incoming pressure
   D. Maximum lay from the water source

3. How much additional water do you anticipate you can supply?

   A. None
   B. 240 gpm
   C. 325 gpm
   D. As much as the pumper supplying you can deliver.

4. Can you supply the additional line?

   A. Yes.
   B. No, the pumper on the hydrant cannot supply any more water.
   C. No, even from a hydrant with support from another pumper, you have reached the practical capacity of your pump.
   D. No, the water supply system has reached its capacity.

Review Questions

When solving the problems below, you should use the following conversion factors when calculating FL 100 for dual lines:

1–2½ in and 1–3 in line	CF = 1.5
1–3 in and 1–3½ in line	CF = 0.062

1. When two or more pumpers are pumping to a Siamese, how much water does each pumper supply?

2. Why is it important to make sure that each line coming off of a wye has the same pressure?

3. What is the first step in the procedure for completing a maximum lay calculation?

4. What is the first step in the procedure for completing a maximum flow calculation?

Activities

Put your knowledge of hydraulics to work by solving the following problems.

1. In **Figure 11-9**, the pumper is pumping through 350 ft of 2½-in hose to a wye. From one discharge of the wye there is a 150-ft, 1½-in hose with a CVFSS nozzle flowing 125 gpm at 100 psi nozzle pressure. From the other discharge there is a 200-ft, 1¾-in hose line using a ¹⁵⁄₁₆-in tip at 50 psi nozzle pressure. Calculate the correct PDP for the pumper.

2. You are the driver on a pumper that responds to a second-alarm fire. When you arrive on the fireground, you are ordered to pump to a ladder pipe that is attempting to put into service a CVFSS nozzle that will flow 1,000 gpm at 100 psi nozzle pressure. The ladder is extended to 70 ft. You lay a 250-ft, 3-in line into the Siamese at the base of the ladder. If there is another engine pumping into the Siamese with a 2½-in line, what is your PDP?

3. You are laying out from a pumper on a hydrant. If you want to be able to deliver 700 gpm through a parallel hose lay, how far can you lay a line? One of the hoses is 2½ in and the other hose is 3 in. The pumper on the hydrant is rated at 1,000 gpm.

4. You have just laid out 500 ft of 4-in hose from a hydrant. If the static pressure on the hydrant is 75 psi, what is the maximum amount of water that you can flow?

5. You are pumping to a 2½-in hand line with a 1⅛-in tip. You are told that the line is 350 ft long. After you calculate the correct PDP and have been pumping to this line for several minutes, you are informed that the line is actually 500 ft long. During the time that you were pumping the pressure required for a 350-ft line, how much water was actually flowing?

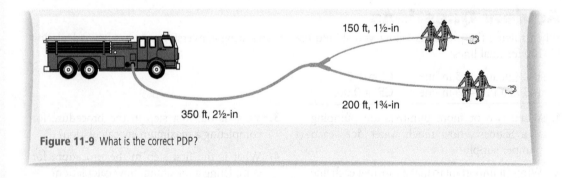

150 ft, 1½-in

200 ft, 1¾-in

350 ft, 2½-in

Figure 11-9 What is the correct PDP?

Challenging Questions

1. Two pumpers are supplying a monitor nozzle. Calculate the PDP for each pumper. Pumper 1 is supplying the monitor nozzle through one line of 3-in hose 350 ft long. Pumper 2 is supplying the monitor nozzle through two lines, one 2½-in hose 300 ft long and a second line of 3-in hose 350 ft long. The monitor nozzle is using a 2-in tip at 80 psi nozzle pressure. ($C = 0.997$)

2. You are pumping through 300 ft of 2½-in line to a wye. From the wye there is one 1¾-in line, 150 ft long with a ⅞-in tip at 45 psi, and a 2½-in line, 100 ft long with a 1-in tip at 50 psi. What is the correct PDP?

3. A 250-ft 2½-in line was operating off a ladder into the third floor. After the fire was knocked down, the nozzle was removed and a wye attached. The wye has two 1½-in, 100-ft lines with CVFSS nozzles that flow 100 gpm at 100 psi nozzle pressure. What is the correct PDP?

4. You must deliver 1,300 gpm to a target hazard 1,500 ft from the closest hydrant. If you place a pumper on the hydrant and all the pumpers are rated at 1,500 gpm, how many pumpers will it take to relay the water? You have parallel supply lines of one 3-in and one 3½-in hose. Remember to include one pumper at the scene.

5. If you were to lay a 4-in supply line from a hydrant that has a static pressure of 70 psi, what would be the maximum distance you could lay and still deliver 700 gpm?

6. You have laid 500 ft of 4-in supply line from a hydrant with a 60 psi static pressure. What

is the maximum amount of water that you can get from the hydrant?

7. You have laid 500 ft of 4-in supply line from a hydrant. Pumper 2, a 2,000 gpm capacity pumper, has hooked up to the hydrant and will pump to you. What is the maximum gpm that pumper 2 can supply to you?

8. As the pump operator, you are pumping to a 200-ft preconnect 1¾-in hand line with a CVFSS nozzle. You pump the correct pressure for a flow of 150 gpm at 100 psi nozzle pressure. Later you discover that a new nozzle was in use that required only 75 psi nozzle pressure. How much water were you actually flowing?

9. You are pumping to a monitor nozzle ($C = 0.997$) with a 1¾-in tip, through what you believe are parallel 2½-in lines, 350 ft long. Later you discover that one of the lines only has 50 ft of 2½-in hose and the other 300 ft is 3-in hose. When you were pumping at the correct pressure for the parallel 2½-in lines and assuming the correct nozzle pressure should have been 80 psi, what were the actual gpm and total friction loss?

10. Two pumpers are supplying a monitor nozzle with a 2-in tip ($C = 0.997$). Pumper 1 is supplying it through one 2½-in line and one 3-in line, each 250 ft long. Pumper 2 is supplying it through one 3-in line, 400 ft long.

    A. What is the PDP of pumper 1 if the nozzle pressure is 75 psi?

    B. What is the PDP of pumper 2 if the nozzle pressure is 75 psi?

11. Calculate the PDP for a pumper supplying two 95 gpm foam eductors. A 200-ft, 3-in line is supplying a wye with 50 ft of 2½-in hose off of each 2½-in discharge. Each 2½-in line then has one of the foam eductors attached. One eductor has 200 ft of 1¾-in hose, and the other eductor has 150 ft of 1½-in hose.

12. What is the maximum lay for a single 3-in supply line with a flow of 550 gpm? The pumper supplying this line is rated at 1,250 gpm.

13. What is the maximum flow through 1,000 ft of 5-in hose?

14. What is the maximum lay for a flow of 1,000 gpm through parallel lines when one line is 3 in and the other line is 2½ in? The pumper is rated at 1,000 gpm.

15. What is the maximum flow through parallel lines, one 3-in line and one 3½-in line? The 3-in line is 500 ft long, and the 3½-in line is 600 ft long.

16. You are supplying two 150-ft, 1¾-in lines from a standpipe on the fourth floor. Each line is supplied from the same riser by use of a wye. What is the correct PDP if each line is flowing 125 gpm at 75 psi and the standpipe is being supplied by a single 50-ft, 3-in line?

17. Go back to Question 16. What would the PDP be if one of the lines were taken off the standpipe system on the fourth floor and the second line were taken off the standpipe system on the third floor, but both nozzles were operating on the fourth floor?

Formula

To find the actual nozzle pressure (INP) when an incorrect pump discharge pressure is used:

$$\frac{CNP}{CPDP} = \frac{INP}{IPDP}$$

Water Supply

LEARNING OBJECTIVES

Upon completion of this chapter, you should be able to:

- Identify the sources of water for firefighting.
- Identify the parts of a municipal water supply system.
- Test a fire hydrant to determine the available gallons per minute (gpm).
- Calculate the efficiency gained by connecting a pumper to a hydrant.
- Identify the needs of a rural water supply.

Case Study

It is late morning as you and the other members of Engine 11 are finishing putting new hose on the engine. An announcement comes over the alerting system that several fire hydrants are out of service in your district because of a water main break.

You and the other members of your company go to the map and locate where the break has occurred. As you identify the hydrants that have been taken out of service, you look for hydrants just outside the affected area that can be used instead of the hydrants placed out of service. Unfortunately, all the most convenient hydrants outside the affected area are painted orange.

1. How would you ensure water for firefighting purposes in the affected area?
2. What is the significance of an orange hydrant?
3. What are the steps of testing a hydrant?

NFPA 1002 Fire Apparatus Driver/Operator Professional Qualifications, 2014 Edition

This chapter addresses the following requisite knowledge elements within sections

5.2.1 and **5.2.2:** low-pressure and private water supply systems, hydrant coding systems, and reliability of static sources.

Introduction

On a planet that is two-thirds covered with water, we are fortunate that water turned out to be such an ideal extinguishing agent. You would think that with such an abundance of water, it would be easy enough to get it to any fire. Unfortunately, getting water to the fire is often as challenging as extinguishing the fire itself.

Water supply sources can be divided into four categories: (1) onboard water; (2) rural water sources (such as ponds, rivers, and lakes); (3) municipal water supply; and (4) private water supply. Every fire department in the world must employ one or a combination of these sources.

As you read this chapter, remember that the water supply is the most important element in fire suppression. If the water is not available, the fire cannot be extinguished. Or if the water is there but not properly used, the fire still may not go out. Even

with a limited water supply, fires can be extinguished successfully, if the water can be delivered from the source of supply to the fire in the most efficient way possible.

Water Supply Requirements

How much water does it take to extinguish a given fire? To answer this question requires knowledge of several factors, including the type of construction, distance between buildings, occupancy, and a factor for the probability of fire spread within the building. When placed in a single formula, these factors determine the needed fire flow (NFF).

There are actually several formulas for calculating the NFF, depending on who is doing the calculations. For preplanning purposes, almost any formula will work to give us a ballpark figure of how much water is needed to extinguish a given fire, including those discussed earlier. For determining how much water is needed to extinguish a specific fire for insurance rating purposes, only one method, the Insurance Services Office, Inc. (ISO) formula, counts. ISO is an independent organization that sets base insurance classifications for fire departments in most of the United States.

It is not the intent of this text to require the calculation of NFF beyond the formulas given earlier. Those calculations are more properly left to texts on firefighting operations. It is, however, necessary to understand the overall concept that governs how much water a municipality should have to meet its needs, including fighting fires. (See Chapter 1 for the details of the formulas used to calculate NFF.)

To determine how much water a municipality needs to be able to meet ISO requirements, we need to know two factors: (1) the **maximum daily consumption (MDC)**, that is, the amount of water the system demands per minute at its peak demand time and (2) the maximum NFF. Together these tell us how many gallons per minute a municipal system must deliver to meet its domestic needs and fight a major fire at the same time. To determine how much water must be in storage, we need to determine how long we need the flow. For structures that require an NFF of up to 2,500 gallons per minute (gpm), the flow must be maintained for up to 2 hours (hr). Where the NFF is 2,501 to 3,500 gpm, the calculated NFF must be maintained for up to 3 hr. And where the NFF exceeds 3,500 gpm, the flow must be maintained for up to 4 hr. For example, at a NFF of 500 gpm, the system should have a minimum of 60,000 gal (500 gpm × 120 min) in storage in addition to domestic water needs. ISO recognizes that there is a practical limit to even a major city's ability to supply water and caps the NFF at 12,000 gpm. Subsequently note that ISO requires a minimum NFF of 500 gpm.

You can estimate the amount of water needed to meet the MDC by using 214.5 gallons (gal) per person.* Simply multiply this figure by the population of the jurisdictions that are served by the water system to find the total gallons of water needed to meet the MDC. Then divide your answer by 1,440 (the number of minutes in 1 day) to find the flow rate. This calculation is represented by the formula MDC = Pop × 214.5/1,440, where MDC is the maximum daily consumption rate in gallons per minute, Pop is the population of the jurisdiction served by the water supply system, 214.5 is the MDC per person, and 1,440 converts the MDC into the MDC needed per minute, or the MDC rate.

$$MDC = \frac{Pop \times 214.5}{1,440}$$

Now we have the gpm needed at the peak demand time (MDC) to meet the domestic needs of the community. Add this to the largest NFF to find the amount of water that the water supply system must have in storage. This translates very neatly into the formula SC = (MDC + NFF) × T, where SC is the storage capacity, MDC and NFF are as previously defined, and T is the amount of time, in minutes, that the flow is needed.

$$SC = (MDC + NFF) \times T$$

If you just wanted to know the flow needed in the system to meet the minimum flow rate, you would only need to add the MDC and NFF. Doing so gives the formula MFR = MDC + NFF, where MFR is the minimum flow rate and MDC and NFF are as defined previously. In this formula all the figures are in gallons per minute.

$$MFR = MDC + NFF$$

---

*In his book, *Fire Suppression Rating Schedule Handbook,* Dr. Harry Hickey states that where adequate records are unavailable, the MDC can be calculated as 150 percent of (or 1.5 times) the average daily consumption (ADC). The American Water Works Association (AWWA) has calculated the ADC to be 143 gal per day per person.

Example 12-1

A small town of 15,000 residents has a target hazard with an NFF of 2,200 gpm. What storage capacity must the system meet in order to satisfy domestic water needs and fight a fire?

Answer

Begin by calculating the MDC.

$$MDC = (Pop \times 214.5)/1,440$$
$$= (15,000 \times 214.5)/1,440$$
$$= 3,217,500/1,440$$
$$= 2,234 \text{ gpm}$$

Add this to the NFF, and then multiply by the time, in minutes, that the flow is needed.

Since the NFF is less than 2,500, in this case the flow must be maintained for 2 hr.

$$SC = (MDC + NFF) \times T$$
$$= (2,234 + 2,200) \times 120$$
$$= 4434 \times 120$$
$$= 532,080 \text{ gal}$$

Example 12-2

Calculate the minimum flow rate to the target hazard for the hypothetical town in Example 12-1.

Answer

$$MFR = MDC + NFF$$
$$= 2,234 + 2,200$$
$$= 4,434 \text{ gpm}$$

During a major fire, this system will need to deliver a minimum of 4,434 gpm.

Sources of Water Supply

Availability of water is a critical factor in fighting a fire. Fortunately, the vast majority of our population lives in areas well served by elaborate and well-maintained water distribution systems. However, although water covers two-thirds of the earth, often water is unavailable or there is no convenient water source for fighting fires. Availability is frequently

environmental, with water being piped in from long distances for cities and areas of high population that are not close to water supplies and have inadequate rain to supply water for the population. In many parts of the country, the only local water is a well that supplies just enough water for individual houses.

The four categories of water supply sources include onboard water, rural water sources, municipal water supply, and private water supply. On most fires, no single source is used exclusively. As an example, in many jurisdictions with municipal water systems, the initial attack is begun with water carried onboard the pumper. Once the driver has made the necessary connections with the supply line, the supply line is charged and the driver *changes over*. <u>Changing over</u> is the process of shutting off water from the onboard tank and opening the intake and admitting water from the charged supply line.

To utilize water supplies to their maximum advantage, it is necessary to understand their particulars. As you study the characteristics of each water source, remember that these sources are often used in conjunction with one another.

Onboard Water

Any firefighting apparatus that has an onboard pump is required to have an onboard water supply Figure 12-1 . The volume of water depends on the

Figure 12-1 The capacity of the onboard water supply depends on the specific type of apparatus.

Courtesy of Robert Selleck/Ferrara Fire Apparatus Inc.

specific kind of apparatus. According to NFPA 1901, *Standard for Automotive Fire Apparatus*, pumper fire apparatus is required to be equipped with a tank that has a minimum capacity of 300 gal. This same standard requires initial attack fire apparatus, or minipumpers, to have a minimum tank capacity of 200 gal.

Because only small fires can be extinguished with this amount of water, many fire departments use much larger-capacity tanks. Tanks of 500 to 1,000 gal are very common. Even these tanks have a limited extinguishing capacity.

The advantage of onboard water is that it is instantly available. The driver only needs to engage the pumps and open the valve to the tank, if it is not kept open, and then water is available. Ideally, the pump/tank valve should always be left open, unless there is some overriding reason for it to be closed. One such example is in freezing weather, when pumps need to be drained to prevent damage to them when they are not in use. Note that if the pump/tank valve is normally closed, it may be necessary to prime the pumps before the pump can develop adequate pressure.

Onboard water is convenient for extinguishing small fires, such as trash fires, small brush fires, and even automobile and small storage shed fires. In rural operations, if used sparingly and efficiently, use of onboard water can allow a primary search or rescue while a water supply is being established. Beyond these examples, extinguishment should generally not be attempted without a secondary water source.

Where a secondary source of water in the form of a municipal water system is not present, fire departments must rely on trucking in their water supply. To do this requires "mobile water supply apparatus" or water tenders. NFPA 1901 defines a *water tender* as any apparatus with an onboard tank of at least 1,000-gal capacity. Considering that we commonly use water at a minimum rate of 100 to 150 gpm, 1,000 gal will only last between 6 minutes (min) 40 seconds (s) and 10 min. This amount is nowhere near adequate. A water tender should be required to carry a minimum of 2,000 gal. At 150 gpm, 2,000 gal will last for 13 min 20 s. Realistically, the actual size of the tender will be governed by several factors, including terrain, bridges, road weight limitations, and funds available for purchase.

Rural Water Sources

When operating beyond the reach of a municipal water system, local fire departments need to develop alternate sources of water supply. Water tenders are the primary source of water. Because structure fires can often require large quantities of water, it may be necessary to employ several tenders to fight a single fire.

When a single tender is not sufficient to fight a fire, multiple tenders are organized to shuttle water to the fire scene. The first tender often deploys a portable drafting basin, a large cistern capable of holding 1,300 to 3,000 gal of water, and then off-loads its water into the drafting basin. Typically, the portable drafting basin will have a capacity of 110 percent of the capacity of the tender. The pumpers on the scene being used to attack the fire then draft water from the basin **Figure 12-2**. The rate of water usage on the fireground determines the frequency at which the basin must be refilled to ensure an uninterrupted water supply. Remember, at just 150 gpm flow, a 1,300-gal basin will need refilling in 8 min 40 s; at 300 gpm it will need refilling in 4 min 20 s; and so on.

Water storage capacity at the scene can be increased by using multiple portable drafting basins in tandem. **Figure 12-3** depicts a 6-inch (6-in) suction sleeve being used to transfer water from the drafting basin in the rear of the photograph to the one in the foreground. Water is being dumped from the rear of

Figure 12-2 Pumper pumping from a portable drafting basin.

© UL LLC. Reprinted with permission.

Figure 12-3 Two portable drafting basins used in tandem. Water is being transferred from the rear basin to the front basin via a jet siphon, not pictured **(A)**, and suction sleeve **(B)**.

Courtesy of Jeff Lohr.

a tender into the rear drafting basin, and a jet siphon in the rear drafting basin is being used in conjunction with the suction sleeve to transfer water. By using multiple portable drafting basins in tandem, you can double and triple the amount of water stored at the scene, while the tender shuttles water from the water source to the scene.

Where a tender shuttle is used to deliver water to the fireground, there must also be a water source to fill the tenders. Sources of water can be a nearby hydrant from a municipal or private water supply system (assuming they will allow you to take their treated water) as well as ponds, lakes, and rivers. Regardless of the source, ideally a pumper with a rated capacity of at least 1,000 gpm needs to be stationed at the source to fill the tenders as they arrive. The idea is to have enough tenders of sufficient capacity that a constant flow can be maintained at the fireground even though the tenders may have to travel several miles to the water source. The farther the source of water from the fire, the more tenders needed to maintain a continuous flow. When a draft site is used to fill tenders, primary consideration should be given to locating the draft pumper as close to the water as possible. Recall that locating a pumper more than 10 feet (ft) above the elevation of the water source will reduce the maximum capacity of the pump. An

unusually long horizontal distance from the water source (more than two suction sleeves) will also reduce the capacity of the pump because of the friction loss in the suction sleeves. (Consult Chapter 8 for more on the relationship between the distance from the water source and pump capacity.)

Fireground Fact

Tender On-load, Off-load Capacity

When tenders are used to supply water for rural firefighting, it is critical that water be on-loaded and off-loaded in minimum time. This point is recognized in NFPA 1901 by requiring an external fill connection that leads directly to the tank. This external fill is required to permit filling at a minimum rate of 1,000 gpm. Additionally, the standard requires that tenders be capable of off-loading 90 percent of their capacity at the rate of 1,000 gpm.

Example 12-3

If we want to maintain a flow of 250 gpm with a 3,000-gal storage capacity at the scene, how many tenders will it take?

Answer

To solve this problem, we must know two pieces of information. First, how long will it take to make a round trip, going from the fire scene to the water source, on-loading a full tank of water, and getting back to the scene? Second, we have to know the capacity of the tenders. For this example, we will make the round-trip time 22 min and set the capacity of the tenders at 2,500 gal. Using water at the rate of 250 gpm, we will exhaust the capacity of a 2,500-gal tender in just 10 min.

2,500 gal of water ÷ 250 gpm = 10 min

With a 22-min round trip, it will take 3 tenders to keep the drafting basin well supplied.

22-min round trip ÷ 10-min capacity per tender = 2.2 or 3 tenders

Some departments employ very large tenders with capacities of as much as 5,000 or 6,000 gal. These tenders are referred to as *nurse tenders.* Rural departments often employ nurse tenders by parking them at a convenient spot on the fireground and then keeping them full. The nurse tender then pumps directly to the pumper(s) on the fireground used to attack the fire, and so a portable drafting basin does not need to be used. NFPA 1142, *Standard on Water Supplies for Suburban and Rural Firefighting,* contains additional information on rural water supplies and getting the most out of the water source.

Whether you are using portable onsite drafting basins or nurse tenders, it is critical that the round-trip time be kept to a minimum. This can be achieved by knowing the location of every possible drafting site or source of water in your jurisdiction. These should be preplanned and, where possible, dry hydrants should be installed Figure 12-4 . A dry hydrant is a 6-in or larger plastic pipe from the water source to a point of access for pumping apparatus.

Figure 12-4 Dry hydrants provide quick and reliable access to water sources.

Courtesy of William F. Crapo.

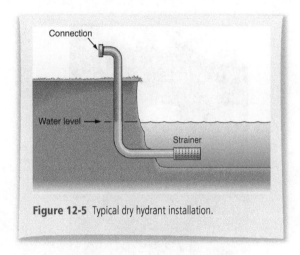

Figure 12-5 Typical dry hydrant installation.

In the water the dry hydrant has a strainer similar to what you would place on the end of your suction hose. At the other end is a 5- or 6-in connection for attaching a length of suction hose from a pumper Figure 12-5 .

The advantage of a dry hydrant is that it saves considerable time in setting up a pumper to draft from a static water source. Dry hydrants can also be placed at convenient spots to allow pumpers better access, such as on harder ground instead of soft earth right next to a pond or lake. Where the dry hydrant must be a relatively long distance from the water source, 8- or 10-in pipe can be used to compensate for the friction loss in the long runs of pipe. When you are using dry hydrants, you must back-flush the hydrant with one-quarter to one-half tank of water before using it. Often marine growth and other debris will try to clog the openings on the strainer. By first back-flushing the hydrant, you will flush out much of this restriction in the strainer, increasing its capacity. (Dry hydrants should be inspected and cleaned annually to minimize obstruction to the strainer.)

Water sources are themselves a separate topic. Water can be obtained from streams, ponds, rivers, swimming pools, and cisterns. Again, these sources should all be preplanned so it will take minimal time

Figure 12-6 A portable pump can be used to draft water if the water source is inaccessible to a fire department engine.

Courtesy of Best Fire Defense LLC.

to establish a continuous, reliable source of water Figure 12-6 . Where water sources are not available, the authority having jurisdiction (AHJ) should require that builders of multihouse developments also supply water for firefighting purposes. Loudoun County, Virginia, for example, requires that all new developments, if not served by municipal water, have a large-capacity cistern installed by the developer. If the cistern is not in the site plans, the county will not approve the development.

Departments or jurisdictions that regularly use tenders and draft sites for fighting fires should have standard operating procedures (SOPs) developed to standardize their use. The SOPs should include everything from requiring a water supply "group" to reserving a separate channel for the water supply group leader to coordinate tenders, fill sites, and dump site.

In addition to preplanning water supply in the rural environment, as mentioned above, preplanning

Fireground Fact

ISO and Rural Water Supply

The section "Water Supply Requirements" demonstrates a method for determining the maximum needed water supply. In rural parts of the country where there is no municipal water supply, this would normally result in an ISO classification of semiprotected ISO rating of 9 if within 5 miles (mi) of the local fire station or of an unprotected ISO rating of 10 if beyond 5 mi. (ISO ratings range from 1 to 10, best to worst.) ISO recognizes that many parts of this country do not have any local water supply for fighting fires and has made allowances for the local fire department to supply water to raise the insurance classification. (Normally the water supply is evaluated independent of the fire department for grading purposes.) To meet a recognized minimum water supply in areas without an adequate municipal supply system, the local fire department must be able to supply 250 gpm for a total of 2 hr. This is a minimum of 30,000 total gal of water over the 2-hr period, without interruption, and it must be started within 5 min of the arrival of the first pumper on the scene.

must include calculating the NFF of every type of building. NFPA 1142 has just such a NFF formula that gives an answer not in terms of rate of flow, but in terms of total water required. In addition, every aspect of hose appliances and pumper connections should be standardized among all jurisdictions that work together, so there is no problem with interoperability. It would be tragic to have a house burn because an odd-size suction sleeve did not allow for the last 5 ft needed to complete a stretch of four sleeves to reach the water.

Municipal Water Supply Systems

A municipal water supply system provides water for firefighting throughout the most densely populated areas of the United States. Understandably, it is the water system most often envisioned when water supply is mentioned.

A municipal water supply system, as it concerns the fire service, is composed of three elements: (1) a storage system, (2) a distribution system, and (3) fire hydrants. The storage system can be in the form of reservoirs, water tanks, or towers. Water to fill these storage elements comes from springs, rivers, lakes, and wells. Exactly which is used to fill the reservoirs and tanks depends upon availability and need. Large towns and cities often have several sources of water to meet their needs.

Pressure for the water supply system is provided by one of two methods: (1) gravity or (2) pumps. Gravity has distinct advantages over pumps for supplying pressure to the water system. Gravity never breaks down or needs routine maintenance, it never needs to be replaced, and it provides a constant and reliable pressure. Simply by placing reservoirs or water tanks on high ground, gravity can be used to provide the pressure needed for the everyday needs of residents and industry and to provide hydrant pressure. Often, where elevated water towers are used, gravity feeds water to the residents during the day at a constant pressure and flow **Figure 12-7** . At night, when demand is lower, a nearby reservoir (higher than the tank) or pump is used to fill the tank again in preparation for the next day's needs.

For some jurisdictions, gravity feed is not possible. In these jurisdictions, pumps are used to supply the pressure needed for the water supply system, which means that pumps must be running constantly, day and night, to maintain the needed flow and pressure around the clock. The disadvantages of pumps are that they can break down, they need routine maintenance, and they can need replacing.

Many municipal water supply systems employ both gravity and pumps to maintain pressure in their systems. It may seem that the pumps are fighting

Figure 12-7 Typical elevated water storage tank/water tower.
Courtesy of William F. Crapo.

gravity to provide pressure, but such is not the case. Large systems are often broken down into many smaller systems. Some of the smaller subsystems can be gravity fed while others, where gravity feed is not feasible, are pumped. These subsystems, regardless of how pressurized, are isolated from one another. Physically there are connections between them that have valves that are kept closed. If an unexpected high demand hits one of the subsystems, such as a major fire, the valves can be opened and water admitted from other subsystems.

Water gets from the point of storage to the point of use by a distribution system **Figure 12-8** . There are three elements of a distribution system: (1) *feeder mains*, (2) *secondary feeder mains*, and (3) *distribution mains*. The terms used here to describe these elements of the water supply system are not universal; different systems may call them

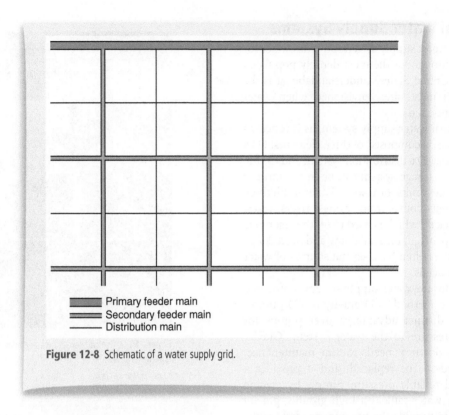

Figure 12-8 Schematic of a water supply grid.

by different names, but they do illustrate the principle. To best describe the role of each element of the distribution system, it is easier to understand if we describe it in reverse order.

The <u>distribution main</u> is the part of the supply system that has the hookup to each individual home, apartment, school, business, factory, or other building. Distribution mains can be whatever size is required to meet the domestic needs of the particular area of the system. This means pipe smaller than 6 in can be used. However, fire hydrants should never be attached to pipe having less than 6-in diameter in residential areas and 8-in diameter in commercial and industrial areas.

<u>Secondary feeder mains</u> supply water from the feeder main to the distribution mains. They are used to effectively break down large areas of the jurisdiction into smaller grids. One secondary feeder may supply water to a residential area while another secondary feeder runs through the commercial area,

feeding the distribution mains that run up and down the blocks feeding the various businesses.

Finally, the secondary feeder main is supplied by the *feeder main*. The <u>feeder main</u> carries water from the point of storage to the secondary feeders. The secondary feeder mains and the distribution mains are designed to form a grid or loop, allowing water to feed each section of the grid from more than one direction. If a break occurs, it can be isolated and a minimum of customers will need to be shut down. This configuration also ensures that a minimum number of fire hydrants are placed out of service if a break occurs.

Private Water Supply

Private water supplies can range from a tank of water to handle the needs of a sprinkler system until the fire department arrives Figure 12-9 , to a small

Figure 12-9 The water tank and pump house (arrow) provide water and pressure for fire protection in this shopping center.

Courtesy of William F. Crapo.

version of a municipal system. In general, they are used where municipal water systems either are not present or are not capable of handling the needs of a particular occupancy.

Just as with a municipal water supply system, it is critical that all elements of the private water supply be maintained. Whereas the responsibility for maintenance of a municipal water supply system rests on the local water authority, the owner of a private system is responsible for the maintenance of that system. The big disadvantage here is that the local fire department usually has no way to verify that these systems are being properly maintained. We can test municipal hydrants, but unless owners of private systems allow us, we cannot test their hydrants.

Water storage for a private water supply system can be in the same forms as a municipal system. Some private systems are simply private extensions of the municipal system where occupants install and maintain their own piping, hydrants, and valves on private property.

Fire Hydrants

A key element of any water supply system used for fighting fires is the fire hydrant. It allows the fire department to gain access to the water system quickly and efficiently. By hooking up to a fire hydrant, a pumper gains access to water in the municipal (private) water system at whatever pressure is in the system. Fire hydrants come in two main types: dry barrel and wet barrel.

To remain reliable, hydrants should be inspected on a regular basis. The inspection should include operation of moving parts, inspection of threads for possible damage, and assurance that caps on each discharge are in place. One final inspection needed is to ensure the threads at each discharge are in place and secure. The threads of the hydrant are not integral and can become loose. If not secure, they can leak and even separate from the hydrant under pressure.

As part of a regular inspection program, some attempt should be made to test flow hydrants on a periodic basis. Once individual hydrant capacity is known, the next logical step is to identify hydrants by capacity. This can be as simple as painting the top of the hydrant according to flow. (Color-coding of hydrants is discussed in the section "Water Supply Testing" later in this chapter.)

Dry Barrel Fire Hydrants

The dry barrel fire hydrant is the hydrant most often used in this country. It is intended to be used in climates where freezing weather is possible. The most distinguishing feature of the dry barrel fire hydrant is the operating stem at the very top of the hydrant that runs down through the hydrant to operate an admission main valve at the bottom of the hydrant **Figure 12-10**. The main valve is located below the frost line to prevent it from freezing. When the hydrant is not in operation, a drain valve, also in the bottom of the hydrant, allows any water in the hydrant to drain out so it cannot freeze. The drain valve closes when the hydrant is open.

When a dry barrel fire hydrant is used, all the outlets receive water anytime the hydrant is turned on, which means that any hose to be connected to the fire hydrant must be connected before the hydrant is turned on or "charged." After the hydrant is charged, if we want to connect another hose to the

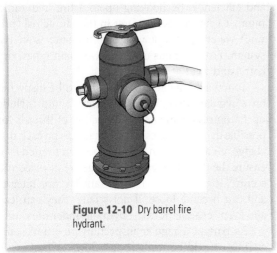

Figure 12-10 Dry barrel fire hydrant.

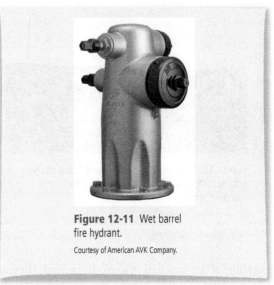

Figure 12-11 Wet barrel fire hydrant.

Courtesy of American AVK Company.

fire hydrant, first we must shut it down. However, if we place a hydrant valve or some sort of gate valve on the hydrant before we charge it, we can charge additional lines as needed.

The exact style and number of outlets of a hydrant can vary according to need and the manufacturer. A typical fire hydrant has two 2½-in outlets and one 4-in, or larger, *steamer connection*. The term **steamer connection** comes from the fact that the first mechanically driven fire engines, called *steamers*, hooked up to the large discharge of hydrants.

Wet Barrel Fire Hydrants

The wet barrel fire hydrant is used only in areas where there is no fear of freezing at any time of the year. Water is constantly in the barrel of the wet barrel fire hydrant and is controlled at each individual discharge 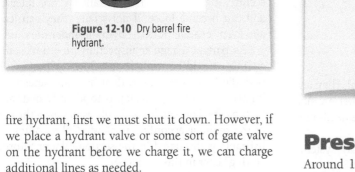. Like the dry barrel fire hydrant, the wet barrel fire hydrant can have one or more 2½-in connections and usually a steamer connection. The primary advantage of the wet barrel fire hydrant is its ability to connect additional lines without having to shut down the hydrant. Its only drawback is that it cannot be used in latitudes where freezing weather may occur.

Pressure Loss in Pipe

Around 1905, Allen Hazen and Gardner Williams performed flow tests on pipes used for water supply and developed the formula known today as the **Hazen–Williams formula**, which is used to determine the pressure (friction) loss in pipe. This formula performs the same function in pipe as the formula introduced earlier for calculating friction loss in hose. However, since the characteristics of pressure loss in pipe are significantly different from those in hose, the formula is slightly different to take these characteristics into account. (See Chapter 6 for a description of calculating friction loss in hose.)

The Hazen–Williams formula is $P_f = (4.52 \times Q^{1.85})/(C^{1.85} \times D^{4.87})$, where P_f is the pressure loss per foot of pipe in pounds per square inch (psi), Q is the number of gallons per minute of flow, C is the roughness coefficient, and D is the internal diameter of the pipe. For this formula to be useful, it is necessary to look up the appropriate values of C and D. Values of C are found in various NFPA publications, such as NFPA 13, 24, and the *Fire Protection Handbook*. In addition, AWWA publications contain charts with values of C. Values of D can be found in various publications from pipe manufacturers.

$$P_f = \frac{4.52 \times Q^{1.85}}{C^{1.85} \times D^{4.87}}$$

Different pipes have different roughness characteristics, and the C factor takes that into consideration. Some of the coefficients for C for a specific kind of pipe even vary with age. It is also important to use the correct coefficient for D because the internal diameter of pipe is not what would be expected. Some pipes actually have an internal diameter larger than the stated pipe diameter, and other pipes are smaller. For example, 8-in class 50 ductile iron pipe has an actual internal diameter of 8.51 in, while class 200 PVC plastic pipe has an actual internal diameter of 7.68 in.

The Hazen–Williams formula looks a little intimidating at first glance; however, it is actually a very basic algebraic expression. The exponentials in the equation are easily calculated with a scientific calculator. Be cautioned that some of the divisors and dividends can get rather large, so pay attention to keep the numbers accurate.

Example 12-4

What is the friction loss for 8-in class 50 ductile iron pipe flowing 500 gpm?

Answer

We already know that D for 8-in class 50 ductile iron pipe is 8.51 and C for ductile iron is 100.

$$\begin{aligned} P_f &= \frac{4.52 \times Q^{1.85}}{C^{1.85} \times D^{4.87}} \\ &= \frac{4.52 \times (500)^{1.85}}{(100)^{1.85} \times (8.51)^{4.87}} \\ &= \frac{4.52 \times 98,422.53}{5,011.87 \times 33,787.57} \\ &= \frac{444,869.84}{169,338,908.46} \\ &= 0.0026 \text{ rounded to } 0.003 \text{ psi/ft friction loss} \end{aligned}$$

By presenting the Hazen–Williams formula here, it is not intended that the average firefighter be able to calculate pressure loss in a municipal water system. Instead it is presented so that the average firefighter can understand the process of calculating pressure loss in a municipal water system and pipe.

The pressure loss as calculated by the Hazen–Williams formula is per foot of pipe, because pipe lengths, unlike hose, can be random lengths. To find the total amount of friction loss, simply multiply by the amount of pipe by the pressure loss per foot as determined by the Hazen–Williams formula. Total pressure loss then becomes $P_t = P_f \times L$, where P_t is the total pressure loss, P_f is the pressure loss per foot as calculated above using the Hazen–Williams formula, and L is the length of pipe in feet.

$$P_t = P_f \times L$$

Example 12-5

Calculate the friction loss from the water tower to the fire hydrant **Figure 12-12**. The pipe is 8-in class 50 ductile iron.

Answer

In Example 12-4, we calculated the friction loss for this pipe to be 0.003 psi/ft.

$$\begin{aligned} P_t &= P_f \times L \\ &= 0.003 \times 2,100 \\ &= 6.30 \text{ psi} \end{aligned}$$

Where elbows and valves are installed in pipes, these fittings are assigned an equivalent in feet to straight pipe. For example, a 90° elbow for 8-in ductile iron pipe is equivalent to 12.84 ft. If you were trying to find the friction loss for 100 ft of 8-in ductile iron pipe with a single 90° elbow in the middle, you would calculate the friction loss for an equivalent of 112.84 ft.

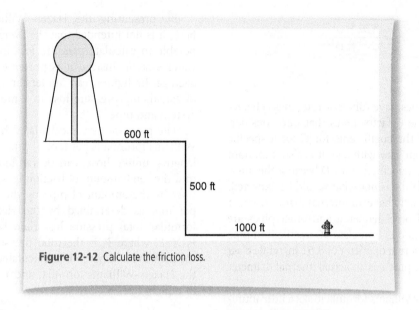

Figure 12-12 Calculate the friction loss.

Example 12-6

Go back to Example 12-5 and recalculate the total friction loss, taking into consideration the two elbows in Figure 12-12.

Answer

We have an equivalent length of 2,100 ft of pipe and two elbows, each equivalent to 12.84 ft, for a total of 2,125.68 ft.

$$P_t = P_f \times L$$
$$= 0.003 \times 2,125.68$$
$$= 6.38 \text{ psi}$$

For those who feel comfortable with the process, a variation of the Hazen–Williams formula can be used that calculates Pt directly. That formula is $Pt = (4.52 \times Q^{1.85} \times L)/(C^{1.85} \times D^{4.87})$. Note that this formula is the Hazen–Williams formula with L included in the formula, so you are solving directly for Pt. When you use this formula, remember to include the equivalent length for each pipe fitting in the system.

$$P_t = \frac{4.52 \times Q^{1.85} \times L}{C^{1.85} \times D^{4.87}}$$

Example 12-7

Calculate the friction loss for the system illustrated in Figure 12-12 with the same 500 gpm, this time for 6-in pipe.

Answer

The following are specific to 6-in class 50 ductile iron pipe:

$$C = 100$$
$$D = 6.4$$

90° elbow is equivalent to 9.98 ft. Total length is 2,100 plus 19.96 ft, or 2,119.96 ft.

$$P_t = \frac{4.52 \times Q^{1.85} \times L}{C^{1.85} \times D^{4.87}}$$

$$= \frac{4.52 \times (500)^{1.85} \times 2,119.96}{(100)^{1.85} \times (6.4)^{4.87}}$$

$$= \frac{4.52 \times 98,422.53 \times 2,119.96}{5,011.87 \times 8,435.22}$$

$$= \frac{943,106,256.70}{42,276,226.06}$$

$$= 22.31 \text{ psi}$$

Water Supply Testing

The ability to determine the amount of water available at a particular fire hydrant is one of the most useful tools a fire department can have. It allows us to determine, before the fire occurs, whether additional water sources will be necessary or extra planning will be required. It can also impact decisions to construct new building or structures. A new factory or apartment complex, for example, cannot be built if the water that would be required to fight a fire is not available. Building planners and their engineer may not always take into consideration the correct amount of water needed. You need the ability to verify needed flows and intelligently challenge any discrepancies.

Testing a water supply system is a process of systematically calculating flows at selected hydrants in the system, and it is usually referred to as *testing hydrants*. Hydrant tests require a minimum of specialized equipment. At minimum, you will need one blind cap with a pressure gauge and drain valve for the test hydrant, a small ruler to measure the hydrant outlet diameter, paper and pen to record data, and a pitot gauge for each flow hydrant.

Note
Testing a water supply system is a process of systematically calculating flows at selected hydrants.

The test procedure consists of placing the blind cap with the pressure gauge on the hydrant you are testing. This hydrant is referred to as the *test hydrant*, and it is the one for which you are determining the flow. Next, flow water from adjacent hydrants, opening enough hydrants that a minimum pressure drop of 10 psi registers on the gauge at the test hydrant. These hydrants are referred to as *flow hydrants*. When you open up the flow hydrant, begin with only one 2½-in discharge open. If this does not give an adequate pressure drop at the test hydrant, open additional discharges on the flow hydrant. To get the

required pressure drop, it may be necessary to open only one hydrant in a small water system or several hydrants in a large municipal water system.

When you are selecting the hydrant to test, it is important to determine the direction of flow in the water mains. If only a single hydrant flow is needed, it is preferable that the flow hydrant be downstream from the test hydrant. However, if it is necessary to flow several hydrants to get the required pressure drop, then the flow hydrants should be between the test hydrant and larger mains. In short, when multiple flow hydrants are used, water should flow past the flow hydrant to the test hydrant **Figure 12-13**. Determination of the direction of flow is not all that difficult. Water always seeks the path of least resistance and flows from larger pipe to smaller pipe. By noting the relationship of the larger pipe and smaller pipe on a water supply map, you can determine the direction of water flow.

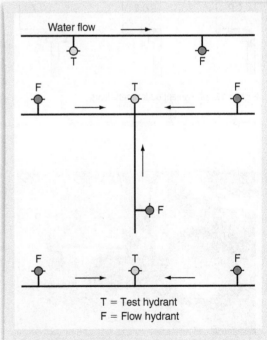

T = Test hydrant
F = Flow hydrant

Figure 12-13 Determine the direction of water flow to the test hydrant.

After we flow water from the flow hydrants, it is necessary to know exactly how much water the hydrants were flowing. To determine this amount, we need to know just how large the inside diameter of each discharge is, to the $\frac{1}{16}$ in, because these can vary. The exact size is obtained by measuring each opening used. We also need to know the discharge coefficient of each outlet, which is obtained by feeling the inside of the hydrant where the outlet meets the hydrant barrel. The coefficients are illustrated in Figure 12-14. The pressure at the flow hydrant must be determined by use of the pitot gauge Figure 12-15. The blade of the gauge should be inserted into the

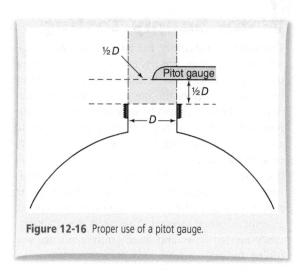

Figure 12-16 Proper use of a pitot gauge.

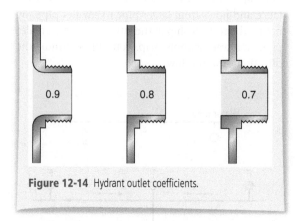

Figure 12-14 Hydrant outlet coefficients.

Figure 12-15 A pitot gauge.
© Jones & Bartlett Learning. Photographed by Glen E. Ellman.

center of the stream, one-half of the diameter of the opening away from the opening. To get an accurate reading, this reading should be taken from an outlet no bigger than 2½ in Figure 12-16.

Where pitot gauges are unavailable, blind caps with pressure gauges can be substituted. The cap with a pressure gauge is put on one of the outlets of the flow hydrant, and the hydrant is charged. A reading taken this way is *approximately* the same as a reading taken with a pitot gauge. The biggest disadvantage to obtaining a flow pressure in this manner is that it is actually a reading of the hydrant residual pressure and not the more accurate *velocity pressure*. **Velocity pressure** is the measure of the kinetic energy of flowing water. Properly done, the velocity pressure is taken at the point of **vena contracta**, the point of maximum stream contraction Figure 12-17.

The following procedure should be followed precisely to get accurate results when testing a hydrant:

1. Place the blind cap with gauge and drain valve on the hydrant selected to be the test hydrant.
2. Open the hydrant all the way. Open the drain valve and bleed off all the air in the hydrant; then close the drain valve. Write down the static pressure.
3. Signal personnel at the flow hydrants to open their hydrant(s), in succession, until a

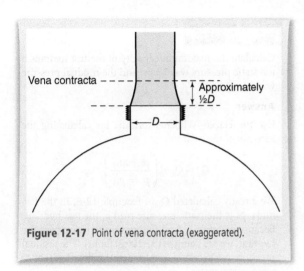

Vena contracta

Approximately ½D

D

Figure 12-17 Point of vena contracta (exaggerated).

minimum of 10 psi drop in pressure from static is obtained at the test hydrant.

4. Signal personnel at the flow hydrants to write down the pressure reading of the flow from their hydrants and write down the residual pressure at the test hydrant.
5. Shut down the flow hydrants slowly in succession to prevent water hammer in the system. Shut down the test hydrant and remove the pressure gauge.
6. Measure and record the exact interior size of all outlets that discharged water during the test as well as the outlet coefficient.
7. Return all the hydrants to a system-ready condition.

Calculating Hydrant Capacity

Calculating the capacity of the test hydrant is a matter of applying the findings of the above test to another version of the Hazen–Williams formula. That formula is $Q_2 = Q_1 \times \left(\dfrac{p_s - p_{r2}}{p_s - p_{r1}} \right)^{0.54}$, where Q_2 is the gallons per minute of flow we are calculating, Q_1 is the gpm flow we calculated as a result of opening the flow hydrants, p_s is the static pressure, p_{r2} is the

pressure we are calculating the flow for, and p_{r1} is the residual pressure obtained from the test. Essentially, $p_s - p_{r2}$ is the pressure drop from static to the theoretical test pressure. The pressure drop $p_s - p_{r1}$ is the drop actually observed during the test.

$$Q_2 = Q_1 \times \left(\frac{p_s - p_{r2}}{p_s - p_{r1}} \right)^{0.54}$$

Most fire department testing is designed to calculate the maximum capacity of the water main, that is, the gpm at 20 psi residual, so p_{r2} is 20. At times we may need to know the capacity of the water main at pressures not intended to produce the maximum capacity. To learn this, simply insert the pressure for which you need a corresponding capacity into the formula as p_{r2}. Then Q_2 will be the flow at pressure p_{r2}.

The first step in calculating the capacity of the test hydrant is to determine how much water was flowing from the flow hydrant(s). This can be done in one of two ways: (1) look up the flow in a chart such as in NFPA 291, *Recommended Practice for Fire Flow Testing and Marking of Hydrants*; or (2) calculate the flow from each hydrant using the formula gpm = $29.84 \times D^2 \times C \times \sqrt{P}$. This is the same formula we used earlier to calculate gpm. Finding the flow in charts is simple and saves time doing the calculations. There is one caution, however: If the chart used has not already taken the hydrant coefficient into account, you must do so by taking the gpm specified in the chart, for the particular outlet size and pressure, and multiplying it by the coefficient. (See Chapter 5 to read the discussion of calculation of gpm.)

By now, use of the formula gpm = $29.84 \times D^2 \times C \times \sqrt{P}$ should be second nature. Here we are going to use it to calculate the flow from a hydrant. Plug in the exact diameter of the outlet(s) flowing water for D, the pressure measured by the pitot gauge for P, and the hydrant coefficient for C. The answer will be the gpm flowing from a particular outlet during the test. If more than one outlet is used during a test, the calculation must be done for each outlet. The pressure obtained at any outlet is the same for all other outlets of the same hydrant. If more than one hydrant was used during the test, the calculations must be repeated for each hydrant. Different hydrants will not normally have the same pressure reading as taken by the pitot gauge, so separate pressures must be taken from each hydrant. (If several hydrants with different-size outlets are used, you quickly understand the value of charts to find gpm.) Once all the calculations are made, the gpm values are added and the result is Q_1.

Tip

Each hydrant will not normally have the same pressure reading as taken by the pitot gauge, so each set of calculations must reflect this.

Example 12-8

When testing a hydrant according to the above procedures, you were required only to flow water from one outlet that has a diameter of 4⁹⁄₁₆ in. The pitot reading gave a pressure of 15 psi, and you determined the hydrant had a coefficient of 0.8. How much water was this hydrant flowing?

Answer

$$\text{gpm} = 29.84 \times D^2 \times C \times \sqrt{P}$$
$$= 29.84 \times (4\tfrac{9}{16})^2 \times 0.8 \times \sqrt{15}$$
$$= 29.84 \times (4.5625)^2 \times 0.8 \times 3.87$$
$$= 29.84 \times 20.82 \times 0.8 \times 3.87$$
$$= 1,923.45 \text{ or } 1923 \text{ gpm}$$

Now that we know how much water was flowing at the flow hydrant, we can determine the amount of water available in the test hydrant at any desired pressure.

Example 12-9

Calculate the maximum capacity of the test hydrant if the static pressure was 78 psi and the residual pressure was 52 psi.

Answer

Use the Hazen–Williams formula for calculating the capacity of the hydrant:

$$Q_2 = Q_1 \times \left(\frac{p_s - p_{r2}}{p_s - p_{r1}} \right)^{0.54}.$$

We already calculated Q_1 in Example 12-8. In the formula, ps is the static pressure read at the test hydrant. Because we want to find the maximum capacity of the hydrant, we are going to use 20 psi for pr_2. The residual pressure at the test hydrant when the flow hydrant(s) was flowing water is pr_1.

$$Q_2 = Q_1 \times \left(\frac{p_s - p_{r2}}{p_s - p_{r1}} \right)^{0.54}$$
$$= 1,923 \times \left(\frac{78 - 20}{78 - 52} \right)^{0.54}$$
$$= 1,923 \times \left(\frac{58}{26} \right)^{0.54}$$
$$= 1,923 \times (2.23)^{0.54}$$

Now find the 0.54 power of 2.23 on your calculator.
$$Q_2 = 1,923 \times 1.54$$
$$= 2,961.42 \text{ or } 2,961 \text{ gpm}$$

Graphing Hydrant Flow

The process just described for calculating the flow from a hydrant at a specific pressure is fairly simple and very accurate. As previously mentioned, when we test a hydrant, we are usually calculating for maximum flow. At times, however, we need to know the amount of water available in the water main at a pressure other than 20 psi. Either we can calculate the pressure at each and every pressure we need a flow for, or we can make just one calculation and graph the capacity of the water supply.

Graphing the capacity of the water supply has the primary advantage of being fast. Its primary disadvantage is that it is not quite as accurate as calculating the flow. Before we can graph the capacity of the hydrant, first we must do one flow test to determine the flow from the flow hydrant(s) and the static and residual pressure at the test hydrant. Once one flow test has been done, we can plot the curve of the hydrant capacity versus pressure.

Graphing the flow of the hydrant requires special graph paper referred to as *1.85 exponential graph paper* Figure 12-18 . The *y*-axis (vertical) of

the paper is calibrated for hydrant pressure, and the *x*-axis (horizontal) is calibrated for hydrant flow. There are usually three scales across the bottom to accommodate a wide range of pressure and flow. The vertical lines are not an equal distance apart but represent the change in pressure versus flow at the 1.85 power.

Using the graph paper to illustrate the capacity of the hydrant over its pressure range is as easy as following these steps:

1. Mark the static pressure of the test hydrant on the *y*-axis at 0 gpm.

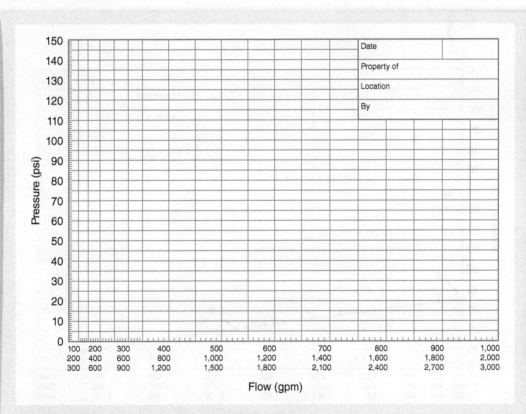

Figure 12-18 Sample of 1.85 exponential graph paper.
Courtesy of National Fire Academy Open Learning Fire Service Program.

2. Determine which scale to use. Choose a scale that spreads out the graph/line of the water flow as much as possible. The graph should end up close to 0 psi when it goes off scale.
3. Mark the gpm flow from the flow hydrant(s) on the x-axis.
4. Extend a line up from the gpm of the flow hydrant, parallel with the reference lines, until you come to the residual pressure of the test hydrant. Mark the spot.
5. Take a straightedge and connect the two points with a line that extends all the way from the x-axis baseline to the y-axis baseline.

Example 12-10

Graph the flow test of a hydrant that has a static pressure of 35 psi and a residual pressure of 20 psi and had one flow hydrant with a flow of 1,300 gpm. (If 1.85 exponential graph paper is not available, copy Figure 12-15.)

Answer

Begin by marking the static pressure of the test hydrant, point A, on the graph Figure 12-19 . Locate the hydrant flow along the bottom of the graph. Find the intersection of 20 psi residual (of the test hydrant) and 1,300 gpm, and mark point B. Finally, draw a straight line connecting these two points and extending across the entire face of the graph.

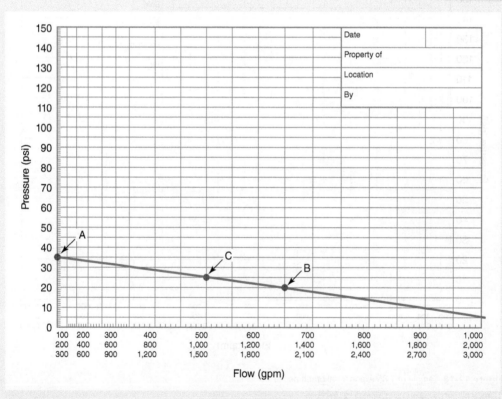

Figure 12-19 Graph of hydrant test.

Courtesy of National Fire Academy Open Learning Fire Service Program.

Once we have the hydrant test graphed, it is easy to find a corresponding flow for any given pressure. For instance, what is the maximum flow from this hydrant at 25 psi? Point C on Figure 12-19 indicates a flow of approximately 1,000 gpm.

Color-Coding of Hydrants

Once you have spent the time and effort to calculate the capacity of individual hydrants, do not just file the results away in a file cabinet or three-ring binder. Mark individual hydrants for field identification according to capacity. **Table 12-1** summarizes the color recommendations for marking hydrants according to NFPA 291. The document recommends that both the top of the hydrant and the individual discharge caps be painted the appropriate color. The use of reflective paint is also recommended for easy identification at night.

Frequently, this information is already in the hands of the local water authority/department, in which case it is simply a matter of obtaining the information and painting the hydrants appropriately. However, where this information does not already exist, coordinate testing with the water authority/department. It may even be willing to assist in or actually perform the testing.

Table 12-1 Marking Hydrants

Hydrant Class	Flow, gpm	Color
Class AA	> 1,500	Light blue
Class A	1,000 – 1,499	Green
Class B	500 – 999	Orange
Class C	< 500	Red

Getting the Most out of the Water Supply

The single most important benefit of learning hydraulics is the ability to make maximum use of the local water supply. Getting the maximum amount of water from the source to the fire in the shortest possible time is the key to successful firefighting.

Four general evolutions are used to get water from the hydrant to the fire: (1) fire to hydrant, (2) hydrant to fire, (3) split lay, and (4) two-pump operation **Table 12-2**. Each is discussed in the following sections with attention to advantages and disadvantages.

Table 12-2 Evolutions Used to Get Water from the Hydrant to the Fire

Evolution	Advantages	Disadvantages
Fire to hydrant	Pumper can pump whatever pressure is needed.	Water supply is limited by length of lay, size of hose, and nozzles left at scene. Any additional lines must be laid from hydrant to fire. Hydraulic balance must be accounted for. All remaining equipment is at the hydrant.
Hydrant to fire	It is easy to pump multiple lines at the proper pressure. Equipment is readily available.	Has a limited water supply (Large-diameter hose [LDH] in single-family residential community may be okay if there is adequate hydrant pressure and supply.)
Split lay	Evolution becomes a two-pump operation once completed.	Two pumpers are required. Water may be delayed.
Two-pump operation	It makes maximum use of water from hydrant. If water is initially limited, second pumper will boost supply through hose already laid.	Two pumpers are required.

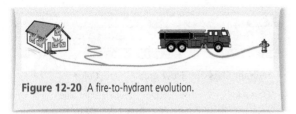

Figure 12-20 A fire-to-hydrant evolution.

Figure 12-21 A hydrant-to-fire evolution.

Fire to Hydrant

In a fire-to-hydrant evolution, the pumper stops in front of the fire building and drops off a skid load of 1½-, 1¾-, or 2½-in hose, and a gated wye, and then lays a line to the hydrant Figure 12-20 . The only advantage to this evolution is that it allows the pumper to connect to the hydrant, which allows you to pump whatever pressure is needed.

There are several disadvantages to this evolution. First, the amount of water supplied is limited by the size of hose and attached nozzles dropped off at the fire. Second, if additional lines are needed, they must be laid from the pumper at the hydrant. Third, unless the hose lines attached to the wye are the same length, require the same nozzle pressure, and flow the same gpm, getting the correct pressure to any line—that is, achieving hydraulic balance—is impossible. Fourth, all the remaining portable equipment is on the pumper, which may be some distance from the fire.

Hydrant to Fire

A hydrant-to-fire evolution involves laying a line from the hydrant to the fire and connecting the supply line directly to the hydrant Figure 12-21 . This procedure has the advantage of placing the pumper at the fire where multiple lines can be easily pumped, each with the correct pressure. All the tools and appliances carried on the pumper are readily available.

The primary disadvantage of this evolution is the limited water supply. With the pumper relying on hydrant pressure to overcome friction loss and provide adequate intake pressure, the water supply available, even with LDH, is limited. To further complicate things, as additional pumpers connect to other hydrants, the pressure in the system will

fall and the total pressure available to supply water will be reduced, reducing the amount of water available.

Example 12-11

How much water can a pumper expect to get from a hydrant if it has laid 400 ft of 4-in hose and the hydrant has a static pressure of 75 psi?

Answer

This is a maximum flow problem. First, we subtract the needed intake pressure of 20 psi from the static pressure of 75 psi, which leaves us with only 55 psi available to overcome friction loss. To find the maximum amount of water, first calculate the allowable friction loss per 100 ft of hose.

$$55 \div 4 = 13.75 \text{ psi} = \text{FL } 100$$

Now find the amount of water a 4-in hose can flow with 13.75 psi of friction loss per 100 ft of hose. We use the formula

$$Q = \sqrt{\text{FL } 100/(\text{CF} \times 2)}$$
$$= \sqrt{13.75/(0.1 \times 2)}$$
$$= \sqrt{13.75/0.2}$$
$$= \sqrt{68.75}$$
$$= 8.29 \text{ or } 829 \text{ gpm}$$

(Consult Chapter 6 for details about this formula.)

While a flow of 829 gpm sounds good, remember that the flow depends on both the static pressure at the hydrant, which can be significantly less than 100 psi, and the length of the lay. LDH also has the disadvantage of forming a roadblock once it is charged. Finally, if the hydrant laid from were to have a capacity of 2,000 gpm, by relying on hydrant pressure alone to move the water, there is no way you could ever get the maximum water from the hydrant. This means that if you had a needed fire flow of 2,000 gpm, even though the hydrant was capable of delivering the water, there would be no way you could ever get the water to the fire.

Split Lay

At times it is not practical for the first-in pumper to lay a line from the hydrant to the fire or from the fire to the hydrant. In these instances, it is necessary for two engine companies to work together to lay the supply line.

To accomplish the split lay, the first-in pumper lays a supply line from a nearby intersection, preferably one with a hydrant on the intersecting street, to the street where the fire is located. A second pumper then completes the lay, from the same intersection to the hydrant or water source Figure 12-22 . Once the second pumper has completed its portion of the lay, the two hose lines are connected. Water can then be delivered from the second-in pumper at the water source to the first-in engine at the fire.

The split lay also has advantages and disadvantages. The biggest disadvantage is that it can take more time getting water in the supply line. The first-in pumper will have to wait for a second pumper to "pick up" its line and complete the lay. However, once the lay is complete, this evolution has the advantage of becoming a two-pump operation.

Two-Pump Operation

A two-pump operation involves the use of two pumpers and gets the most water from the hydrant Figure 12-23 . The first pumper lays hose from the

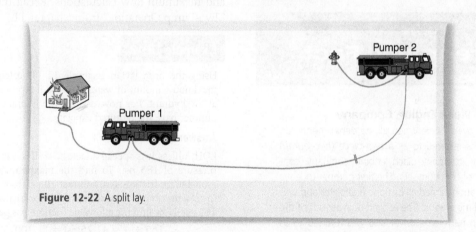

Figure 12-22 A split lay.

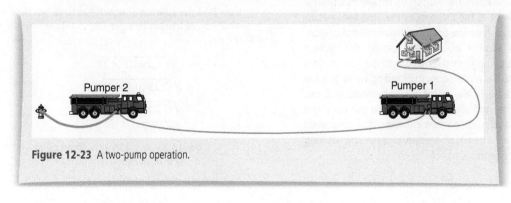

Figure 12-23 A two-pump operation.

hydrant to the fire, and the second connects to the hydrant and pumps to the first. In a possible variant, the first pumper stops at the fire, and the second pumper lays hose from the fire to the hydrant, connects to the hydrant, and pumps to the first pumper. Both are textbook examples of two-pump operations. Since they are technically relay operations, pump discharge pressure (PDP) for the pumper at the hydrant is calculated as described in Chapter 10 in the section "Relay Formula."

The only disadvantage of a two-pump operation is that it requires two pumpers. In all other ways, this is the most efficient way to get the maximum water out of the hydrant. With the proper use of a hydrant valve, water will not be delayed. Even if water is limited initially, once a second pumper arrives, it is then connected to the hydrant through the hydrant valve, thereby increasing the water supply in the hose already laid.

Fireground Fact

Two-Piece Engine Company

There was a time when all fire departments ran what is referred to as a *two-piece* (*two-pump*) engine company. Each engine company consisted of two pumpers. The first pumper to leave the station carried all but one firefighter and was called the *wagon*. The second pumper out of the station had just a driver and was referred to as the *pumper*. The wagon would lay out from a hydrant to the fire, and all the hose lines necessary to fight the fire were taken off the wagon. The pumper would then "catch" the hydrant and pump from the hydrant to the wagon. This system is very efficient in providing water to a fire because it allows for fire-to-hydrant, hydrant-to-fire, and split lays with a single company in a time-efficient manner. However, it is considered today to be an unnecessary expense and has long since become extinct.

What gives this evolution the advantage over the hydrant-to-fire evolution is that you are no longer restricted to hydrant pressure. The pumper at the hydrant can pump as much water as is in the system as long as the intake pressure stays above 5 psi and the pumper at the hydrant has sufficient capacity. Even as other pumpers hook up to the system and the residual pressure in the system is reduced, the pumper at the hydrant can compensate by adding more pressure with the pump. (The needed discharge pressure can be maintained because the pump will do more work and contribute more of the pressure.)

This evolution does have some limiting factors that must be considered when calculating PDP. These factors are capacity of the hydrant, capacity of the pump, maximum capacity of the hose, and maximum pressure of the hose. In addition to straightforward relay calculations, the advantage of this evolution becomes very evident in maximum lay and maximum flow calculations. Recall the section "Two-Pump Operation" from Chapter 11.

Example 12-12

Using the hose lay in Example 12-11, calculate the maximum amount of water possible if a pumper were at the hydrant. For now, assume the hydrant and the pumper have the required capacity.

Answer

LDH, unless it is special attack hose, has a maximum pressure of 185 psi. To find the maximum pressure available for friction loss, subtract 20 psi from 185 psi, to get a maximum pressure allowable for friction loss of 165 psi. Now find the allowable friction loss per 100 ft.

$$165 \div 4 = 41.25 \text{ psi} = \text{FL } 100$$

Now find the amount of water 4-in hose can flow with 41.25 psi of friction loss per 100 ft of hose. We use the formula

$$
\begin{aligned}
Q &= \sqrt{\text{FL } 100/(\text{CF} \times 2)} \\
&= \sqrt{41.25/(0.1 \times 2)} \\
&= \sqrt{41.25/0.2} \\
&= \sqrt{206.25} \\
&= 14.36 \text{ or } 1{,}436 \text{ gpm (remember,} \\
&\quad Q \text{ is in hundreds)}
\end{aligned}
$$

By placing a pumper on the hydrant in Example 12-12, we have increased the water supply from the hydrant by 73 percent. The pumper on the hydrant also allows us to compensate for residual pressure losses as other hydrants are used.

As mentioned previously, the only disadvantage of this system is that it requires two pumpers. This, however, is not a major disadvantage. Most structural fire responses mandate a response of at least two engine companies. By directing the second-in engine to "catch" the hydrant, the two-pump operation can be accomplished universally. Where it is desired to have the first two engines secure their own hydrants, *hydrant valves* should be used.

Hydrant valves allow later-responding engines to hook up to the steamer connection of the hydrant and boost pressure or water supply to the pumper at the fire, without interrupting the water flow. As additional pumpers respond, they can be directed to hydrants that are already in use to boost the pressure, water supply, or both. Personnel from pumpers on hydrants can utilize hose and equipment from pumpers at the scene. Regardless of which of the four evolutions is used to supply water to a working structure fire or major incident, you should never rely on a single source of water. If a single supply line is supplying the incident and it breaks, there will be no water for anyone.

Chapter Summary

- Water supply sources can be divided into four categories: (1) onboard water, (2) rural water sources, (3) municipal water supply, and (4) private water supply.
- The maximum daily consumption (MDC) is the amount of water a system requires to meet domestic water needs.
- A municipal water system should have enough reserve capacity to store an adequate amount of water to meet the needed fire flow (NFF): a 2-hr supply for up to 2,500 NFF; a 3-hr supply for 2,501 to 3,500 NFF; and a 4-hr supply for over 3,500 NFF.
- NFPA 1901 requires pumper fire apparatus to carry a minimum of 300 gal of water.
- The primary water source for rural operations is onboard water and water tenders.
- Water tenders need to be able to on-load and off-load water at 1,000 gpm.
- Rural water sources include streams, ponds, rivers, swimming pools, and cisterns.
- Municipal water supply systems are composed of three elements: (1) storage systems, (2) distribution systems, and (3) fire hydrants.
- Pressure for municipal water systems can come from elevation, pumps, or a combination of the two.
- Fire hydrants can be either dry barrel or wet barrel.
- The Hazen–Williams formula is used to calculate the friction loss in pipe.
- The formula for calculating flow from a hydrant is a variation of the Hazen–Williams formula.
- Once the flow from a hydrant at a given pressure has been calculated, the flow at any pressure can be found by graphing the flow on 1.85 exponential graph paper.
- There are four general evolutions to get water from the hydrant to the fire: (1) fire-to-hydrant lay, (2) hydrant-to-fire lay, (3) split lay, and (4) two-pump operation.

Key Terms

changing over The process of shutting off water from the onboard tank and opening the intake to admit water from the charged supply line.

distribution main The element of the distribution system that has the hookup to each individual home, apartment, school, business, factory, or other building.

feeder main The element of a distribution system that carries water from the point of storage to the secondary feeders.

Hazen–Williams formula The formula used to determine pressure (friction) loss in pipe.

hydrant valves Valves that allow later-responding pumpers to hook up to the steamer connection of the hydrant and boost pressure or water supply to the pumper at the fire.

maximum daily consumption (MDC) The amount of water a system demands, per minute, at its peak demand time.

secondary feeder mains The element of a distribution system that supplies water from the feeder main to the distribution mains.

steamer connection The large outlet on a hydrant whose name comes from the fact that the first mechanically driven fire engines, called *steamers*, hooked up to the large discharge of hydrants.

velocity pressure The measure of the kinetic energy of flowing water.

vena contracta The point of maximum stream contraction.

Case Study

The Recycle Facility Fire

On one otherwise nondescript afternoon, as this author was leaving work to go home, a dispatch for fire at a recycle facility was transmitted. He immediately turned around and responded to the incident.

Upon arriving at the scene, he found several large buildings, used for the processing and storage of recycled material, fully involved. A second alarm had already been requested by the battalion chief, and the fire chief was in the process of assuming command and requesting a third alarm.

Throughout the progress of the evening, water became a major challenge. Fire fighters were able to contain the fire, but extinguishing it became a problem because of access difficulties and lack of water. Even with the arrival of pumpers on the fourth alarm, water was still in short supply.

After the fire, the fire chief went to the water department and inquired about the availability of water for firefighting that night. After some analysis, it was determined that more water was available; in fact, while we were fighting the fire, a water storage tank on the east side of the city was being filled at the rate of 1,000 gpm. Furthermore, the water that was used to fill the tank was being supplied through a water main that went right past the fire scene.

At the time, it was this particular fire department's policy that their engines (four of them) should establish independent water supplies. Additional engines responding on mutual aid also typically established an independent water supply unless specifically directed to send personnel to the scene.

1. Based on this information, what could have been done to improve the water supply after the first-in units had committed?

   A. Additional responding engines should have been organized so that for every two pumpers responding, one pumper could pump the hydrant and the other pumper could lay to the fire.

   B. Once the first-in engines had established independent water supplies, nothing more could have been done to improve overall water usage.

   C. The water department should have been dispatched to shut down mains leading away from the fire.

   D. A public service or reverse 9-1-1 message should have been initiated to request that citizens temporarily cease water use.

2. What operational procedure(s) would you implement to prevent this from happening again?

   A. Require that all third-alarm pumpers be organized into relays to get water from hydrants farther from the fire.

   B. Equip all engine companies with hydrant valves so that if fires go beyond the initial response, later-responding pumpers can hook up to the hydrant and pump additional water from the hydrant.

   C. Require that when engines respond to fires in industrial and commercial structures, for every two engines, one will pump the hydrant and the other will lay to the fire.

   D. Both B and C.

3. When you are trying to deliver large volumes of water, as required by fires such as this one, what other steps can be added to preplans?

A. Put additional pumpers on the initial alarm.

B. Carry water supply maps to identify which mains carry the most water.

C. Put tenders on the assignment if adequate mains are not in the area.

D. All the above.

4. Under a situation such as this, wherein pumpers might be in danger of running out of water, what precautions should be taken?

A. Ensure that all unnecessary or ineffective streams are shut down.

B. Switch to smaller tips.

C. Use only master streams.

D. Lower the discharge pressure to reduce flow.

Review Questions

1. What are the two methods of creating pressure for a municipal water supply system?

2. How can the amount of water on scene be increased at a rural fire where no hydrants are available?

3. What is the name given to a large tender that stays on scene and supplies water to the pumper pumping the attack lines?

4. What is the maximum amount of time it should take to load a 3,000-gal tender?

5. What kind of hydrant has water constantly in the barrel?

6. When you are calculating the capacity of a hydrant, what is the target pressure at maximum capacity?

7. What is the advantage of a two-pump operation?

8. What is the pressure limitation placed on LDH?

Activities

Use your knowledge of water supply to solve the following problems.

1. You are the fire chief of a small department in a town of 2,500 residents. What is the MDC?

2. If the community in Activity 1 had a single target hazard with an NFF of 1,500 gpm, what is the MFR?

3. How much minimum storage capacity must this small town have in order to meet domestic needs and fire suppression requirements?

4. One commercial building in this small town has a sprinkler system. For the sprinkler system to work properly, it needs a minimum pressure of 35 psi at the entrance to the building. With the pipe layout shown in **Figure 12-24**, is it possible to deliver the required water if the sprinkler system is designed to flow a maximum of 300 gpm? The pipe is 8-in Class 50 ductile iron with a C factor of 100 and $D = 8.51$. Each 45° bend is equivalent to 9 ft of pipe and the 90° turn is equivalent to 18 ft of pipe. The hydrant on the public water system, at the point where the pipe enters the private system, has a static pressure of 65 psi.

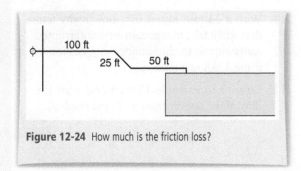

Figure 12-24 How much is the friction loss?

53 psi. If the flow hydrant was flowing one outlet with an actual diameter of 2½ in and had a flow pressure of 47 psi, how much water was it flowing? How much water will the test hydrant have available at 25 psi? The hydrant has a *C* factor of 0.9. Graph this problem on a sheet of 1.85 exponential graph paper.

6. Using the Hazen–Williams formula, verify the flow at 25 psi for the hydrant in Activity 5.

7. Go back to Example 12-11 and calculate the maximum lay possible for 829 gpm with 4-in LDH with a two-pump operation.

5. As the result of a hydrant test, you have determined that the test hydrant has a static pressure of 63 psi and a residual pressure of

Challenging Questions

1. You live in a community of 25,000 people; what is the MDC of your community?

2. If the community in Question 1 has a target hazard that requires a fire flow of 2,500 gpm, what gpm flow must the water system be able to supply?

3. What is the minimum amount of water that needs to be in storage for this community's total water needs (the community in Questions 1 and 2)?

4. What is the maximum pressure loss that will be realized in the piping from the town water tank to the water supply grid in your town (Questions 1–3) during a fire at the target hazard? The piping from the tank is 24-in cast iron and has a straight run of 400 ft. (*C* = 100, *D* = 24.34)

5. During a hydrant test, the test hydrant has a static pressure of 88 psi and a residual pressure of 75 psi. The flow hydrant had a residual

pressure of 54 psi and a *C* factor of 0.9, and the opening was measured to be 2⁹⁄₁₆ in.

A. How much water is flowing at the flow hydrant?
B. How much water will the test hydrant flow at a residual of 20 psi?
C. Now chart the flow curve on a piece of 1.85 exponential paper.

6. With a 5-in supply line 700 ft long connected directly to the hydrant, what is the maximum gpm that can be expected with a hydrant pressure of 75 psi and the pumper 20 ft above the hydrant?

7. With a 4-in supply line, what is the maximum length of hose lay possible if you want to deliver 1,000 gpm and you have a 1,250-gpm-rated pumper connected to the hydrant?

8. During a hydrant test you have a residual pressure at the flow hydrant of 38 psi while flowing water from a single 2½-in discharge that has a

C factor of 0.9. The test hydrant has a static pressure of 65 psi and a residual pressure of 40 psi.

A. How much water is flowing from the flow hydrant?

B. How much water will the test hydrant flow at 20 psi?

9. Using water at the rate of 250 gpm, how often will it be necessary to fill a 3,000-gal drafting basin?

10. With a 23-min round trip, how many tenders will it take to maintain an uninterrupted water supply to the pumper in Question 9, using 1,500-gal tenders?

11. Go back to Example 12-9 and calculate the flow of the test hydrant at 35 psi residual.

Formulas

To calculate a community's maximum daily consumption:

$$MDC = \frac{Pop \times 214.5}{1,440}$$

To calculate a community's needed storage capacity:

$$SC = (MDC + NFF) \times T$$

To find the minimum flow rate:

$$MFR = MDC + NFF$$

Hazen–Williams formula:

$$P_f = \frac{4.52 \times Q^{1.85}}{C^{1.85} \times D^{4.87}}$$

To find the total pressure loss for a system:

$$P_t = P_f \times L$$

To directly calculate the total pressure loss:

$$P_t = \frac{4.52 \times Q^{1.85} \times L}{C^{1.85} \times D^{4.87}}$$

Hydrant capacity formula:

$$Q_2 = Q_1 \times \left(\frac{p_s - p_{r2}}{p_s - p_{r1}} \right)^{0.54}$$

References

Cote, Arthur E., P.E., Editor-in-Chief, *Fire Protection Handbook*, 20th ed. Quincy, MA: National Fire Protection Association, 2008.

Hickey, Harry E., *Fire Suppression Rating Schedule Handbook*. U.S.A.: Professional Loss Control Foundation, 1993.

NFPA 291, *Recommended Practice for Fire Flow Testing and Marking of Hydrants*. Quincy, MA: National Fire Protection Association, 2016.

NFPA 1142, *Standard on Water Supplies for Suburban and Rural Fire Fighting*. Quincy, MA: National Fire Protection Association, 2012.

NFPA 1901, *Standard for Automotive Fire Apparatus*. Quincy, MA: National Fire Protection Association, 2009.

Standpipes, Sprinklers, and Fireground Formulas

LEARNING OBJECTIVES

Upon completion of this chapter, you should be able to:

- Better understand the principles of installation and operation of a standpipe system.
- Understand the basic concepts of the design and installation of sprinkler systems.
- Test a sprinkler system for proper performance.
- Verify the water supply needs for a sprinkler system.
- Calculate the hydrant capacity at a fire.
- Understand the use and weaknesses of fireground formulas.

Driving is starting to become familiar to you. For the last few months, whenever the regular driver is off, the captain lets you drive so you can get practical experience to go along with your study of hydraulics.

You have been on the scene of a fire on the sixth floor of a commercial building for over an hour. The building is equipped with a combination standpipe/sprinkler system. You have been pumping two lines into the fire department connection (FDC) and were just informed that the 2½-inch (2½-in) hand line you are pumping to is switching from a smooth-bore tip to a fog tip. You have also been informed that the stream you are pumping to the aerial ladder can be shut down.

1. Why are there restrictions of maximum pressure on combination sprinkler/standpipe systems?
2. Is it wise to use the same pumper to pump a sprinkler system and other lines?
3. What role do hand formulas play as a component of fireground formulas?

NFPA 1002 Fire Apparatus Driver/Operator Professional Qualifications, 2014 Edition

This chapter addresses the following requisite knowledge elements within sections

5.2.1 and **5.2.4:** hydraulic calculations for friction loss and flow using both written formulas and estimation methods; safe operation of the pump; calculation of pump discharge pressure; hose layouts; location of fire department connection; operating principles of sprinkler systems as defined in NFPA 13, NFPA 13D, and NFPA 13R; and operating principles of standpipe systems as defined in NFPA 14.

Introduction

Fire suppression systems such as standpipes and sprinkler systems are so common in fire suppression today that we often take them for granted and we often misuse them. In this chapter you will learn about basic design requirements and testing of both standpipe and sprinkler systems. This will not qualify you to design such systems, but it will give you a better understanding of the systems so you can effectively support them at fires and test them.

In addition to standpipe and sprinkler systems, this chapter explores the subject of fireground formulas. Some firefighters use fireground formulas exclusively instead of learning hydraulics. This practice may be adequate for someone who only operates the pump at drills, but for actual fireground operation we must be as exact as possible. Some fireground formulas can actually be useful, but most just substitute for learning the correct way.

As you study this chapter, a word of caution is in order. The specifications for both standpipe and sprinkler systems mentioned in this chapter come from their respective National Fire Protection Association (NFPA) standards. These standards do not always agree with model or local codes. In fact, in many places NFPA standards are flexible and allow the authority having jurisdiction (AHJ) to modify the standard to meet local preferences. Where the AHJ has modified the standard, the modifications must be followed.

Standpipe Systems

Standpipe systems are traditionally thought of as vertical piping in the stairways of buildings to which firefighters can attach hose for firefighting. Standpipe systems, however, do not need to be either vertical or in buildings. In buildings that cover large expanses of property, standpipe systems can be helpful even when the building is only one story. Instead of firefighters dragging several hundred feet of hose in from the street, a horizontal standpipe system allows them to hook up their standpipe hose to the system at the outlet closest to the fire.

Systems can be installed to gain access to difficult-to-reach portions of any structure Figure 13-1. Several unique applications of standpipe systems include the following:

- A townhouse community built on the roof of another building. There is no vehicle access to the town homes, so a standpipe system was required with risers spaced throughout the courtyard.

Figure 13-1 A manual-dry standpipe system installed to protect a parking garage.

© A. Maurice Jones, Jr./Jones & Bartlett Learning.

- Bridges with standpipe systems intended to get water up to the bridge or down to the highway below.
- A most ingenious system that runs through some woods and up a hill to provide water to an elementary school on top of the hill.
- Standpipe systems installed on long piers.
- Standpipe systems on the outside of a grain silo.

Each of the mentioned systems, as well as more traditional standpipe systems, is designed to solve a common problem: They allow firefighters to get water from an FDC (point *A*) to a riser closer to the fire (point *B*). A **standpipe system** is therefore defined as a water distribution system designed to more efficiently get water for firefighting purposes from point *A* to point *B*.

The following information about standpipe system design and testing is *general information* only. NFPA 14, *Standard for the Installation of Standpipe and Hose Systems,* contains specific criteria for the installation and maintenance of standpipe systems and should be consulted before the design review, inspection, or testing of any system is carried out.

Classes of Systems

There are three classes of standpipe systems:

- Class 1 standpipe systems are intended solely for use by fire department personnel. Each outlet is gated and is fitted with a 2½-in hose connection.
- Class 2 standpipe systems are intended for use by building occupants and are fitted only with 1½-in gated outlets and are usually fitted with 100 feet (ft) of 1½-in hose and nozzle.
- Class 3 standpipe systems are also intended for use by building occupants, but are fitted with 2½-in outlets reduced to 1½-in with 1½-in hose and nozzle, which allows fire department personnel to use the system as they would a Class 1 system as well as allowing the occupants to use the system.

The primary difference between these systems is the water supply requirements. Class 1 and Class 3

Class	Intended Uses	Water Supply	Size Outlet	Hose
1	FD personnel	30 min	2½ in	None
2	Occupants	30 min	1½ in	100 ft, 1½-in with nozzle
3	Both	30 min	2½ in reduced to 1½ in	100 ft, 1½-in with nozzle

Table 13-1 Summary of Standpipe Classes

systems require a minimum of 500 gallons per minute (500 gpm) for the first riser and 250 gpm for each additional riser, to a maximum of 1,250 gpm, with 100 pounds per square inch (psi) residual. A Class 2 system only requires a flow of 100 gpm with 65 psi residual. Where a water supply is required for any class system, it must be for a minimum of 30-minute (30-min) duration. Table 13-1 summarizes the classes of standpipe systems by intended use, water supply, size outlet, and hose requirements.

Types of Systems

Standpipe systems can be subdivided into the following seven types:

1. *Automatic-wet.* This system is connected to a water source and is capable of automatically supplying the system demand.
2. *Manual-wet.* This system has water in it but is unable to meet system demand. The water is there to indicate if a leak develops and eliminate the need to discharge large amounts of air before water begins to flow. Water is supplied by the fire department.
3. *Automatic-dry.* This system is dry, under air pressure, causing a dry-pipe valve to activate when an outlet is opened. The water supply for this system must meet system demand.
4. *Semiautomatic-dry.* This system is similar to automatic-dry but may or may not have pressurized air in the system. This system has a remote device at the hose connection that activates a valve, most likely a deluge valve, which then allows water into the system. The water supply has to meet system demand.

5. *Manual-dry.* This system is always dry and needs water from the fire department to meet demand.
6. *Combination.* This type of standpipe system supplies both hose connections and automatic sprinklers.
7. *Wet standpipe.* This type of standpipe system has piping containing water at all times.

Standpipe Design

Standpipe systems are designed with one of two overall design concepts: *actual length* or *exit location*. The actual-length concept is used only for Class 2 systems. **Actual-length concept** stipulates that hose connections be located so that 100 ft of hose and a 30-ft stream are able to reach any area of the building. The 100 ft is not a straight-line measurement but an actual travel distance measurement. This means distance must be measured around obstructions so that every square foot of the building can be reached with 100 ft of hose and a 30-ft stream.

The **exit location concept** allows hose connections to be located in exit stairwells, horizontal exits, and exit passageways. This method is used for Class 1 and Class 3 systems. The idea is that in a properly designed building, exit structures will be adequately placed to provide sufficient spacing and number of 2½-in outlets. NFPA 14, however, allows for additional 2½-in hose connections under specific situations. The size of pipe depends on the type of system, height of the standpipe, and required flow, as determined by hydraulic calculation. Generally, however, the minimum size pipe for a Class 1 or Class 3 system is 4 in, the minimum size pipe for a Class 2 system is 2 in, and combination systems have a minimum size of 6 in.

With **hydraulic design**, the size of pipe in a standpipe system is determined by computer programs to provide the minimum design flow at the required pressure. To hydraulically design a system, the designer must know the minimum pressure and flow at the hydraulically most remote outlet. From there, the design is calculated back to the water supply, using hydraulic calculations (the Hazen–Williams formula) to determine the size of the pipe.

Water Supply

The water supply for standpipes comes from one of two sources: The fire department pumps into the system, or the system has its own supply. On systems of type 2 and type 5, the FDC is the only source of water for the system. Therefore, before the system can be built, the local water supply must be tested to make sure the fire department can supply the system requirements.

System types 1, 3, 4, and 6 must be provided with an independent source of water. This source can be as simple as a connection to a municipal or private system, or more complicated, such as water tank and pump. Pressure tanks and elevated water tanks also serve as water supplies for standpipes.

Regardless of the source of water for a standpipe system, each system must also be fitted with at least one FDC. This FDC should be conveniently located, ideally with one located only a few feet away from the main entrance to the building for use by the first-in engine company. Additional FDCs should be located for convenient use by additional responding engine companies. At no time should contractors ever be allowed to put FDCs in the rear of buildings; they *must* be located where it is *convenient for the fire department* to use them. FDCs for supplying or supplementing standpipe systems can consist of a single 2½-in female coupling, multiple 2½-in female couplings (traditionally referred to as Siamese), or single-port quick-connect couplings.

Testing Standpipe Systems

Testing of standpipe systems is a simple process divided into three phases: (1) a pretest, (2) a hydrostatic

test, and (3) a flow test. Additional specifics of design and testing of standpipe systems are found in NFPA 14.

Pretest

The pretest is nothing more than a physical check of the system. This visual inspection should begin with a review of the approved plans to verify that the installation complies with all plan details, including pipe size and location of valves, and that proper methods were followed to connect pipe. You need to ensure that every outlet is properly capped and closed off. Each outlet must have a working valve and a handle to operate the valve. Check the threads on the system; it is not unheard of for contractors to install connections with the wrong threads. Look at the method used to tie the pipe to the building; it needs to be secure to prevent the pipe from moving. Finally, while you are conducting the pretest, check the overall physical condition of the pipe and all components. The system should give the impression that it was installed by professionals with neat welds (where required) and joints that line up.

Hydrostatic Test

Once the pretest is completed, the system is filled with water and pressurized for 2 hours (hr). The minimum test pressure is 200 psi, or 50 psi above the minimum design pressure if the minimum pressure is greater than 150 psi. While the system is under pressure, walk the system while you search for leaks—not just at valves and obvious points, but at joints and along long runs of pipe.

The contractor should be able to provide a pump for the hydrostatic test. Once the system is pressurized, the pump is shut off and should be disconnected to prevent the contractor from adding pressure if the system is leaking. After 2 hr, if the system has not lost any pressure, it passes.

Before you begin the hydrostatic test, make sure *all* air is exhausted from the system. To do this, it is necessary to flow water from the topmost outlet of each riser. Once you are certain the air is exhausted, the test time begins.

Flow Test

The flow test is conducted after the system has passed the hydrostatic test. The flow test determines whether the system is capable of flowing the required gallons per minute. Remember, the most remote outlet must flow 500 gpm, and each additional riser must flow 250 gpm. There is no time period on the flow test. Once you can verify the flow, the test is over.

Verifying the flow is a relatively simple process. Use a smooth-bore tip and take a reading with a pitot gauge. By now you should be adept at calculating the flow when the nozzle pressure and tip diameter are known. To make things easier at the test site, calculate in advance the minimum pressure needed to get the required flow with the size tip used. As long as you meet or exceed the calculated nozzle pressure, you will have the required minimum flow. This procedure must be done at each riser while all risers are flowing water.

Example 13-1

Calculate the nozzle pressure necessary to get a flow of 500 gpm from a 1⅜-in tip. Assume a C factor of 0.97.

Answer

Use the formula $P = \left(\dfrac{\text{gpm}}{29.84 \times D^2 \times C} \right)^2$

$$P = \left(\frac{\text{gpm}}{29.84 \times D^2 \times C} \right)^2$$

$$= \left(\frac{500}{29.84 \times (1.375)^2 \times 0.97} \right)^2$$

$$= \left(\frac{500}{29.84 \times 1.89 \times 0.97} \right)^2$$

$$= \left(\frac{500}{54.71} \right)^2$$

$$= (9.14)^2$$

$$= 83.54 \text{ or } 84 \text{ psi}$$

Verifying the residual pressure is a bit more complicated. One method is to put an in-line gauge on the outlet with an in-line valve after the gauge. With water flowing, if the in-line gauge does not read 100 psi, the valve can be closed down until a pressure of 100 psi is read on the gauge. With a minimum of 100 psi on the gauge there must be sufficient pressure at the nozzle, measured with a pitot gauge, to verify a 500-gpm flow.

Tip

If additional outlets are allowed to flow more water than the minimum required, they will reduce the pressure available to the system by creating excessive friction loss.

Another way to ensure the minimum residual pressure is to simply use enough hose and properly sized smooth-bore tip to ensure a minimum pressure of 100 psi at the outlet. For example, 100 ft of 2 ½-in hose has a friction loss (FL 100) of 50 psi when flowing 500 gpm. Take one-half of the 50 psi, using only 50 ft of hose, and add it to the 84 psi required nozzle pressure from Example 13-1, and you now have a required pressure of 109 psi. With this setup, as long as you have a pitot reading of 84 psi at the tip, you are assured of having both the minimum residual pressure and the minimum flow.

If there is more than one riser, you only have to verify the residual pressure at the most remote outlet. Remember that pressure throughout should be equal in a properly designed system. Make sure additional outlets are flowing at least 250 gpm, but no more than 250 gpm. The valve on the outlet can be closed off a bit to restrict the flow if necessary. If additional outlets are allowed to flow more water than the minimum required, they will reduce the pressure available to the system by creating excessive friction loss.

Note

When you are flow-testing automatic and semiautomatic standpipe systems, the flow test should include a test of the entire system, including the dry-pipe or deluge valves. When you are testing a Class 2 system, the flow and residual pressure requirements should be reduced appropriately.

Unusual Fireground Operations

Earlier we covered calculation of the pump discharge pressure (PDP) for a standpipe system under normal circumstances. However, the fireground often throws the unexpected at us, such as pumping to a foam line at the top of a 120-ft grain silo. What would the PDP be when you are using a 125-gpm nozzle? What restrictions should concern you? Is this even possible given the restrictions? The answer to this real-life problem can be found in Activity 3 at the end of the chapter. (See Chapter 10 for discussion of calculating the PDP for a standpipe system under normal circumstances.)

Finally, this safety note is important: When you are doing flow tests from the roofs of buildings, either nozzles should be anchored or some sort of diffusion device should be used. Opening and closing nozzles creates a reaction that has the potential to knock a firefighter off balance and off a building if he or she is too close to the edge.

When you are doing flow tests from the roofs of buildings, either nozzles should be anchored or some sort of diffusion device should be used. Also make sure that water discharge is managed and does not create a hazard on the roof or on the ground or damage the roof.

Sprinkler Systems

The single most important fire protection invention has been the sprinkler system. When used in a residential application, the sprinkler system has the ability to cut loss of life to approximately one-third the rate of unprotected homes. As an added advantage, it reduces property damage and can significantly reduce insurance rates.

Useful applications for sprinkler systems are virtually unlimited. Systems can be designed to protect the largest buildings imaginable or a single apartment. Transformer vaults and piers can also be protected by sprinkler systems.

The installation and testing of sprinkler systems is governed by one of the following three NFPA standards:

- NFPA 13, *Standard for the Installation of Sprinkler Systems*
- NFPA 13D, *Standard for the Installation of Sprinkler Systems in One- and Two-Family Dwellings and Manufactured Homes*
- NFPA 13R, *Standard for the Installation of Sprinkler Systems in Low-Rise Residential Occupancies*

Types of Systems

There are four types of sprinkler systems. Specific elements of each type of system can be altered to some extent, but they always fall into one of the following types.

Wet-Pipe System

A wet-pipe sprinkler system is designed to have water, under pressure, in it at all times, which allows water to flow immediately when the sprinkler head opens. It is the simplest type of system to design and maintain. The wet-pipe system is the most common system and should be used whenever possible. The only disadvantage of the wet-pipe system is that it must be kept at a temperature of at least 40°F to prevent freezing.

The wet-pipe system is designed with a wet-pipe valve between the incoming water supply and the system. The primary purpose of the valve is to prevent backflow when the system operates. In some instances the valve also activates the flow alarm.

Dry-Pipe System

The dry-pipe system is found in environments subject to freezing conditions. Normally the dry-pipe system has compressed air in the piping. When a sprinkler head activates, the air escapes and water enters the system. The dry-pipe valve is designed to cause the air in the sprinkler system to hold back the water in the supply system. The dry-pipe valve is often referred to as a differential valve, because it is designed so that a small amount of air pressure can hold back a larger amount of water pressure. Thus, a differential in pressure exists. A dry-pipe valve can require as little as 1 psi of air pressure on the system side for every 5 or 6 psi of water pressure on the supply side.

The dry-pipe system has several disadvantages. Because the system is used where it can freeze, the sprinkler valve must be placed in a heated room. Additionally, because compressed air is required in the system, a means to maintain the air pressure is needed. Finally, because the system is full of air, water will not flow immediately. However, NFPA 13 requires that a dry system be designed so that water will flow from the most remote outlet within 60 seconds (s).

Preaction System

The preaction sprinkler system is not a single system, but three variations of the same system. In general, the preaction system is very similar to the dry-pipe system. The primary difference is that a preaction valve replaces the dry-pipe valve. A preaction valve is activated by a remote fire sensor and is independent of the operation of the sprinkler heads. Another difference is that the air in the preaction system may or may not be pressurized.

In the first type of preaction system, when the fire detection system detects a fire, it automatically trips the preaction valve, allowing water to flow into the system. When the sprinkler heads closest to the fire activate, water is already well on the way, reducing the time it takes to get water to the fire.

The second mutation of the preaction system maintains pressurized air in the system. This variation requires that the fire detection system trip the preaction valve, *and* the sprinkler head(s) must operate before water is admitted to the system. For example, if the preaction valve is tripped by the fire

detection system before the sprinkler heads activate, water will not flow. Once one of the heads activates, the system must evacuate the air before the system will operate.

The final mutation of the preaction system is most similar to the second. There is pressurized air in the system, but in this system water begins to flow when *either* the preaction valve is tripped or a sprinkler head operates.

Deluge System

A deluge system operates most similarly to the first variation of the preaction system. A fire detection system senses the fire and activates the deluge valve, allowing water to flow. All sprinkler heads on a deluge system are open, allowing water to flow from every head in the system.

Sprinkler System Design

Today, sprinkler systems are hydraulically calculated by computers, but a word of caution is in order here: Do not assume a computer-generated sprinkler system design is correct. The computer does only what its operator tells it to do. Qualified reviewers should review all sprinkler plans before any system is built.

Tip

Do not assume a computer-generated sprinkler system design is correct. Qualified reviewers should review all sprinkler plans before any system is built.

Sprinkler systems are not intended to extinguish a fire if the building is fully involved. They are only intended to hold a fire in check if they are unable to extinguish the fire in its incipient stage. In large buildings, enough water cannot be delivered to supply every sprinkler head if they all activate, so the sprinkler system is separated into design areas. A **design area/zone** is a portion of a larger building wherein the combined flow from all sprinklers in that area is used to calculate the required flow of the system. This fact is primarily important in determining the required gallons per minute. You need only enough water to supply the sprinkler demand in the largest design area. Design areas are also referred to as *zones*.

Figure 13-2 is a photograph of a typical wet sprinkler system. Note the three system specification plates (*A*, *B*, and *C*) hanging from the system.

Each plate specifies the water flow requirement and residual pressure for each of three zones supplied by this one valve. Also, notice the outside stem and yoke (OS&Y) valve. When the stem is protruding, as it is in this picture, the valve is open. If the stem is not showing, the valve is closed. This is a good visual reference that the valve is open or closed. Finally, note the FDC and how it feeds into the system.

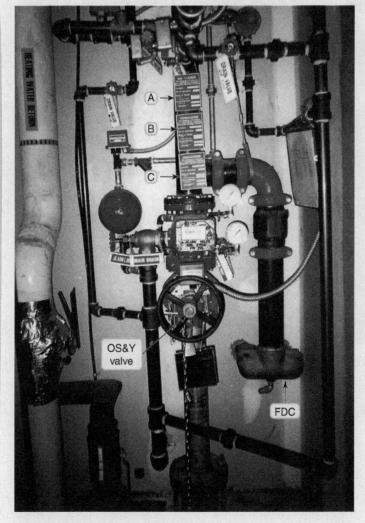

Figure 13-2 A typical wet sprinkler system.
Courtesy of William F. Crapo.

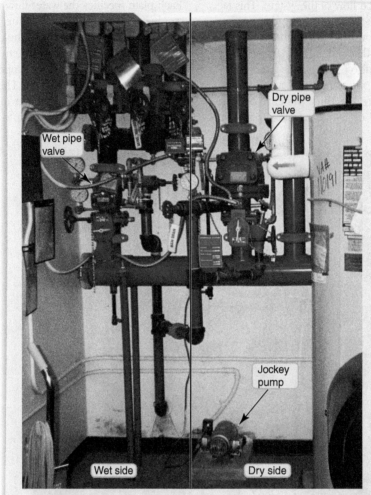

Figure 13-3 A split sprinkler system: the left side is wet and the right side is dry.

Courtesy of William F. Crapo.

Figure 13-3 is a photograph of a combination wet and dry system. The dry-pipe valve feeds only one zone of the system that protects unheated portions of the building. The remainder of the system, to the left of the dry-pipe valve, serves three zones and is a wet system. Note the air compressor, also known as a *jockey pump*, is used to maintain the air pressure on the dry side of the system.

The basic philosophy in calculating the number of heads for any given design area is to achieve a specified water delivery *density*. The **density** is the amount of water delivered per square foot of surface or floor area. The required density is dependent on the type of fuel, fuel geometry, distance the sprinkler head is from the fuel, fire load, and heat release of the fuel. These factors are summarized into three basic **hazard levels**: light, ordinary, and extra hazard.

To achieve a specified density depends on the heads used and the spacing between heads. The same sprinkler heads can be used to achieve different densities simply by varying the spacing between them. When heads are spaced closer together, a higher density is achieved, and when heads are spaced farther apart, a lower density is obtained.

Sprinkler Heads

The heart of any sprinkler system is the sprinkler head itself. Sprinkler heads come in one of three versions: upright, pendant, and sidewall Figure 13-4. There are sprinkler heads intended for special applications, but they are just special applications of one of these three versions. Each head has a different kind of fusible link, or operating mechanism. The upright head has a mechanism held together by a **eutectic metal**, or solder sensitive to a specific temperature. The pendant head has a mechanism that contains a temperature-sensitive chemical pellet. Finally, the sidewall head has a glass bulb with a

Figure 13-5 Sprinkler heads and their year of manufacture. The basic design has remained essentially unchanged for nearly 100 years.

Courtesy of William F. Crapo.

temperature-sensitive liquid called a *frangible bulb*. With the exception of the deflector design (pendant, upright, or sidewall) and a more modern look, the basic design of the sprinkler head has remained essentially unchanged since the early twentieth century Figure 13-5.

The operation of a sprinkler head is very basic. It is simply a frame that holds a deflector in place. To prevent water from flowing when it is not needed, a fusible link holds a valve cap in place. Figure 13-6 illustrates a typical sprinkler head with the fusible link removed. The gasket prevents water from leaking prior to operation. When the heat of a fire activates the fusible link, water pressure forces the valve cap off and allows water to flow.

Testing Sprinkler Systems

Sprinkler system testing is divided into three different parts, just as with standpipe systems: pretest, pressure test, and flow test. Since the details of testing of sprinkler systems can be specific, it is recommended that the appropriate NFPA standard be consulted.

During the following tests, it is not necessary to have sprinkler heads in place. The outlets for the heads can be capped if the heads will prevent proper

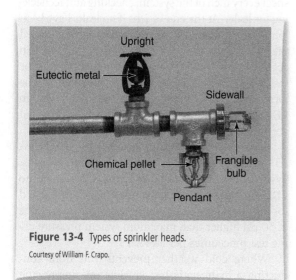

Figure 13-4 Types of sprinkler heads.

Courtesy of William F. Crapo.

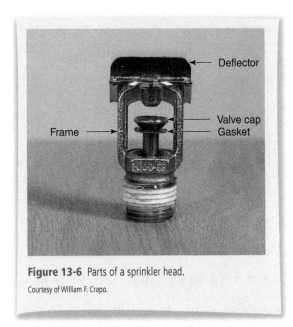

Figure 13-6 Parts of a sprinkler head.

Courtesy of William F. Crapo.

installation of a ceiling or wall later. For obvious reasons, deluge systems cannot have heads in place to conduct a hydrostatic test. However, when the system is made of plastic pipe, all heads *must* be in place before any test is conducted, because the plastic pipe may not be able to take the stress of putting on caps, removing the caps, and then putting on the sprinkler heads. A word of caution is appropriate here concerning plastic pipe: Some contractors like to put the sprinkler head on the threaded connection before gluing the connection in place; *this must not be allowed under any circumstances.* Glue from the joint can flow out of the joint when it is put together and flow onto the inside of the sprinkler head, rendering the head inoperable.

Pretest

Before a sprinkler system is tested, a walk-through inspection should be made of the entire system. During this visual part of the inspection, the inspector should be looking for obvious defects, making sure the proper hangers were used and verifying that *all* piping is visible. It is absolutely necessary that the inspector be able to see every inch of pipe to make sure it is not leaking during the hydrostatic test.

Pressure Test

Once the pretest has been completed, the entire sprinkler system must be put under a hydrostatic test for 2 hr. After the pressure has been achieved, the pump is disconnected as in the standpipe test. The pressure during the test is 200 psi or 50 psi over the maximum operating pressure of the system if the maximum pressure is greater than 150 psi. A word of caution is in order here: Sprinkler systems that are put together using steel pipe and joined using rubber gaskets and split compression fittings at joints can cause a problem. These rubber gaskets can actually seal leaks at high pressure, but not at low pressure. The test pressure in these systems may be sufficiently high to prevent high-pressure leaks, while low-pressure leaks will not be evident.

Tip

Rubber gaskets can actually seal leaks at high pressure. If the test pressure in these systems is too high, low-pressure leaks will not be evident.

During the 2 hr of the test, the inspector must inspect every inch of the system, checking and rechecking for leaks. At the end of 2 hr if no leaks have been found, the system can be certified as having passed the hydrostatic test. Sprinkler systems installed in accordance with NFPA 13D are only required to be tested for 15 min at the pressure provided by the local water supply unless they are equipped with an FDC. If equipped with an FDC, they must be tested according to NFPA 13 requirements.

Systems designed and installed by the NFPA 13R standard also have special testing requirements. Systems that have both fewer than 20 heads and no FDC must pass a hydrostatic pressure test performed for the aboveground piping system. This test is done at 50 psi higher than maximum system pressure using test procedures per NFPA 13.

Where cold weather prevents you from conducting a hydrostatic test, an interim air pressure test can be accepted. The air pressure test is for 24 hr with 40 psi pressure. Once the weather permits a hydrostatic test, it must be completed before final approval can be given to the sprinkler system. When air is used to test wet systems under these circumstances, under no circumstances should the test be performed on sprinkler systems that contain plastic pipe.

Dry sprinkler systems require an additional pressure test. Since these systems are required to hold back the flow of water with air pressure, it is necessary to test the system for air leaks. To do so requires a 24-hr pressure test at 40 psi of air pressure. Once the test is begun, the system cannot be pumped up for the remainder of the test. At the end of the 24-hr period, the pressure in the system cannot have dropped more than 1.5 psi. If a drop of more than 1.5 psi is observed, the system must be repaired and retested before being placed in service.

Flow Test

The flow test of a sprinkler system actually consists of two tests. The first test is conducted at the *inspector's test connection*. The **inspector's test connection** simulates water flowing from the most remote head on a dry sprinkler system or from any downstream head on a wet system. When the inspector's test connection is opened, you are looking for two functions. First, you want to make sure water flows. At this point there is no need to measure the flow; you just want to verify that water will flow when the test connection is opened. When you are conducting water flow tests on a dry system, water must flow from the most remote outlet within 60 s. In the second test, you want to determine that the water flow alarm is working. The water flow alarm should sound the alarm within 5 min of system activation.

Main Drain Test

The final test of the hydraulics of the sprinkler system involves verifying that the system has adequate water. Verification is done with a main drain test that is part of the flow test. The main drain is located

at the sprinkler system valve on the supply side. The proper procedure for testing is to first record the static pressure, then open the main drain valve fully and record the residual pressure. The pressure drop should be in line with the anticipated pressure drop as specified on the system specification plate, located at the sprinkler valve riser. Main drains typically have a 2-in diameter.

> **Note**
>
> When you are conducting water flow tests on a dry system, water must flow from the most remote outlet within 60 s.

Water Supply

In general, water supply sources are the same as for standpipe systems. Again, it is worth repeating that FDC should be located for ease of use by the fire department, not the convenience of the contractor. Also, as with the standpipe system, the FDC can be a single 2½-in connection, multiple 2½-in connections (Siamese), or a single-port quick-connect.

The hazard level of the occupancy governs the amount of water required by a sprinkler system. The desired density as determined by the hazard level and number of sprinkler heads in the design area gives the gpm required. The sprinkler design engineer should supply this information.

Determining the amount of water flowing from a sprinkler head is not a difficult task for anyone who has progressed this far in this text. Recall the formula for calculating gpm is gpm = $29.84 \times D^2 \times C \times \sqrt{P}$. You can calculate the gpm from a sprinkler head simply by inserting the diameter of the opening and the C factor for the head. Since this might have to be done over and over with heads of the exact same type for a particular design area, it makes sense to shorten the formula. Instead of having to use the entire formula for each head in a sprinkler system, simply multiply $29.84 \times D^2 \times C$ for each type and size of head. When multiplied, these factors create what is known as a *K* factor for a particular sprinkler head. The formula for calculating gpm from a sprinkler head then becomes gpm $= K \times \sqrt{P}$. The *K* factors are typically supplied on sprinkler plans for each type and size of head. Some sprinkler heads actually have the *K* factor stamped on the deflector. (See Chapter 5 for more details on the formula for calculating the gpm.)

$$\text{gpm} = K \times \sqrt{P}$$

Example 13-2

What is the flow from a sprinkler head at 7 psi if it has a *K* factor of 5.3?

Answer

$$\begin{aligned}
\text{gpm} &= K \times \sqrt{P} \\
&= 5.3 \times \sqrt{7} \\
&= 5.3 \times 2.65 \\
&= 14.05 \text{ or } 14 \text{ gpm}
\end{aligned}$$

With this formula, calculating the gpm for each individual head requires minimal effort.

Where sprinkler systems, or even standpipe systems, must be connected to the municipal water supply, a dedicated fire main will be run to the system. The fire main must have sufficient pressure and capacity to supply the needs of the system. One requirement is that once the pipe is run to the building but before the final connection is made, the fire main must be flushed out in the presence of a fire inspector. This ensures that there are no restrictions to the flow of water that would interfere with the operation of the system or debris that would later cause problems. The requirement that the fire main be flushed before final connection should not be taken lightly. Figure 13-7 is an example of the type and size of debris that was flushed from a fire main prior to final connection.

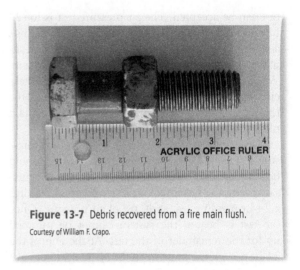

Figure 13-7 Debris recovered from a fire main flush.
Courtesy of William F. Crapo.

Inspection of Water-Based Suppression Systems

Inspections and testing of water-based suppression systems should be conducted on a regular basis. These systems should be inspected and tested in conjunction with their appropriate NFPA standard, that is, NFPA 13, NFPA 14, and NFPA 25, *Standard for the Inspection, Testing, and Maintenance of Water-Based Fire Protection Systems.*

When sprinkler systems are inspected, special attention must be given to the condition of sprinkler heads. For a sprinkler head to operate, it must be in the same condition as it was on the day it was installed. Several defective heads are displayed in Figure 13-8. Sprinkler head A has been painted and will not work as intended. Even if the chemical pellet is melted at the correct temperature, the paint will likely prevent the mechanism from releasing. Sprinkler head B is a frangible bulb with some of the liquid missing. If this head operates at all, it will be at the wrong temperature. Finally, head C is also a frangible bulb, and all the liquid has leaked out, rendering it as effective as a plug on the end of a pipe.

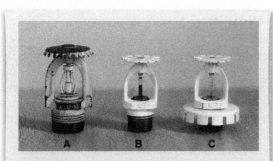

Figure 13-8 Defective sprinkler heads: **A.** Painted head, **B.** Frangible bulb with some liquid missing, **C.** Frangible bulb is empty.

Courtesy of William F. Crapo.

Fireground Formulas

By now you understand the principles of hydraulics as they pertain to the fire service. You can use any number of formulas in order to calculate gpm, friction loss, nozzle reaction, and so forth. However, even if you are armed with all these formulas, using them on the fireground is impossible. For this reason it is necessary to develop a means to apply your knowledge at the scene of a fire without getting bogged down with calculations.

Calculating the correct PDP on the fireground must be a simple, yet accurate process. It should involve simply adding nozzle pressure, adding friction loss, adding or subtracting elevation, and adding appliance loss. The most complex of these factors is friction loss. On the fireground we do not have the luxury of pencil, paper, and a calculator to calculate FL 100. Rather, we need to know the correct gpm and FL 100 for every tip size carried. For example, if your officer calls you on the radio and says your company has extended the 1¾-in line by 100 ft, you should automatically know that the line is flowing 150 gpm and the FL 100 is 35 psi. Instinctively, you increase the PDP by 35 psi— it should be performed about as quickly as it takes to read this sentence.

When you are calculating total friction loss on the fireground, even with the FL 100 memorized, it may be necessary to round off some numbers.

But just as when you do calculations with a pencil, paper, and calculator, try not to round off until the total friction loss is calculated. If you round off the FL 100 before calculating the total friction loss, remember that you will probably have to round off the final number again. This is true because, from a practical perspective, gauges can only be read in increments of 5 psi. In general, it is better to be a few psi high than a few psi low. With a little attention to detail, calculations on the fireground can be almost as accurate as those done with pencil, paper, and calculator.

There are several methods, other than memorizing FL 100 data, of finding the correct friction loss for various hose lines and nozzles. One such method relies on knowing how many fingers you have on a hand and assigning numbers to them that represent gpm. These rules of thumb or "hand methods" of calculating friction loss should be avoided. Although hand methods of calculating friction loss can be relatively accurate for some hose sizes, if you have two or three different sizes of hose, you will need two or three different hand methods. They will each require about nine different numbers that must be memorized in a specific sequence. Another reason to avoid using hand methods is that they require a great deal of mental calculation that, to be accurate, involves decimals and fractions. These are calculations that must be done in your head— there are no calculators on the fireground while you are surrounded by chaos.

The only acceptable method that the author recommends of simplifying PDP calculations on the fireground involves a two-step approach. First, personnel who drive and operate pumping apparatus should be required to learn, through memorization, the gpm each nozzle will deliver. They should also know the FL 100 and nozzle pressure associated with the correct gpm for all appliances carried on the apparatus. On the fireground, it should take only a moment to mentally recalculate the correct PDP if a tip is changed or if the length of a line is altered. This does require memorization, but you have to memorize the tip sizes, flows, and nozzle pressures even if you use a hand method. The only additional memorization will be the FL 100 for each tip flow. This additional memorization required to learn the FL 100 for tips and appliances carried on your apparatus is less that what is required for even a single hand method, and it doesn't change from one circumstance to another.

The second step is to provide a *friction loss chart*, accessible to the operator, for quick reference. The friction loss chart can be custom-designed to reflect your department's hose sizes, flow requirements for different nozzles, and special appliance recommendations. The biggest advantage of the friction loss chart is that friction loss in supply lines can be calculated at any flow. It takes a little time to calculate the FL 100 values for a friction loss chart such as in Figure 13-9 , but in the long run it is far preferable to relying on a rule of thumb to find the friction loss. Calculations for the sample friction loss chart in Figure 13-8 took less than 1 hr, using formulas presented in this text. On the fireground, once you have looked up the correct FL 100, you must still mentally calculate the correct PDP. But this process removes the step of having to first mentally calculate FL 100, eliminating one step of mental calculations under stressful conditions.

In Figure 13-9, the first column is the appliance and its required nozzle pressure. The second column is the gpm for each appliance in the first column and at various increments of flow. The remaining columns are the FL 100 values for a single 2½-in hose, a single 3-in hose, two 3-in hoses, and a 5-in hose. Allowances for special appliances are also included as well as gpm, nozzle pressure, and FL 100 for various 1¾-in appliances. This chart can be customized to your department, to include only the hose sizes and nozzles/tips that you carry. This friction loss chart can be protected in a piece of plastic and taped inside a cabinet convenient to the pump panel, or silk-screened onto a piece of aluminum and attached to the apparatus in a spot convenient to the pump operator. In making this chart, the only concession to simplicity was the rounding off of all smooth-bore tip flows. They were rounded to the nearest 5 gpm, which means the largest deviation from exact flow in the chart is less than 2 gpm.

Elevation

Previously whenever elevation was calculated, we used the formula $P = 0.433 \times H$. On the fireground, calculating elevation to this extent is impossible. A good standard practice is to calculate elevation on the fireground at ½ psi per foot (psi/ft) of elevation. If you are pumping uphill or downhill to another pumper or a nozzle, just estimate the elevation and use one-half of it as the elevation allowance. When you are dealing with buildings, it is accepted practice to allow 5 psi for each floor above and below the position of the pumper. Both of these concepts were previously mentioned in Chapter 10. Also recall from prior discussion that when the extension of an aerial device is known, use half the extension, or ½ psi/ft of elevation. (See Chapters 10 and 11 for calculation of elevation using the formula $P = 0.433 \times H$.)

Example 13-3

You are the pump operator of the first-in engine company operating at a second alarm residential fire. You have already charged two preconnect lines and have them set at the correct pressure. The officer from Engine 28, which responded on the second alarm, asks you to supply a hand line for his company. The hand line will be 150 ft of 2½-in hose with a 1⅛-in tip. What will be the correct PDP for this line? Round friction loss to the nearest 5 gpm.

Answer

Without the use of a calculator, pencil, or paper, you need to determine the correct PDP. Use numbers from Figure 13-9.

$$PDP = NP + FL\,1 + FL\,2 \pm EP + AL$$
$$= 50 + 20 + 0 \pm 0 + 0$$
$$= 70 \text{ psi}$$

Calculating Available Water on the Fireground

One of the most valuable tools a fire chief can have on the fireground is knowledge of how much water is available at the time of the fire. If the department has been well organized, it has already made

Appliance/Pressure	GPM	2½ in	3 in	2–3 in	5 in
	100	2	1		
	150	5	2		
	200	8	3		
1-in tip @ 50 psi	205	8	3		
	225	10	4		
CVFSS @ 100 psi	240	12	5		
1⅛-in tip @ 50 psi	260	14	5		
	300	18	7		
1¼-in tip @ 50 psi	320	20	8		
	350	25	10		
	375	28	11		
	400	32	13	3	1
1¼-in tip @ 80 psi	415	34	14	4	1
	425	36	14	4	1
	450	48	16	4	2
	500	50	20	5	2
1¾-in tip @ 80 psi	505	50	20	5	2
	525		22	6	2
	550		24	6	2
	575		26	7	3
1½-in tip @ 80 psi	600		29	7	3
	625		31	8	3
	650		34	8	3
	675		36	9	4
	700		39	10	4
1⅝-in tip @ 80 psi	705		40	10	4
	725		42	11	5
	750		45	11	5
	775		48	12	5
	800		51	13	5
1¾-in tip @ 80 psi	815			13	5
	850			14	6
	900			16	6
	950			18	7
	1,000			20	8
	1,050			22	9
	1,100			24	10
	1,150			26	11
	1,200			29	12
	1,250			31	13
	1,300			34	14
	1,350			36	15
	1,400			39	16
	1,450			42	17
	1,500			45	18
	1,550			48	19
	1,600			51	20

Appliance Loss

Standpipe	25 psi
Monitor nozzle	20 psi
Sprinkler system	150 psi
Combination system	175 psi maximum
Wye/Siamese	10 psi
Foam eductor	200 psi

1¾-in

Nozzle	GPM	NP	FL 100
CVFSS	125	100 psi	24 psi
CVFSS	150	100 psi	35 psi

Figure 13-9 Sample friction loss chart.

flow tests and has labeled each fire hydrant with its expected flow. While this is very useful, it is not a perfect solution. The primary weakness with flow-testing hydrants is that you only test one hydrant at a time. Subsequently, the water that is calculated as being available in that hydrant is dependent on other hydrants not being used. At a major fire, using only one hydrant is a virtual impossibility. We need to develop a means of determining how much water is available while the fire is burning.

Determining the amount of water available at a hydrant during a fire requires that a pumper be attached to the hydrant. (Here is another argument in favor of a two-pump operation.) The pump operator should first connect the soft sleeve to the hydrant and then charge the hydrant. After the hydrant has been charged, the operator must note the static pressure. Knowing the static pressure is critical in determining the amount of water left in the hydrant.

When it becomes necessary to estimate the amount of water left in the hydrant, the current residual pressure must be noted. The amount of water left in the hydrant is determined by the difference between the static pressure observed when the hydrant was first charged and the current residual pressure. In addition, the gpm flow of the pumper must be known. The percentage drop in pressure, from static to residual, corresponds to the amount of water available at that hydrant Table 13-2. A drop of up to 10 percent indicates that the hydrant can supply three times more water than it is already supplying. A pressure drop of more than 10 percent but not more than 15 percent means the hydrant can supply two times more water than it is already flowing. Finally, if the pressure drop is more than 15 percent but not more than 25 percent, the hydrant is capable of supplying one time the amount of water it is already supplying.

Example 13-4

You are operating a pumper that has hooked up to a hydrant near the fire. Your static pressure was 70 psi and your residual pressure is 55 psi. You are supplying a ladder pipe that is flowing 500 gpm. Assuming the capacity of the pumper is not an issue, how much more water can you supply?

Answer

$$70 - 55 = 15 \text{ psi}$$

Because 7 is 10 percent of 70 and you have a 15 psi drop, your total drop is just over 20 percent. This puts you in the over 15 percent, but not over 25 percent, category; you can supply another 500 gpm.

Two final points are relevant to this method of determining hydrant capacity. First, this method is only an approximation. Unfortunately, there is no way to determine precisely how much water is available on the fireground in real time. Second, to be even close, the residual pressure must be taken at the time you want to know the available water remaining. As additional pumpers hook up to the water main, they take water and pressure away from you. For example, when you first hook up to the hydrant and are the only pumper flowing water, you may have a 10 percent drop while supplying 500 gpm. After two or three other pumpers hook up to hydrants and begin flowing water, you may discover you now have a 25 percent drop in pressure. If the water supply officer were to ask you how much water you had left and you based your calculation on your initial residual pressure, you would have mistakenly told him you had 1,500 gpm more. In reality you only have approximately another 500 gpm left.

High-Rise Operations

High-rise firefighting can be the most challenging of all firefighting operations, as it tests both firefighters

Table 13-2 Calculating Hydrant Capacity

Pressure Drop	Hydrant Capacity
Up to 10%	3 × current flow
Over 10% but not over 15%	2 × current flow
Over 15% but not over 25%	1 × current flow

and equipment. Our concern is how it will challenge the ability to pump to hose lines on upper floors.

Typically high-rise buildings can require unusually high pressures. The maximum output of a typical pumper is 300 psi. In high-rise operations it is easy to require pump pressures close to the maximum. For instance, a 150-ft, 2½-in line flowing 322 gpm from a 1¼-in tip and operating on the 30th floor requires a PDP of 261 psi. There are two points to remember here: First, the test pressure of the standpipe system may be reached, and second, the test pressure of the hose used to supply the standpipe becomes a factor. This makes it imperative that all hose be tested annually according to NFPA 1962, *Standard for the Care, Use, Inspection, Service Testing, and Replacement of Fire Hose, Couplings, Nozzles, and Fire Hose Appliances*. In short, regardless of your best efforts, it may be necessary to push your equipment, as well as the firefighters, to the limit.

Chapter Summary

- A standpipe system can be thought of as a system designed to get water, for firefighting purposes, from point A to point B.
- NFPA 14 governs the design and installation of standpipe systems.
- There are three classes of standpipe systems: Class 1, intended solely for use by FD personnel; Class 2, intended solely for use by building residents; Class 3, which has hose for use by building residents, but also has a 2½-in outlet for use by FD personnel.
- Standpipe systems can be further classified into seven types of systems: automatic-wet, manual-wet, automatic-dry, semiautomatic-dry, manual-dry, combination, and wet.
- The water supply for standpipe systems can come from one of two sources; either it has its own supply, or the FD supplies the system.
- When you are flow-testing a standpipe system, the most remote outlet must be able to flow 500 gpm, and each additional outlet must be able to flow 250 gpm.
- There are three NFPA standards governing the design and installation of sprinkler systems: NFPA 13, NFPA 13D, and NFPA 13R.
- There are four types of sprinkler systems: wet-pipe, dry-pipe, preaction, and deluge.
- All sprinkler systems are hydrostatically tested for 2 hr at 200 psi, or 50 psi over the operating pressure, whichever is higher.
- In addition to the hydrostatic test, dry systems and certain preaction systems must also be tested with 40 psi air pressure for 24 hr.
- Water-based fire protection systems are inspected in accordance with NFPA 25.
- Where necessary on the fireground, estimate elevation at ½ psi per foot of elevation.
- Water available on the fireground can be estimated based on the pressure drop when water is supplied from a hydrant; 0 to 10 percent drop has three times as much water available; 11 to 15 percent drop has two times as much water available; and 16 to 25 percent drop has one time as much water available.
- NFPA 1962 specifies test procedures and pressures for the annual hose test.

Key Terms

Actual-length concept A standpipe system design concept that stipulates that hose connections must be located so that 100 ft of hose and a 30-ft stream are sufficient to reach any area of the building.

density The amount of water delivered per square foot of surface or floor area.

design area/zone A portion of a larger building wherein the combined flow from all sprinklers in that area is used to calculate the required flow of the system.

eutectic metal Solder sensitive to a specific temperature.

exit location concept A standpipe system design concept that allows hose connections to be located in exit stairwells, horizontal exits, and exit passageways.

hazard levels The three classifications—light, ordinary, and extra hazard—into which occupancies are categorized based on the level of hazard present.

hydraulic design A method in which pipe sizes are determined by computer programs to provide the minimum design flow at the required pressure.

inspector's test connection In a sprinkler system flow test, the valve that stimulates water flowing from the most remote head in a dry system or any downstream head in a wet system.

standpipe system A water distribution system designed to get water for firefighting purposes from point A to point B.

Case Study

One Meridian Plaza

On February 23, 1991, a fire broke out on the 22nd floor of One Meridian Plaza in Philadelphia, Pennsylvania. Before it was extinguished, the fire claimed floors 21 through 29 and, sadly, the lives of three Philadelphia firefighters.

This tragic fire was set up by several factors in a building that by even the most basic standards should never have burned. The building was a 38-story, fire-resistive construction high-rise, built to code and with fire safety in mind. The standpipe system was fully operational with pressure-reducing valves at each outlet. On February 23, however, the building and codes were put to the test and failed.

The fire started on the 22nd floor. Extension was possible through improperly protected vertical and horizontal openings. This allowed fire to drop down to the 21st floor and to extend vertically to the 29th floor.

Firefighting efforts were hampered by the improperly set pressure-reducing valves that had been installed on the standpipe system. These valves are designed to prevent excessive pressure that would create extra work for the firefighting crews on the attack lines. This way, crews on multiple floors would not have to deal with higher pressures necessary for crews on upper floors. However, these pressure-reducing valves had been set too low, and it was impossible to get acceptable fire streams from the system.

After about 11 hr, the Philadelphia Fire Department pulled all personnel from the building. The fire continued unabated until it reached the 30th floor. The 30th floor was the first fully sprinklered floor. Upon reaching the 30th floor, the fire was stopped by the activation of just 10 sprinkler heads. While it is not certain just how much water was flowing from these sprinkler heads, if they had each been flowing a maximum of 25 gpm, a total flow of just 250 gpm controlled the fire. Had the entire building been sprinklered, three firefighters would still be alive and there would be no articles about One Meridian Plaza.

This fire has provided many lessons on the construction of high-rise buildings and fire protection. What it proves most of all is that when the water is (1) supplied in sufficient quantities, (2) in the right form, (3) and at the right spot, fire can be stopped. This is exactly what sprinkler systems do.

1. What factor did not contribute to fire spread at One Meridian Plaza?

   A. Improperly protected vertical and horizontal openings.
   B. Improperly set pressure-reducing valves.
   C. Fire beyond the reach of aerial devices.
   D. Inadequate water supply.

2. How were manual firefighting efforts hampered?

   A. The fire was too high in the building to get adequate pressure for hose lines.
   B. Improperly set pressure-reducing valves.
   C. Improper sprinkler system for the building.
   D. All the above.

3. How can recurrences of fires such as this one be prevented?

   A. Equip all such buildings with sprinklers.
   B. Inspect all construction daily.
   C. Test all sprinkler/standpipe systems as they become operational.
   D. Answers A and C.

4. What NFPA standard covers the inspection, testing, and maintenance of water-based fire protection systems?

   A. NFPA 13
   B. NFPA 13D
   C. NFPA 14
   D. NFPA 25

Review Questions

1. Define a standpipe system.

2. What NFPA standard specifically addresses standpipe systems?

3. How many classes of standpipe system are there?

4. Is it possible for a standpipe system to be maintained without water but operate automatically when needed?

5. What is the maximum flow for which a standpipe system is designed?

6. How many types of sprinkler systems are there?

7. What is the design area when you are referring to a sprinkler system?

8. Are all sprinkler systems required to have a 24-hr air pressure test?

9. What is the purpose of flowing water from the inspector's test connection?

10. What NFPA standard addresses inspection, testing, and maintenance of water-based fire protection systems?

11. Why should rules of thumb be avoided?

12. If you are operating a pumper connected to a hydrant, what do you need to know in order to estimate how much more water the hydrant can supply?

13. Why is it critical to flush fire mains before they are connected to sprinkler/standpipe systems?

14. What would the test pressure be for a standpipe system that has a working pressure of 175 psi?

15. How is the K factor for a sprinkler head determined?

Activities

1. What pressure would be required on a 1-in tip to obtain a flow of 250 gpm?

2. How many gallons per minute will flow from a sprinkler head with a K factor of 5.6 and operating at 10.33 psi?

3. How would your department respond to a situation like the one outlined in the Fireground Fact "Unusual Fireground Operations"?

   A. What restrictions concern you?
   B. What would the PDP be when you are using a 125 gpm nozzle?
   C. Is pumping to the top of the silo even possible, given the restrictions?

Challenging Questions

1. You are the pump operator pumping through 100 ft of 3-in hose into a standpipe system of a high-rise building. Your company is operating with 150 ft of 2½-in hose and an automatic nozzle flowing 225 gpm on the 30th floor. After operating with 225 gpm for a short time, the officer notifies you that a second company has arrived and is operating with the same hose and nozzle configuration. Since the company that has just joined up with your officer is from the same engine that is your water supply company, you will have to supply both lines.

   A. How can you do this without changing your PDP?

   B. What is the minimum capacity pump needed to do this?

2. You are operating a pumper connected to a hydrant. When you first charge the hydrant, you have a static pressure of 90 psi. After charging a supply line to another pumper and supplying 700 gpm, you have a residual of 50 psi. How much water is left in the hydrant?

3. In Figure 13-5, the sprinkler head manufactured in 2003 has a K factor of 25.2. What would the flow be at 10 psi?

Formula

To calculate gpm from a sprinkler head:

$$\text{gpm} = K \times \sqrt{P}$$

References

NFPA 13, *Standard for the Installation of Sprinkler Systems*. Quincy, MA: National Fire Protection Association, 2013.

NFPA 13D, *Standard for the Installation of Sprinkler Systems in One- and Two-Family Dwellings and Manufactured Homes*. Quincy, MA: National Fire Protection Association, 2016.

NFPA 13R, *Standard for the Installation of Sprinkler Systems in Low-Rise Residential Occupancies*. Quincy, MA: National Fire Protection Association, 2013.

NFPA 14, *Standard for the Installation of Standpipe and Hose Systems*. Quincy, MA: National Fire Protection Association, 2013.

NFPA 25, *Standard for the Inspection, Testing, and Maintenance of Water-Based Fire Protection Systems*. Quincy, MA: National Fire Protection Association, 2014.

NFPA 1962, *Standard for the Care, Use, Inspection, Service Testing, and Replacement of Fire Hose, Couplings, and Nozzles, and Fire Hose Appliances*. Quincy, MA: National Fire Protection Association, 2013.

Math Review

To have a complete understanding of the rules and principles of hydraulics, one must understand the math involved. This appendix reviews the basic principles necessary to be successful with the mathematical calculations in this text.

Arithmetic

Math in its most rudimentary form is called *arithmetic*. The rules that govern arithmetic are basic and learned in our first few years of grade school. Therefore, we will not spend time reviewing all these, but will mention only a few rules that will become very important later in the review of algebra.

Arithmetic involves four basic mathematical operations: addition, subtraction, multiplication, and division. The order in which you carry out these functions can have a bearing on the outcome of the problem; thus, you must strictly adhere to this order.

Addition

Addition can be done in any order, and the answer will always be the same.

> **Example**
>
> $1 + 2 = 3$ is the same as $2 + 1 = 3$.

This principle is referred to as *commutativity*. Simply stated, commutativity is interchangeability in the order of an operation.

In addition to commutativity, addition can be done in any order, regardless of groupings, without affecting the answer.

> **Example**
>
> $2 + (3 + 5) = 10$ is the same as $5 + (2 + 3) = 10$.

This principle is referred to as *associativity*. Thus, addition is both commutative and associative.

Subtraction

To be correct, subtraction must be done in a specific order. Subtraction is neither commutative nor associative.

> **Example**
>
> $5 - 3 = 2$ is not the same as $3 - 5 = -2$.

Thus, subtraction is not commutative.

Also, groupings are important in getting the correct answer.

> **Example**
>
> $25 - 13 - 6 = (25 - 13) - 6 = 12 - 6 = 6$

If 6 is subtracted from 13 first, the answer becomes $25 - (13 - 6) = 18$, which is incorrect.

Thus, subtraction is not associative.

Multiplication

Multiplication and addition are first cousins, in that multiplication is actually repeated additions.

> **Example**
>
> $4 + 4 + 4 = 12$ is the same as $3 \times 4 = 12$.

This is convenient because, just as with addition, multiplication is both commutative and associative.

> **Example**
>
> $2 \times (4 \times 5) = 40$ is the same as $4 \times (5 \times 2) = 40$.

Also, there are four ways to indicate multiplication: ×, ˙, *, or no sign at all, such as 2Q for 2 times quantity. Whenever a number and either a letter or symbol come together to indicate multiplication, the number is always expressed first. As in the example above, 2Q indicates 2 times quantity and would never be expressed as Q2.

Multiplication and Addition

Some mathematical operations require both multiplication and addition. Here, sequence can vary within defined parameters.

Example

$$2 \times (4 + 3) = (2 \times 4) + (2 \times 3) = 14$$

Example

$$(2 + 3) \times (4 + 5) = [2 \times (4 + 5)] + [3 \times (4 + 5)]$$
$$= (2 \times 4) + (2 \times 5) + (3 \times 4) +$$
$$(3 \times 5) = 45$$

This ability to vary the multiplication or addition, as illustrated in the examples above, is referred to as the *distributive law of multiplication over addition*.

Division

Division is the inverse of multiplication, but it is neither commutative nor associative; the sequence counts.

Example

$$4 \div 2 = 2 \text{ is not the same as } 2 \div 4 = \frac{1}{2}.$$

It is possible to indicate that one number (the dividend) is being divided by another (the divisor) in more than one way. It can be expressed as $4 \div 2$ or as a fraction such as $\frac{1}{2}$.

Exponentials

Exponentials and roots are important to the study of hydraulics. The exponential is expressed as a number N to the x power, or N^x. The number N

can be any number, and x is the exponential, or power, we want to find. The power is the number of times we use that number as a factor to find the appropriate power.

Example

5^2 is 5 to the power 2, or second power (commonly called the square). To find 5 to the second power, we multiply 5 times itself, using 5 as a factor 2 times. Or $5 \times 5 = 25$. Thus, 5 squared, denoted by 5^2, is 25.

The third power is commonly referred to as the *cube* of a number. So 5^3 is $5 \times 5 \times 5 = 125$.

Root

Roots of numbers are the opposite of the exponential. Typically the most common root is the *square root*. It is the opposite of squaring a number.

Example

As shown above, the square of 5, or 5^2, is 25. Therefore, the square root of 25, denoted by $\sqrt{25}$, is 5.

The symbol for indicating the need to find the square root is $\sqrt{\ }$, also referred to as the *radical sign*. It is possible in more advanced formulas to need to find the third root, also called the *cube root*, or roots of even higher powers. However, the square root is the single most common root and is all that is needed for this text. The actual calculation to find the square root of a number can be tedious, especially when the answer is not a whole number. Therefore, the easiest way to find the square root of a number is with a small handheld calculator.

Algebra

Beyond arithmetic, the formulas and calculations in this book require an understanding of the basic principles of algebra. Algebra is nothing more than basic arithmetic using symbols or letters, in place of numbers, to create formulas. These formulas then become the means to solve problems. The algebraic formulas in this text are mostly specific to the study of hydraulics; however, some are specific to geometry.

Area

Area is the amount of space, in square units, that is taken up by a flat, or planar, object. To find the area of an object, we take the given measurements of the object and insert them into the appropriate formula for calculating area.

In our study of fire science, we frequently need to find the area of square or rectangular objects such as a wall or floor, so that we may calculate the needed fire flow. To calculate the area in this instance, we multiply the length of the object by its width or height. This gives us the formula for the area of a square or rectangle:

Area = length × width (or height)

In the form of an algebraic equation, it is written

$$A = L \times W \text{ (or } H)$$

Both L and W must have the same unit of measurement. If L is in feet (ft), then W must also be in ft. Units of measurement cannot be mixed. Remember that the answer will then be in units squared. If L and W are in ft, the answer is in square feet (ft^2). If L and W are in inches (in), the answer is in square inches (in^2), and so forth.

Example

Find the area of a floor that is 10 ft long by 15 ft wide.

$$A = L \times W$$
$$= 10 \times 15$$
$$= 150 \text{ ft}^2$$

The formulas for calculating the area of other common objects are given at the end of this appendix.

Volume

Another critical function to know in the study of hydraulics and fire science is how to calculate volume,

or the amount of space contained in a solid object. When you are finding volume, the answer is given in units cubed, such as cubic inches (in^3) or cubic feet (ft^3).

Whether it is used to calculate the volume of an onboard water tank or the volume of a room, this is one formula that every firefighter must know:

Volume = length × width × height

Expressed as an algebraic equation, it is

$$V = L \times W \times H$$

Again, all units must be the same to get a valid answer.

Example

Calculate the volume of water in a portable drafting basin if the basin is 10 ft wide by 10 ft long and water in the basin is 2 ft deep.

$$V = L \times W \times H$$
$$= 10 \times 10 \times 2$$
$$= 200 \text{ ft}^3$$

To find the volume in gallons when the volume in cubic feet is given, we use the formula Gallons = $V \times$ 7.48. See Chapter 1 for an introduction to this formula.

$$\text{Gallons} = V \times 7.48$$
$$= 200 \times 7.48$$
$$= 1{,}496$$

The formulas for calculating the volume of other common objects are given at the end of this appendix.

Units of Measurement

As important as it is to keep all dimensions and other operators in similar units, you will encounter instances where numbers are given with mixed units. For example, the dimensions of a floor may be given as 12 ft 10 in long and 13 ft 8 in wide. In this instance, it will be necessary to either convert the feet to inches (multiply by 12) or convert the inches to feet (divide by 12). Although the conversion of feet to inches is indeed workable, it will result in extremely large numbers and will usually require a reconversion to feet when finished. The simpler solution is to convert inches to feet by dividing inches by 12. If inches are converted to feet, the units of feet will be a decimal that represents a portion of a full foot, such as 0.5 for one-half of a foot, 0.75 for three-quarters of a foot, and so forth.

Example

To calculate the area of a floor that is 12 ft 10 in long and 13 ft 8 in wide, first we perform unit conversions. Given the number 12 ft 10 in (also written as 12'10"), to convert 10 in to feet, divide 10 by 12 (10/12), which gives 0.8333. The entire number can now be written as 12.8333 ft. Next convert 13 ft 8 in to a single measurement of feet: 8/12 = 0.6667. The entire number is then 13.6667 ft. Now we can put these into our formula to calculate the area of the floor.

Rounding

Before we calculate the area, however, we must round off the numbers (eliminate unnecessary or unwanted numbers) to a precision of only two decimal places to the right of the decimal point. (The more places there are to the right of the decimal, the higher the precision of the answer.)

The rule for rounding is first to determine the precision of your number. By specifying that we are going to round our numbers to two places to the right of the decimal point, we are saying that the answer will be 0.XX, instead of the current 0.XXXX. This, however, does not mean that we disregard any number beyond the second decimal place. We have to look at the next number and decide whether we need to round the second number up by 1. This is the rule: if the number beyond the level of precision you are seeking is no greater than 4, you leave the number without rounding up. The number 0.8333 will then be 0.83 when rounded to a precision of just two decimal places. If the next number is 5 or larger, you round up. The number 0.6667, rounded to a precision of two places, is then 0.67, because the third number in the original number is greater than or equal to 5.

When you are rounding, only the next number beyond the desired precision counts. In a number such as 0.8347, if you had chosen to round to two decimal places, you would not first round 4 to 5 because the 7 is larger than 5. You would only look at the 4, and the result would be 0.83.

Once the numbers have been properly rounded to the required precision, put them in the formula.

Example

$$A = L \times W$$
$$= 12.83 \times 13.67$$
$$= 175.3861 \text{ ft}^2$$

When rounded to the chosen level of precision, the area is 175.39 ft².

Order of Operations

When a single equation or formula requires multiple operations, there is a specific sequence of operations. This sequence must be adhered to: multiplication, division, addition, and subtraction. Also any operation within fences, such as parentheses and square brackets, must be carried out first, and any exponential or root must be calculated before carrying out the remaining operations.

Consider the equation $D = \sqrt{\text{gpm}/(29.84 \times C \times \sqrt{P})}$ Since the entire problem is inside a radical sign, once the interior set of operations is worked out, the last step is to find the square root.

Inside the radical sign, first the square root is calculated. Then the three numbers enclosed in the parentheses are multiplied (multiplication precedes division). Next the gpm is divided by the product 29.84 × $C \times \sqrt{P}$. Finally, the square root is calculated, and we have the answer. Refer to Chapter 5 for the introduction to this equation.

By following the guidelines given above, you will be able to find the area and volume in a minimum amount of time. Furthermore, these guidelines will both help you work every formula in this book easily and yield precise answers.

Example

Find the diameter of a nozzle if it is discharging 812 gpm at 80 psi tip pressure and the C factor is 0.997. To solve this problem, substitute 812 for gpm, 80 for P, and 0.997 for C.

$$D = \sqrt{\text{gpm} / (29.84 \times C \times \sqrt{P})}$$
$$= \sqrt{812 / (29.84 \times 0.997 \times \sqrt{80})}$$
$$= \sqrt{812 / (29.84 \times 0.997 \times 8.94)}$$
$$= \sqrt{812/265.97}$$
$$= \sqrt{3.05}$$
$$= 1.75" \text{ or } 1\frac{3}{4} \text{ in}$$

Formulas for Calculating the Area and Volume of Selected Shapes

The formulas designed to find the area and volume for the most common geometric shapes are provided below.

Circle

$$\text{Area} = 0.7854 \times D^2$$

Cylinder

$$\text{Volume} = 0.7854 \times D^2 \times H$$

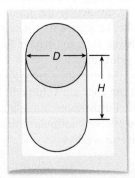

Ellipse

$$\text{Area} = 3.14 \times A \times B$$

Elliptical Tank

$$\text{Vol.} = 3.14 \times A \times B \times H$$

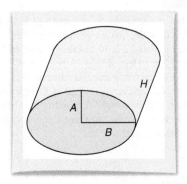

Parallelogram

$$\text{Area} = W \times L$$
$$\text{Volume} = W \times L \times H$$

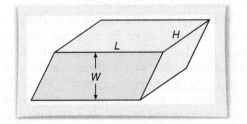

Sphere

$$\text{Area} = 12.56 \times r^2$$
$$\text{Volume} = 4.188 \times r^3$$

Square or Rectangle

$$\text{Area} = L \times W$$
$$\text{Volume} = L \times W \times H$$

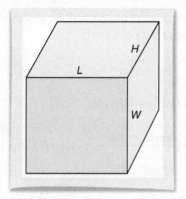

Trapezoid

$$\text{Area} = W\left(\frac{L_1 + L_2}{2}\right)$$

$$\text{Volume} = W\left(\frac{L_1 + L_2}{2}\right) \times H$$

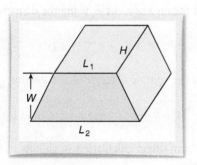

Triangle

$$\text{Area} = W \times \frac{H}{2}$$

$$\text{Volume} = W \times \frac{H}{2} \times L$$

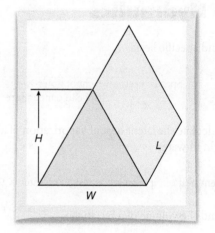

Formulas

Chapter 1: Introduction to Hydraulics

To find specific heat:

$$\text{Specific heat} = \text{weight of water} \times \text{change in temperature}$$

To calculate the latent heat of vaporization of a given weight of water:

$$\text{Latent heat of vaporization} = \text{weight of water} \times 970.3$$

To find the volume of a rectangle:

$$V = L \times W \times H, \text{ where } V \text{ is the volume, } L \text{ is the length, } W \text{ is the width, and } H \text{ is the height}$$

To find out how much water is needed to fill a room with steam:

$$CFW = V \div 1,700$$

To calculate how much water is in a container:

$$\text{Gallons} = V \times 7.48$$

To find the gallons of water needed to fill a room with steam:

$$\text{Gallons} = V \div 227$$

The Iowa State rate-of-flow formula:

$$\text{Gallons per minute (gpm)} = V \div 100, \text{ where } V \text{ is the volume in cubic feet}$$

To find out how many Btu will be absorbed by the number of gallons of water:

$$\text{Btu absorbed} = 9,343 \times \text{gallons}$$

The NFA fire flow formula:

$$NFA = \left(\frac{\text{length} \times \text{width}}{3} + \text{exposure charge} \right) \times \% \text{ involvement}$$

To find the gallons of water when the volume is in cubic inches:

$$\text{Gallons} = V \div 231$$

To find the volume of a cylinder in cubic inches or cubic feet:

$$V = 0.7854 \times D^2 \times H \text{ (or } L)$$

To find the area of a circle, using the diameter:

$$\text{Area} = 0.7854 \times D^2$$

To find the volume of a cylinder in gallons:

$$V = 7.48 \times 0.7854 \times D^2 \times H \text{ (or } L)$$

To perform direct calculation of gallons of water in a cylinder:

$$V = 5.87 \times D^2 \times H \text{ (or } L)$$

To find the weight of water in a container:

$$W = V \times 8.34$$

To preform direct calculation of the weight of water in a cylinder:

$$W = 49 \times D^2 \times H \text{ (or } L)$$

Chapter 2: Force and Pressure

To find pressure when force and area are known:

$$P = \frac{F}{A}$$

To find pressure when the height (elevation) of water is known:

$$P = 0.433 \times H$$

To find the height of water when the pressure it creates is known:

$$H = 2.31 \times P$$

To find absolute pressure:

$$\text{Absolute pressure} = \text{relative pressure} + \text{atmospheric pressure}$$

To find force when pressure and area are known:

$$F = P \times A$$

To find force when the height of the water and the area it acts on are known:

$$F = 0.433 \times H \times A$$

Chapter 3: Bernoulli's Principle

Conservation of energy:

$$\text{Total PE} = \text{total KE}$$

Bernoulli's equation:

$$PH_1 + EH_1 + VH_1 = PH_2 + EH_2 + VH_2$$

Torricelli's equation:

$$v = \sqrt{2g(y_2 - y_1)}$$

To find velocity when height is known with Torricelli's equation simplified:

$$v = 8.02\sqrt{H}$$

Revised Bernoulli equation:

$$\frac{P_1}{0.433} + Z_1 + \frac{v_1^2}{64.35} = \frac{P_2}{0.433} + Z_2 + \frac{v_2^2}{64.35}$$

Chapter 4: Velocity and Flow

To find velocity when pressure is known:

$$v = \sqrt{2 \times g \times 2.31 \times P}$$

To find velocity when pressure is known, simplified:

$$v = 12.19\sqrt{P}$$

To find pressure when velocity is known:

$$P = \left(\frac{v}{12.19}\right)^2$$

To compare the area and velocity in two sizes of hose for the same flow:

$$A_1 \times v_1 = A_2 \times v_2$$

To find the diameter of a circle when area is known:

$$D = \sqrt{\frac{A}{0.7854}}$$

Basic quantity flow formula:

$$Q = A \times v$$

To find flow from a circular opening:

$$Q = 0.7854 \times D^2 \times v$$

To find flow from a circular opening when the diameter is in inches:

$$Q = 0.7854 \times D^2 \times \frac{1}{144} \times v$$

To calculate flow in 1 min:

$$Q = 0.7854 \times D^2 \times \frac{1}{144} \times v \times 60$$

Chapter 5: GPM

To calculate gpm when diameter and velocity are known:

$$gpm = 2.448 \times D^2 \times v$$

Basic gpm formula:

$$gpm = 0.7854 \times D^2 \times \tfrac{1}{144} \times 12.19 \times \sqrt{P} \times 60 \times 7.48$$

To calculate gpm when diameter and pressure are known:

$$gpm = 2.448 \times D^2 \times 12.19 \times \sqrt{P}$$

Freeman's formula with the coefficient of discharge:

$$gpm = 29.84 \times D^2 \times C \times \sqrt{P}$$

AIA formula:

$$gpm = 29.83 \times D^2 \times C \times \sqrt{P}$$

To find the diameter when gpm and pressure are known:

$$D = \sqrt{\frac{gpm}{29.84 \times C \times \sqrt{P}}}$$

To find the pressure when gpm and diameter are known:

$$P = \left(\frac{gpm}{29.84 \times D^2 \times C}\right)^2$$

To find the velocity when gpm and diameter are known:

$$v = \frac{gpm}{2.448 \times D^2 \times C}$$

Chapter 6: Friction Loss

To find the friction loss multiplier:

$$F_m = \left(\frac{v_2}{v_1}\right)^2$$

To find the conversion factor:

$$CF = \frac{D_1^5}{D_2^5}$$

To find the friction loss per 100 ft of 2½-in hose:

$$FL\ 100 = 2Q^2$$

To find the friction loss per 100 ft of hose other than 2½-in hose:

$$FL\ 100 = CF \times 2Q^2$$

To find the conversion factor when the friction loss and quantity of flow are known:

$$CF = \frac{FL\ 100}{2Q^2}$$

To use the abbreviated friction loss formula:

$$FL\ 100 = CF \times Q^2$$

To use the equivalent-length formula:

$$L_2 = L_1 \times \left(\frac{CF_1}{CF_2}\right)$$

To use the equivalent friction loss formula:

$$FL_2\ 100 = FL_1\ 100 \times \left(\frac{CF_2}{CF_1}\right)$$

To find the quantity flow when the friction loss and conversion factor are known:

$$Q = \sqrt{\frac{FL\ 100}{CF \times 2}}$$

Chapter 7: Pump Theory and Operation

To find net pump pressure:

$$\text{Discharge} - \text{intake} = \text{net pump}$$
$$\text{pressure} \quad \text{pressure} \quad \text{pressure}$$

Chapter 8: Theory of Drafting and Pump Testing

To find pressure from inches of mercury:

$$P = 0.489 \times Hg$$

To find the height of water from inches of mercury:

$$H = 1.13 \times Hg$$

Reverse lift calculation:

$$Hg = 0.882 \times H$$

To find friction loss on intake side of pump at draft:

$$FL = (\text{dynamic reading} - \text{static reading}) \times 0.489$$

Chapter 9: Fire Streams

To find a horizontal range of streams:

$$HR = \frac{1}{2}NP + 26*$$

To find a vertical range of streams:

$$VR = \frac{5}{8}NP + 26*$$

To calculate the nozzle reaction:

$$NR = 1.57 \times D^2 \times P$$

To find the nozzle reaction when gpm and pressure are known:

$$NR = 0.0543 \times \text{gpm} \times \sqrt{P}$$

---

*Add 5 for each ⅛ in of nozzle diameter more than ¾ in.

To calculate the area of coverage AC for foam:

$$AC = \text{gpm (of the nozzle)}/AR$$

Chapter 10: Calculating Pump Discharge Pressure

Pump discharge pressure formula:

$$PDP = NP + FL\,1 + FL\,2 \pm EP + AL$$

To find the total friction loss:

$$FL = FL\,100 \times L$$

Relay formula:

$$PDP = 20 + FL \pm EP$$

Chapter 11: Advanced Problems in Hydraulics

To find the actual nozzle pressure (INP) when an incorrect pump discharge pressure is used:

$$\frac{CNP}{CPDP} = \frac{INP}{IPDP}$$

Chapter 12: Water Supply

To calculate a community's maximum daily consumption:

$$MDC = \frac{Pop \times 214.5}{1,440}$$

To calculate a community's needed storage capacity:

$$SC = (MDC + NFF) \times T$$

To find the minimum flow rate:

$$MFR = MDC + NFF$$

Hazen–Williams formula:

$$P_f = \frac{4.52 \times Q^{1.85}}{C^{1.85} \times D^{4.87}}$$

To find the total pressure loss for a system:

$$P_t = P_f \times L$$

To directly calculate the total pressure loss:

$$P_t = \frac{4.52 \times Q^{1.85} \times L}{C^{1.85} \times D^{4.87}}$$

Hydrant capacity formula:

$$Q_2 = Q_1 \times \left(\frac{p_s - p_{r2}}{p_s - p_{r1}}\right)^{0.54}$$

Chapter 13: Standpipes, Sprinklers, and Fireground Formulas

To calculate gpm from a sprinkler head:

$$\text{gpm} = K \times \sqrt{P}$$

Answers to Wrap-Up Questions

Chapter 1: Introduction to Hydraulics

Case Study

1. C
2. A
3. D
4. B

Review Questions

1. 8.56 lb
2. 62.4 lb
3. 7.48 gal
4. 970.3 Btu
5. 227 ft³

Activities

1. $V = L \times W \times H$
   $$= 100 \text{ ft} \times 50 \text{ ft} \times 10 \text{ ft}$$
   $$= 50{,}000 \text{ ft}^3$$

   To convert to gallons:

   Gallons $= 50{,}000 \times 7.48$
   $$= 374{,}000$$

2. This problem can be solved in two ways. One method is to find how many cubic feet of water are necessary to fill 10,000 ft³ with steam and then to find the equivalent volume of water.

   Cubic feet of water $= V \div 1{,}700$
   $$= 10{,}000 \div 1{,}700$$
   $$= 5.88 \text{ ft}^3$$

   Now find how many gallons of water are in 5.88 ft³ of water.

   Gallons $=$ cubic feet of water $\times 7.48$
   $$= 5.88 \times 7.48$$
   $$= 43.98$$

Or we can use this formula:

Gallons $= V \div 227$
$$= 10{,}000 \div 227$$
$$= 44.05$$

(The difference in the results of the two methods is due to rounding.)

3. Weight of water $= 2{,}000 \text{ gal} \times 8.34 \text{ lb/gal of water}$
   $$= 16{,}680 \text{ lb}$$

4. Multiply 100 by the number of Btu absorbed per gallon of water:

   Btu $= 100 \times 9{,}343$
   $$= 934{,}300$$

Challenging Questions

1.
   A. $19{,}182 \div 8.34 = 2{,}300$ gal
   B. $2{,}300 \div 7.48 = 307.49$ ft³

2. $V = 4.1888 \times r^3$
   $$= 4.1888 \times (10)^3$$
   $$= 4.1888 \times 1{,}000$$
   $$= 4{,}188.80 \text{ gal}$$

3. $W = 49 \times D^2 \times L$
   $$= 49 \times (4.5)^2 \times 10$$
   $$= 49 \times 20.25 \times 10$$
   $$= 9{,}922.50 \text{ lb}$$

4. $V = 0.7854 \times D^2 \times L$
   $$= 0.7854 \times (4.5)^2 \times 10$$
   $$= 0.7854 \times 20.25 \times 10$$
   $$= 159.04 \text{ ft}^3$$

   Now find the weight of 159.04 ft³ of salt water.

   $W = 159.04 \times 64$
   $$= 10{,}178.56 \text{ lb}$$

5. 210 + 265 = 475 total gpm

   475 ÷ 2 = 237.5 gpm being added to dead weight of building.

   237.5 × 30 = 7,125 gal of water added to dead weight of building in 30 min.

   7,125 × 8.34 = 59,422.5 lb of water added to dead weight of building in 30 min.

6. 2 ft^3 of water weighs 2 × 62.5 = 125 lb

   1 lb of water will absorb 200 − 62 = 138 Btu going from 62°F to 200°F

   Total Btu absorbed is 125 × 138 = 17,250 Btu

7. 125 gpm for 30 s is 62.5 gal

   227 ft^3 of steam from 1 gal of water × 62.5 = 14,187.5 ft^3 of steam/volume of room

8.
   A. $V = L \times W \times H$
      = 25 ft × 10 ft × 13 ft
      = 3,250 ft^3 but a single pool is one-half of this volume, or 1,625 ft^3

   B. Gallons = $V \times 7.48$
      = 1,625 × 7.48
      = 12,155

9. 7.48 × 2 = 14.96 gal

10. Rate-of-flow formula:

    gpm = $V \div 100$
       = $(L \times W \times H) \div 100$
       = (20 × 12 × 8) ÷ 100
       = 1,920 ÷ 100
       = 19.20 gpm for 30 s, or 9.6 gal

    Straight conversion:

    Gallons = $V \div 227$
       = 1,920 ÷ 227
       = 8.46

11. The Iowa formula has a built-in margin of safety.

Chapter 2: Force and Pressure
Case Study

1. C
2. C
3. A
4. A

Review Questions

1. Force is pressure times the area it is acting on. Pressure is force per unit of area.
2. Piezometric plane
3. Static pressure
4. Residual pressure

Activities

1. $P = 0.433 \times H$
   = 0.433 × 123
   = 53.26 or 53 psi

2. $H = 2.31 \times P$
   = 2.31 × 63
   = 145.53 or 146 ft above the hydrant

3. $F = P \times A$

   The formula for the area of a circle is $A = 0.7854 \times D^2$, where A is the area and D is the diameter. (This formula is more convenient than the conventional $A = \pi \times r^2$ because we universally use the diameter in the fire service. Either one will work, so use what you feel comfortable with.)

   First find the area:

   $A = 0.7854 \times D^2$
      = 0.7854 × (4.5)^2
      = 0.7854 × 20.25
      = 15.9 in^2

Now find the force:

$F = P \times A$
$= 10 \times 15.9$
$= 159$ psi of force on the cap

This amount of force reveals why it is so diffi-cult to operate gates (valves) or remove blind caps under pressure.

4. First find the area of the bottom of the container:

$A = L \times W$
$= 5 \times 5$
$= 25$ in²

Now find the pressure:

$P = F/A$
$= 25/25$
$= 1$ psi

5. $H = 2.31 \times P$
$= 2.31 \times 14.7$
$= 33.96$ ft

We now know that atmospheric pressure at sea level will raise a column of water 33.96 ft. This fact is critical when we discuss drafting. See Chapter 8.

Challenging Questions

1. $P = 0.433 \times H$
$= 0.433 \times 60$
$= 25.98$ psi

2. An area of 2 ft² $= 288$ in².

$F = P \times A$
$= 25.98 \times 288$
$= 7,482.24$ lb of force

3. $P = 0.433 \times H$
$= 0.433 \times 35$
$= 15.16$ psi

Remember: The amount of pressure exerted by a column of water of a given height is the same amount of pressure needed to lift water to that height.

4. $H = 2.31 \times P$
$= 2.31 \times 70$
$= 161.7$ ft

5.

A. First find the area of the bottom of the tank.

$A = 0.7854 \times D^2$
$= 0.7854 \times (10)^2$
$= 0.7854 \times 100$
$= 78.54$ ft²

To convert square feet to square inches, multiply by 144.

$A = 78.54 \times 144$
$= 11,309.76$ in²

Now find the pressure if the tank is half empty.

$P = 0.433 \times H$
$= 0.433 \times 20$
$= 8.66$ psi

Finally, find the force.

$F = P \times A$
$= 8.66 \times 11,309.76$
$= 97,942.52$ lb of force

B. $P = 0.433 \times H$
$= 0.433 \times 40$
$= 17.32$ psi

C. First find the pressure exerted by 30 ft of water.

$P = 0.433 \times H$
$= 0.433 \times 30$
$= 12.99$ psi

Since there is 100 psi pressure added in this scenario, we must remember to apply Pascal's principle and increase the pressure at the base by 100 psi, making total pres-sure 112.99 psi.

We already know the area is 11,309.76.

$F = P \times A$
$= 112.99 \times 11,309.76$
$= 1,277,889.78$ or $1,277,890$ lb of force from weight of water and added 100 psi.

This is the force on the bottom of the tank.

6. First find the area of the cap:

$$A = 0.7854 \times D^2$$
$$= 0.7854 \times (4.5)^2$$
$$= 0.7854 \times 20.25$$
$$= 15.90 \text{ in}^2$$

Now find the force on the cap:

$$F = P \times A$$
$$= 65 \times 15.90$$
$$= 1{,}033.5 \text{ lb of force}$$

7. $H = 2.31 \times P$
$$= 2.31 \times 65$$
$$= 150.15 \text{ ft above the hydrant}$$

8. If absolute pressure includes 14.7 psi for sea level, relative pressure is simply absolute pressure minus 14.7, or $140 - 14.7 = 125.3$ psi relative pressure.

9. First find the area of the 2½-in cap on the hydrant.

$$A = 0.7854 \times D^2$$
$$= 0.7854 \times (2.5)^2$$
$$= 0.7854 \times 6.25$$
$$= 4.91 \text{ in}^2$$

Now find the pressure on the cap.

$$P = 0.433 \times H$$
$$= 0.433 \times 83$$
$$= 35.94 \text{ psi}$$

Finally, calculate the force:

$$F = P \times A$$
$$= 35.94 \times 4.91$$
$$= 176.47 \text{ lb of force}$$

10. The 100 lb of force/pressure on the piston rod that is driving a piston of 10 in² is exerting a pressure of 100/10, or 10 psi. So 10 psi will be added to the pressure in the tank.

11. $P = 0.433 \times H$
$$= 0.433 \times 10$$
$$= 4.33 \text{ psi due to height of water}$$

Total pressure is $4.33 + 50 = 54.33$ psi.

Chapter 3: Bernoulli's Principle

Case Study

1. C
2. A
3. D
4. B

Review Questions

1. 1738
2. As one increases, the other decreases.
3. Energy.
4. Examples of potential energy include static pressure, a drawn bow, water in a reservoir or/tank, and a stretched elastic band.
5. One-half the potential energy and one-half the kinetic energy
6. Yes, friction will reduce the output of a system.
7. Elevation.

Activities

1. $PH_1 + EH_1 + VH_1 = PH_2 + EH_2 + VH_2$

$$100 \text{ psi} + 25 \text{ ft} + 0 = (100 \text{ psi} + 10.83 \text{ psi}) + 0 + 0$$
$$(100 \times 2.31) + 25 = (110.83 \times 2.31)$$
$$231 \text{ ft} + 25 \text{ ft} = 256 \text{ ft}$$

2. $PH_1 + EH_1 + VH_1 = PH_2 + EH_2 + VH_2$

$$0 + EH_1 + 0 = 35 \text{ psi} + 0 + 0$$
$$EH_1 = 35 \times 2.31$$
$$EH_1 = 80.85 \text{ ft}$$

3. Use the abbreviated form of Torricelli's equation, $v = 8.02 \sqrt{H}$.

$$v = 8.02 \sqrt{H}$$
$$= 8.02 \sqrt{50}$$
$$= 8.02 \times 7.07$$
$$= 56.70 \text{ or } 57 \text{ fps}$$

4. Use the formula $v = 8.02 \times \sqrt{H}$.

$$v = 8.02 \times \sqrt{H}$$
$$= 8.02 \times \sqrt{120}$$
$$= 8.02 \times 10.95$$
$$= 87.82 \text{ or } 88 \text{ fps}$$

5. Using the revised form of Bernoulli's equation, we get

$$P_1/0.433 + Z_1 + v_1^2/64.35 = P_2/0.433 + Z_2 + v_2^2/64.35$$
$$54.33 \text{ psi}/0.433 \text{ psi} + (-10 \text{ ft}) + 0 = 0 + 0 + (85.98 \text{ fps})^2/64.35 \text{ fps}^2$$
$$125.47 \text{ ft} - 10 \text{ ft} = 7,382.56 \text{ fps}^2/64.35 \text{ fps}^2$$
$$115.47 \text{ or } 115 \text{ ft} = 114.88 \text{ or } 115 \text{ ft}$$

Challenging Questions

1. $H = 2.31 \times P$
   $= 2.31 \times 35$
   $= 80.85 \text{ ft}$

2. The gravitational acceleration constant

3. It eliminates the need to make separate calculations for pressure, elevation, and velocity.

4. $H = 2.31 \times P$
   $= 2.31 \times 6.06$
   $= 14.00 \text{ ft}$

Chapter 4: Velocity and Flow

Case Study

1. D
2. A
3. B
4. D

Review Questions

1. 32.174 fps per second (the rate at which all objects accelerate due to gravity).
2. Per Torricelli's theorem, the discharge velocity is equivalent to water being dropped the same distance as the elevation head.
3. $\sqrt{2 \times g}$
4. It converts square feet to square inches.

Activities

1. We learned that the shape and volume of a tank are irrelevant in determining pressure at the base. When we calculate the velocity from an opening at the base of a tank, these principles hold true. The shape and volume have no effect on velocity; only the water level does. To calculate the velocity of water at the point of discharge, the only fact you need to know is the elevation of the water above the discharge. In this problem it is 125 ft. See Chapter 2 for more about the shape, volume, and pressure of a tank.

   $v = 8.02 \sqrt{H}$
   $= 8.02 \times \sqrt{125}$
   $= 8.02 \times 11.18$
   $= 89.66 \text{ or } 90 \text{ fps}$

2. First calculate the pressure created by an elevation of 125 ft.

   $P = H \times 0.433$
   $= 125 \text{ ft} \times 0.433$
   $= 54.13 \text{ psi}$

   Now that you know the pressure created by 125 ft of elevation, calculate the velocity of the water.

   $v = 12.19 \times \sqrt{P}$
   $= 12.19 \times \sqrt{54.13}$
   $= 12.19 \times 7.36$
   $= 89.72 \text{ or } 90 \text{ fps}$

   This answer is within 0.06 fps of the answer arrived at in Activity 1. This difference is easily explained by the rounding of numbers in both problems.

3. First find the velocity for the 1⅛-in tip at 50 psi.

$$v = 12.19 \times \sqrt{P}$$
$$= 12.19 \times \sqrt{50}$$
$$= 12.19 \times 7.07$$
$$= 86.18 \text{ or } 86 \text{ fps}$$

Now calculate the velocity of water in the 2½-in hose, using the formula $A_1 \times v_1 = A_2 \times v_2$, where A_1 = area of the 1⅛-in tip, v_1 = velocity from the 1⅛-in tip, A_2 = area of the 2½-in hose, and v_2 = velocity of water in the 2½-in hose.

$$A_1 \times v_1 = A_2 \times v_2$$
$$0.7854 \times (1.125)^2 \times 86.18 = 0.7854 \times (2.5)^2 \times v_2$$
$$0.7854 \times 1.27 \times 86.18 = 0.7854 \times 6.25 \times v_2$$
$$85.96 = 4.91 \times v_2$$
$$v_2 = 85.96/4.91$$
$$= 17.51 \text{ or } 18 \text{ fps}$$

4. To verify the answer to Activity 3, we need only compare the area of the 2½-in hose to the area of the 1⅛-in nozzle and the velocity of water in the hose to the velocity of water leaving the nozzle.

The area of the 2½-in hose compared to the area of a 1⅛-in tip is 4.9/0.994, or 4.93, times the area of the 1⅛-in tip. The velocity of the water in the 1⅛-in tip compared to the velocity of the water in the 2½-in hose is 86.18/17.51, or 4.92, times the velocity of the 2½-in hose. The fact that both the ratio of the areas and the ratio of the velocities are the same verifies that the answer in Activity 3 is correct. Just remember that the relationship of area to velocity is an inverse one.

5. To find how much water is flowing, use the formula $Q = A \times v$.

$$Q = A \times v$$
$$= 1.5 \times 3$$
$$= 4.5 \text{ cfs}$$

6. When you are finding the area of the 2-in opening, remember to convert it to square feet.

$$Q = A \times v$$
$$= 0.7854 \times D^2 \times \frac{1}{144} \times v$$
$$= 0.7854 \times (2)^2 \times \frac{1}{144} \times 108.76$$
$$= 0.7854 \times 4 \times 0.0069 \times 108.76$$
$$= 2.37 \text{ cfs}$$

7. First find the velocity of the water.

$$v = 8.02 \times \sqrt{H}$$
$$= 8.02 \times \sqrt{85}$$
$$= 8.02 \times 9.22$$
$$= 73.94 \text{ fps}$$

Now calculate the amount of water flowing in 1 min.

$$Q = 0.7854 \times D^2 \times \frac{1}{144} \times v \times 60$$
$$= 0.7854 \times (2.5)^2 \times \frac{1}{144} \times 73.94 \times 60$$
$$= 0.7854 \times 6.25 \times 0.0069 \times 73.94 \times 60$$
$$= 151.23 \text{ or } 151 \text{ cfm}$$

Challenging Questions

1. $$v = 8.02 \times \sqrt{H}$$
$$= 8.02 \times \sqrt{37}$$
$$= 8.02 \times 6.08$$
$$= 48.76 \text{ fps}$$

2. $$A_1 \times v_1 = A_2 \times v_2$$
$$1.5 \times v_1 = 1 \times 1$$
$$v_1 = 1/1.5$$
$$v_1 = 0.67$$

The velocity of the water in hose 1 is 0.67, or two-thirds, that of the water in hose 2.

3. $$A_1 \times v_1 = A_2 \times v_2$$
$$0.7854 \times D^2 \times v_1 = A_2 \times v_2$$
$$0.7854 \times (2.5)^2 \times 10 = A_2 \times 20.42$$
$$0.7854 \times 6.25 \times 10 = A_2 \times 20.42$$
$$49.09 = A_2 \times 20.42$$
$$A_2 = 49.09/20.42$$
$$= 2.40 \text{ in}^2$$

4. $$Q = 0.7854 \times D^2 \times \frac{1}{144} \times v$$
$$= 0.7854 \times (1.75)^2 \times \frac{1}{144} \times 109$$
$$= 0.7854 \times 3.06 \times 0.0069 \times 109$$
$$= 1.82 \text{ cfs}$$

5. First find the velocity from the pressure:

$$v = 12.19 \times \sqrt{P}$$
$$= 12.19 \times \sqrt{80}$$
$$= 12.19 \times 8.94$$
$$= 108.98 \text{ cfs}$$

Now solve for the quantity flow:

$$Q = 0.7854 \times D^2 \times \tfrac{1}{144} \times v \times 60$$
$$= 0.7854 \times (1\tfrac{5}{8})^2 \times \tfrac{1}{144} \times 108.98 \times 60$$
$$= 0.7854 \times 2.66 \times 0.0069 \times 108.98 \times 60$$
$$= 94.87 \text{ cfm}$$

6. $Q = 0.7854 \times D^2 \times \tfrac{1}{144} \times v$
$$= 0.7854 \times (3)^2 \times \tfrac{1}{144} \times 35$$
$$= 0.7854 \times 9 \times 0.0069 \times 35$$
$$= 1.72 \text{ cfs}$$

7. $v = 12.19 \times \sqrt{P}$
$$= 12.19 \times \sqrt{80}$$
$$= 12.19 \times 8.94$$
$$= 108.98 \text{ fps}$$

8. $Q = 0.7854 \times D^2 \times \tfrac{1}{144} \times v \times 60$
$$= 0.7854 \times (1.5)^2 \times \tfrac{1}{144} \times 108.7 \times 60$$
$$= 0.7854 \times 2.25 \times 0.0069 \times 108.7 \times 60$$
$$= 80.04 \text{ cfm}$$

Chapter 5: GPM

Case Study

1. C
2. A
3. C
4. D

Review Questions

1. $0.7854 \times D^2$ is the area of a circle.
   $\tfrac{1}{144}$ converts square feet to square inches.
   v is the velocity of discharge.
   60 is the flow in 1 min.
   7.48 converts cubic feet of flow to gallons.

2. The number 2.448 is a constant derived from $0.7854 \times \tfrac{1}{144} \times 60 \times 7.48$.

3. Substitute $12.19 \times \sqrt{P}$ for v.
4. John Freeman used 62.5 lb as the weight of 1 ft³ of water instead of 62.4 lb.

Activities

1. gpm $= 0.7854 \times D^2 \times \tfrac{1}{144} \times v \times 60 \times 7.48$
   $$= 0.7854 \times (1.75)^2 \times \tfrac{1}{144} \times 109 \times 60 \times 7.48$$
   $$= 0.7854 \times 3.06 \times 0.0069 \times 109 \times 60 \times 7.48$$
   $$= 816.45 \text{ or } 816 \text{ gpm}$$

2. gpm $= 29.84 \times D^2 \times C \times \sqrt{P}$
   $$= 29.84 \times (\tfrac{15}{16})^2 \times 0.97 \times \sqrt{50}$$
   $$= 29.84 \times 0.88 \times 0.97 \times 7.07$$
   $$= 180.08 \text{ or } 180 \text{ gpm}$$

3. The C factor from Table 5-1 is 0.997.

$$D = \sqrt{\text{gpm}/(29.84 \times C \times \sqrt{P})}$$
$$= \sqrt{650/(29.84 \times 0.997 \times \sqrt{80})}$$
$$= \sqrt{650/(29.84 \times 0.997 \times 8.94)}$$
$$= \sqrt{650/265.97}$$
$$= \sqrt{2.44}$$
$$= 1.56 \text{ or } 1\tfrac{9}{16}\text{-in tip.}$$

4. $P = [\text{gpm}/(29.84 \times D^2 \times C)]^2$
   $$= \{300/[29.84 \times (1.25)^2 \times 0.97]\}^2$$
   $$= [300/(29.84 \times 1.56 \times 0.97)]^2$$
   $$= (300/45.15)^2$$
   $$= (6.64)^2$$
   $$= 44.09 \text{ or } 44 \text{ psi}$$

5. $v = \text{gpm}/(2.448 \times D^2 \times C)$
   $$= 600/[2.448 \times (1.25)^2 \times 0.97]$$
   $$= 600/(2.448 \times 1.56 \times 0.97)$$
   $$= 600/3.7$$
   $$= 162.16 \text{ or } 162 \text{ fps}$$

6. The C factor from Table 5-1 is 0.95.

$$\text{gpm} = 29.84 \times D^2 \times C \times \sqrt{P}$$
$$= 29.84 \times (^{17}\!/_{32})^2 \times 0.95 \times \sqrt{15}$$
$$= 29.84 \times 0.28 \times 0.95 \times 3.87$$
$$= 30.72 \text{ or } 31 \text{ gpm}$$

Challenging Questions

1. $\text{gpm} = 2.448 \times D^2 \times v$
$$= 2.448 \times (1.125)^2 \times 81.5$$
$$= 2.448 \times 1.27 \times 81.5$$
$$= 253.38 \text{ or } 253 \text{ gpm}$$

(Note: Since this formula was developed before the introduction of the C factor, this answer is off by the correct value of C. If we correct for a C of 0.97, the more accurate flow becomes $253.38 \times 0.97 = 245.78$ or 246 gpm.)

2. $\text{gpm} = 29.84 \times D^2 \times C \times \sqrt{P}$
$$= 29.84 \times (1.875)^2 \times 0.997 \times \sqrt{70}$$
$$= 29.84 \times 3.52 \times 0.997 \times 8.37$$
$$= 876.52 \text{ or } 877 \text{ gpm}$$

3. $D = \sqrt{\text{gpm}/29.84 \times C \times \sqrt{P}}$

$$= \sqrt{247.69/29.84 \times 0.97 \times \sqrt{45}}$$

$$= \sqrt{247.69/29.84 \times 0.97 \times 6.71}$$

$$= \sqrt{247.69/194.22}$$

$$= \sqrt{1.28}$$

$$= 1.13 \text{ in}$$

4. $P = [\text{gpm}/(29.84 \times D^2 \times C)]^2$
$$= \{181.5/[29.84 \times (0.94)^2 \times 0.97]\}^2$$
$$= [181.5/(29.84 \times 0.88 \times 0.97)]^2$$
$$= (7.13)^2$$
$$= 50.84 \text{ or } 51 \text{ psi}$$

5. $v = \text{gpm}/(2.448 \times D^2 \times C)$
$$= 210/[2.448 \times (1)^2 \times 0.97]$$
$$= 210/(2.448 \times 1 \times 0.97)$$
$$= 210/2.37$$
$$= 88.61 \text{ fps}$$

6. $P = [\text{gpm}/(29.84 \times D^2 \times C)]^2$
$$= \{20/[29.84 \times (0.5)^2 \times 0.75]\}^2$$
$$= [20/(29.84 \times 0.25 \times 0.75)]^2$$
$$= (20/5.6)^2$$
$$= (3.57)^2$$
$$= 12.74 \text{ psi}$$

7. $\text{gpm} = 2.448 \times D^2 \times v$
$$= 2.448 \times (1.5)^2 \times 85.98 \times C$$
$$= 2.448 \times 2.25 \times 85.98 \times 0.997$$
$$= 472.16 \text{ or } 472 \text{ gpm}$$

8. $\text{gpm} = 29.84 \times D^2 \times C \times \sqrt{P}$
$$= 29.84 \times (1.5)^2 \times 0.997 \times \sqrt{50}$$
$$= 29.84 \times 2.25 \times 0.997 \times 7.07$$
$$= 473.26 \text{ or } 473 \text{ gpm}$$

9. $v = \text{gpm}/(2.448 \times D^2 \times C)$
$$= 325/[2.448 \times (1.25)^2 \times 0.97]$$
$$= 325/(2.448 \times 1.56 \times 0.97)$$
$$= 325/3.7$$
$$= 87.84 \text{ fps}$$

10. $\text{gpm} = 29.84 \times D^2 \times C \times \sqrt{P}$
$$= 29.84 \times (0.5)^2 \times 0.75 \times \sqrt{15}$$
$$= 29.84 \times 0.25 \times 0.75 \times 3.87$$
$$= 21.65 \text{ or } 22 \text{ gpm}$$

11. $\text{gpm} = 29.84 \times D^2 \times C \times \sqrt{P}$
$$= 29.84 \times (1.375)^2 \times 0.97 \times \sqrt{80}$$
$$= 29.84 \times 1.89 \times 0.97 \times 8.94$$
$$= 489.07 \text{ or } 489 \text{ gpm}$$

Chapter 6: Friction Loss

Case Study

1. A
2. C
3. C
4. D

Review Questions

1. The conversion of useful energy into nonuseful energy due to friction
2. Resistance to flow of a liquid

3. The smooth, orderly flow of water, with layers, or cores, of water gliding effortlessly over the next layer of water.

4. The flow of water is disorganized and random.

5.
   A. As the velocity of flow increases, the friction loss will increase by the square of the velocity increase. Since the flow in this example is doubled, the friction loss will be increased by $(2)^2$, or 4.
   B. $4.5 \times 4 = 18$ psi

6. The velocity of the water creates friction loss.

7. $FL\ 100 = CF \times 2Q^2$

Activities

1. The conversion factor for 2-in hose in Table 6-1 is 4.

   $$FL\ 100 = CF \times 2Q^2$$
   $$= 4 \times 2 \times (2)^2$$
   $$= 4 \times 2 \times 4$$
   $$= 32\ psi$$

2. $CF = FL\ 100/2Q^2$
   $$= 18/[2 \times (3)^2]$$
   $$= 18/(2 \times 9)$$
   $$= 18/18$$
   $$= 1$$

   A conversion factor of 1 indicates 2½-in hose.

3. From Table 6-1, the conversion factor for 1¾-in hose is 7.76.

   $$L_2 = L_1 \times (CF_1/CF_2)$$
   $$= 500 \times (1/7.76)$$
   $$= 500 \times 0.129$$
   $$= 64.5\ \text{rounded down to a single 50-ft length of hose}$$

4. From Table 6-1, the conversion factor for 3-in hose is 0.4, and the conversion factor for 3½-in hose is 0.17.

   $$FL_2\ 100 = FL_1\ 100 \times (CF_2/CF_1)$$
   $$= 20 \times (0.17/0.4)$$
   $$= 20 \times 0.425$$
   $$= 8.5\ psi$$

5. From Table 6-1, the conversion factor for 3-in hose is 0.4, and for 4-in hose it is 0.1. Remember to use 1 as FL_1.

   $$FL_2\ 100 = FL_1\ 100 \times (CF_2/CF_1)$$
   $$= 1 \times (0.4/0.1)$$
   $$= 1 \times 4$$
   $$= 4$$

   The answer tells us that 3-in hose has 4 times the friction loss of 4-in hose.

6. The conversion factor for 3½-in hose is 0.17.

   $$Q = \sqrt{FL\ 100/(CF \times 2)}$$
   $$= \sqrt{45/(0.17 \times 2)}$$
   $$= \sqrt{45/0.34}$$
   $$= \sqrt{132.35}$$
   $$= 11.5\ \text{or 1,150 gpm (remember, } Q \text{ is in hundreds)}$$

7. Calculate the friction loss for a given flow in both 1½-in hose and 1¾-in hose. Then multiply the FL for the 1½-in hose by 1.5 and the FL for the same flow in the 1¾-in hose by 2.325.

   First, calculate the friction loss for 125 gpm in 1½-in hose.

   $$FL\ 100 = CF \times 2Q^2$$
   $$= 12 \times 2 \times (1.25)^2$$
   $$= 12 \times 2 \times 1.56$$
   $$= 37.44\ psi$$

   Total friction loss is $37.44 \times 1.5 = 56.16$, rounded to 56 psi.

   Now calculate the friction loss for 125 gpm in 1¾-in hose.

   $$FL\ 100 = CF \times 2Q^2$$
   $$= 7.76 \times 2 \times (1.25)^2$$
   $$= 7.76 \times 2 \times 1.56$$
   $$= 24.21\ psi$$

   Total friction loss is $24.21 \times 2.325 = 56.29$, rounded to 56 psi.

   This proves that Example 6-7 is correct.

8. We already know that at 11.25 psi the friction loss 3-in hose will flow 375 gpm. First calculate how much water will flow in 3½-in hose with 11.25 psi friction loss.

$$Q = \sqrt{FL\ 100/(CF \times 2)}$$
$$= \sqrt{11.25/(0.17 \times 2}$$
$$= \sqrt{11.25/0.34}$$
$$= \sqrt{33.09}$$
$$= 5.75\ or\ 575\ gpm$$

So 375 gpm for 3-in hose + 575 gpm for 3½-in hose = 950 gpm.

$$CF = FL\ 100/2Q^2$$
$$= 11.25/[2 \times (9.5)^2]$$
$$= 11.25/(2 \times 90.25)$$
$$= 11.25/180.5$$
$$= 0.062$$

Challenging Questions

1. $CF = D_1{}^5/D_2{}^5$
   $$= (1)^5/(2)^5$$
   $$= \frac{1}{32}$$

You will have about 0.03 (1 ÷ 32 = 0.03) times as much friction loss in the larger hose as in the smaller hose.

2. $FL\ 100 = CF \times 2Q^2$
   $$= 7.76 \times 2 \times (1.5)^2$$
   $$= 7.76 \times 2 \times 2.25$$
   $$= 34.92\ or\ 35\ psi$$

3. $FL\ 100 = 2 \times (3)^2$
   $$= 2 \times 9$$
   $$= 18\ psi$$

4.
   A. First find FL 100.

      125 − 89 = 36 lb total friction loss

      36 ÷ 5 = 7.2 psi friction loss is FL 100

Next calculate the flow, then the correction factor.

$$gpm = 29.84 \times D^2 \times C \times \sqrt{P}$$
$$= 29.84 \times (1.5)^2 \times 0.997 \times \sqrt{80}$$
$$= 29.84 \times 2.25 \times 0.997 \times 8.94$$
$$= 598.43\ or\ 598\ gpm$$

$$CF = FL\ 100/2Q^2$$
$$= 7.2/[2 \times (5.98)^2]$$
$$= 7.2/(2 \times 35.76)$$
$$= 7.2/71.52$$
$$= 0.1$$

   B. A correction factor of 0.1 tells us it is 4-in hose

5. $L_2 = L_1 \times (CF_1/CF_2)$
   $$= 300 \times (7.76/4)$$
   $$= 300 \times 1.94$$
   $$= 582\ or\ 550\ ft\ (hose\ comes\ in\ 50\text{-}ft\ lengths)$$

You will get an additional 250 ft of hose for the same total friction loss.

6. $FL_2\ 100 = FL_1\ 100 \times (CF_2/CF_1)$
   $$= 20 \times (7.75/1)$$
   $$= 20 \times 7.75$$
   $$= 155\ psi$$

7. $Q = \sqrt{FL\ 100/(CF \times 2)}$
   $$= \sqrt{155/(7.76 \times 2)}$$
   $$= \sqrt{155/15.52}$$
   $$= \sqrt{9.99}$$
   $$= 3.16\ or\ 316\ gpm$$

8. $Q = \sqrt{FL\ 100/(CF \times 2)}$
   $$= \sqrt{50/(7.76 \times 2)}$$
   $$= \sqrt{50/15.52}$$
   $$= \sqrt{3.22}$$
   $$= 1.79\ or\ 179\ gpm$$

Chapter 7: Pump Theory and Operation

Case Study

1. B
2. C
3. C
4. C

Review Questions

1. 1912
2. Positive displacement
3. Positive displacement pumps can pump liquids and gas; nonpositive displacement pumps can only pump liquids.
4. Impeller
5. Eye
6. To accommodate the ever-increasing volume of liquid as the discharge is approached, and to constrain the water and convert velocity to pressure.
7. 140 psi
8. Reduce the throttle
9. The formation and subsequent collapse of steam bubbles in the pump due to excessive low pressure.
10. Positive displacement
11. Testing pumps at different pressures and flows and the conversion of velocity to pressure in volute
12. Water running contrary to the direction of the main flow
13. Net pump pressure
14. Water enters through the intake to the stage 1 impeller. From there, the water is directed to the stage 2 impeller then directed out of the pump through the discharge manifold.
15. End thrust
16.
    A. Gauge
    B. Flowmeter
17. Making pumps ready, getting water, pumping, and shutting down.
18. To have a reserve of water should something go wrong
19. Less friction loss
20. Making pumps ready and getting water
21. Each time the driver changes; at a minimum, once a day.

Activities

1. Because the pump is in the series position, each impeller adds one-half of the pressure difference between the intake pressure and the discharge pressure.

   Discharge pressure − intake pressure = pressure generated by pump (net pump pressure)

   $$235 - 65 = 170 \text{ psi}$$

   Each impeller adds one-half of the total pressure generated by the pump.

   $$170 \div 2 = 85 \text{ psi}$$

   Each impeller generates 85 psi.

2. The pump is in the parallel configuration, so each impeller will develop the total pressure added by the pump.

   Discharge pressure − intake pressure = pressure generated by pump (net pump pressure)

   $$150 - 45 = 105 \text{ psi}$$

   Each impeller develops 105 psi.

3. The following experiment will prove the direction of the force that accelerates water in the impeller of a radial flow pump. Take a piece of string about 2 ft long and tie a small weight to one end. Twirl the string around just fast enough to keep the weight at the end of the string, scribing a circle. When the weight is at the top of the circle, let go of the string. In what direction did the weight go Figure C-1 ?

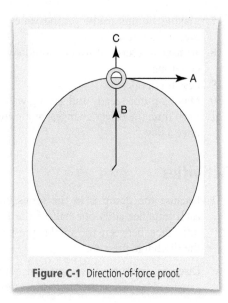

Figure C-1 Direction-of-force proof.

In Figure C-1, you can see the circle scribed by the weight at the end of the string. When the weight is released, it will fly in the direction of force on the weight. The weight will fly out in the direction indicated by A.

It was mentioned toward the beginning of this chapter that the nonpositive displacement pumps are often referred to as *centrifugal pumps*. At one time it was mistakenly believed that a force (a center-fleeing force, or centrifugal force) pushed objects out from the center of a circle, as indicated by arrow B in Figure C-1. However, if a centrifugal force were acting in the direction of B, our weight would go straight out in the direction of C when the string was released. Instead the weight follows the path indicated by A. This force is *tangential* to the direction of rotation of the weight at the end of the string. The actual force that imparts energy to water in the radial flow pump is everyday, garden variety, inertia.

4. The tank is filled at the end of step 2, getting water. The purpose is to ensure an adequate supply of onboard water in case of an emergency, such as a broken supply line.

5. When you are operating in a relay or two-pump operation, 20 psi is generally the minimum intake pressure considered safe. However, from a hydrant where you want to get every last drop of water, 5 psi is considered the minimum.
6. Bourdon tube gauges are 100 percent mechanical and are connected to the pump by thin copper or plastic tubing. Electronic gauges have pressure-sensing devices, at points of discharge, which are connected to the gauge by wires.

Challenging Questions

1. Net pump pressure = discharge pressure – intake pressure
   = 150 – 30
   = 120 psi

2. Each impeller will generate one-half of the discharge pressure, or 60 psi.
3. Each impeller will generate the entire pressure, or 120 psi.
4. 1,500 gpm at 200 psi

Chapter 8: Theory of Drafting and Pump Testing
Case Study
1. C
2. D
3. B
4. A

Review Questions

1. Atmospheric pressure pushes water into the pump.
2. The process of removing water from a static source by the influence of the atmosphere.

3. Temperature, atmospheric pressure, pump condition, weather, and friction loss on the intake side.
4. Required operational clearances around the impeller shaft have to be loose enough that the shaft can turn without restriction, yet tight enough to prevent air and water leaks.
5. 0.882 inHg
6. 20 ft
7. Hotter water will tend to vaporize as a vacuum is created.
8. To test the pump at 100 percent capacity, 70 percent capacity, and 50 percent capacity
9. 22 inHg
10. 30 psi
11.
    A. 5 min
    B. No more than 10 inHg

Activities

1. The atmospheric pressure will be 1.5 psi less than at sea level.

$$Lift = 2.31 \times pressure$$
$$= 2.31 \times 13.2$$
$$= 30.49 \text{ ft}$$

2. From Table 8-1, 1 inHg is equal to 0.489 psi.

$$Pressure = 0.489 \times Hg$$
$$= 0.489 \times 30.36$$
$$= 14.8 \text{ psi atmospheric pressure}$$

3. From Table 8-1, 1 inHg is equivalent to 1.13 ft of lift.

$$Lift = 1.13 \times Hg$$
$$= 1.13 \times 10$$
$$= 11.3 \text{ ft}$$

4. Find how much pressure it would take to push the water 6 ft up the tube and subtract it from the atmospheric pressure. Recall the formula $P = 0.433 \times H$. See Chapter 2 for a discussion of this formula.

$$P = 0.433 \times H$$
$$= 0.433 \times 6$$
$$= 2.60 \text{ psi}$$

$$\text{Internal pressure} = \text{atmospheric pressure} - 2.60 \text{ psi}$$
$$= 14.7 \text{ psi} - 2.60 \text{ psi}$$
$$= 12.10 \text{ psi}$$

5. $FL = (\text{dynamic reading} - \text{static reading}) \times 0.489$
$$= (18 - 11) \times 0.489$$
$$= 7 \times 0.489$$
$$= 3.42 \text{ psi}$$

6. The difference is that the single-stage pump cannot change configuration from full capacity to 50 percent capacity. However, with the two-stage pump, the full-capacity test and 70 percent capacity test were both done in the parallel position, but the 50 percent capacity test was done in the series position. Obviously, the two-stage pump is more efficient, pumping a little water at high pressure rather than a lot of water at lower pressure.

Challenging Questions

1. If 1 psi = 2.037 inHg, then 2.037 × 0.5 = 1.02 inHg.

2. $H = 1.13 \times Hg$
$$= 1.13 \times 7.5$$
$$= 8.48 \text{ ft of lift}$$

3. At 4,000 ft there will be 2 psi less atmospheric pressure, or 14.7 − 2 = 12.7 psi atmospheric pressure.

$$H = 2.31 \times P$$
$$= 2.31 \times 12.7$$
$$= 29.34 \text{ ft of lift}$$

4. $FL = (\text{dynamic reading} - \text{static reading}) \times 0.489$
$$= (23 - 5.5) \times 0.489$$
$$= 17.5 \times 0.489$$
$$= 8.56 \text{ psi}$$

5. $Hg = 0.882 \times H$
   $= 0.882 \times 15.5$
   $= 13.67$ inHg

6. Remaining pressure = atmospheric pressure
   $- (0.433 \times H)$
   Remaining pressure $= 14.7 - (0.433 \times 15.5)$
   Remaining pressure $= 14.7 - 6.71$
   Remaining pressure $= 7.99$ psi

7. First, find the velocity (v_2) in a device from which we can calculate the flow. In this case, a 1½-in tip will work.

   Area A of a 2½-in line $= 0.7854 \times D^2$
   $= 0.7854 \times (2½)^2$
   $= 0.7854 \times 6.25$
   $= 4.91$ in^2

   Area of a 1½-in tip is

   $A = 0.7854 \times D^2$
   $= 0.7854 \times (1½)^2$
   $= 0.7853 \times 2.25$
   $= 1.77$ in^2

   $A_1 \times v_1 = A_2 \times v_2$

   $4.91 \times 35 = 1.77 \times v_2$
   $171.85 = 1.77 \times v_2$
   $v_2 = 171.85/1.77$
   $= 97.09$ fps

   Find the pressure if the velocity is 97.09 fps.

   $P = (v/12.19)^2$
   $= (97.09/12.19)^2$
   $= (7.96)^2$
   $= 63.36$ psi

   Now find out how much water will flow from a 1½-in tip at 63.36 psi.

   gpm $= 29.84 \times D^2 \times C \times \sqrt{NP}$
   $= 29.84 \times (1½)^2 \times 0.997 \times \sqrt{63.36}$
   $= 29.84 \times 2.25 \times 0.997 \times 7.96$
   $= 532.83$ or 533 gpm maximum flow through 2½-in hose during pump test

8. $P = 0.489 \times Hg$
   $= 0.489 \times 15$
   $= 7.34$ psi

9. First convert the dynamic reading to psi.

   $P = 0.489 \times Hg$
   $= 0.489 \times 13.5$
   $= 6.60$ psi

   The pressure showing on the discharge gauge will be 200 psi − 6.60 psi, or 193.40 psi.

Chapter 9: Fire Streams
Case Study

1. C
2. B
3. A
4. C

Review Questions

1. To absorb maximum Btu
2. To achieve the best conversion of water to steam
3. Nine-tenths of the volume
4. To break the stream into small drops
5. Larger-diameter streams will put out more water, which will have greater range
6. 32°
7. 45°
8. The third law
9. Find an equivalent for D^2.
10. 30°
11. Forward pressure and back pressure are the result of elevation; friction loss is the result of water rubbing against the lining of the hose or appliance.
12. If it is back pressure, it is added to the PDP; if it is forward pressure, it is subtracted from PDP.
13. Devices such as monitor nozzles, ladder pipes, Siamese, wyes, and standpipe systems.

Activities

1. To achieve the desired 45° angle, the nozzle is placed a distance from the building equal to the desired reach. In this case the nozzle should be about 20 ft from the building.

2. Add 5 for each ⅛ in that the nozzle diameter is greater than ¾ in. (If it is not a full ⅛ in, do not add 5.)

$$HR = ½NP + 26*$$
$$= ½(50) + 31$$
$$= 25 + 31$$
$$= 56 \text{ ft}$$

3. Add 5 for each ⅛ in that the nozzle diameter is greater than ¾ in.

$$VR = ⅝NP + 26*$$
$$= ⅝(50) + 31$$
$$= 31.25 + 31$$
$$= 62.25 \text{ ft, rounded to } 62 \text{ ft}$$

4. $$NR = 1.57 \times D^2 \times P$$
$$= 1.57 \times (1.5)^2 \times 80$$
$$= 1.57 \times 2.25 \times 80$$
$$= 282.6 \text{ or } 283 \text{ lb}$$

5. $$NR = 0.0543 \times gpm \times \sqrt{P}$$
$$= 0.0543 \times 205 \times \sqrt{31.14}$$
$$= 0.0543 \times 205 \times 5.58$$
$$= 62.11 \text{ or } 62 \text{ lb}$$

6. We only need to overcome enough back pressure to get our water up to the fifth floor. Because the line will be operating at or close to the floor on the fifth floor, calculate for a full 4 floors of height, or 48 ft.

$$P = 0.433 \times H$$
$$= 0.433 \times 48$$
$$= 20.78 \text{ or } 21 \text{ psi}$$

7. First, find out how many gallons of concentrate will be used each minute.

$$60 \times 0.06 = 3.6 \text{ gpm of concentrate}$$

Second, find how long a 5-gal container of concentrate will last.

$$5/3.6 = 1.39 \text{ min or } 1 \text{ min } 23 \text{ s}$$

Finally, find how many 5-gal containers of concentrate will be needed.

$$15/1.39 = 10.79, \text{ or } 11 \text{ containers}$$

However, this problem is not yet complete. We are using fluoroprotein foam, which

requires an application rate of 0.16 gpm/ft². We still need to find how much a single line will cover. To do that, we use the formula AC = gpm/AR.

$$AC = gpm/AR$$
$$= 60/0.16$$
$$= 375 \text{ ft}^2$$

With a spill/fire area of 900 ft² and a single line only able to cover an area of 375 ft², you will need three hand lines and 33–5-gal containers to cover this fire.

Challenging Questions

1. $$HR = ½NP + 26*$$
$$= (½ \times 50) + 26*$$
$$= 25 + 36$$
$$= 61 \text{ ft}$$

2. $$VR = ⅝NP + 26*$$
$$= (⅝ \times 50) + 26*$$
$$= 31.25 + 36$$
$$= 67.25 \text{ or } 67 \text{ ft}$$

3. $$NR = 1.57 \times D^2 \times P$$
$$= 1.57 \times (⅞)^2 \times 40$$
$$= 1.57 \times 0.77 \times 40$$
$$= 48.36 \text{ or } 48 \text{ lb}$$

4. Foam used per minute = gpm × percent concentration
$$= 340 \times 0.03$$
$$= 10.2 \text{ gal of concentrate in } 1 \text{ min}$$

5. $$AC = gpm/AR$$
$$= 340/0.1$$
$$= 3,400 \text{ ft}^2 \text{ coverage}$$

6. We know from Example 9-9 that we have 30 psi available for friction loss before the elevation pressure is factored in. If the nozzle is 15 ft below the pumper, we need to add 0.433 × 15 = 6.50 psi to the allowable friction loss. We now have 36.50 psi available for the friction loss. Our hose can be 36.50/14 = 2.61, rounded down to 250 ft long.

*Add 5 for each ⅛ in of nozzle diameter greater than ¾ in.

Chapter 10: Calculating Pump Discharge Pressure

Case Study

1. D
2. C
3. D
4. C

Review Questions

1. 20 psi
2. 150 AL + 10 EP = 160 psi
3. 20 psi
4. When both lines have the same total friction loss

Activities

1. First, find the gpm for a 1-in tip at 45 psi.

   $$gpm = 29.84 \times D^2 \times C \times \sqrt{P}$$
   $$= 29.84 \times (1)^2 \times 0.97 \times \sqrt{45}$$
   $$= 29.84 \times 1 \times 0.97 \times 6.7$$
   $$= 193.93 \text{ or } 194 \text{ gpm}$$

   Now find the FL 100 for 194 gpm in 2½-in hose.

   $$FL\ 100 = CF \times 2Q^2$$
   $$= 1 \times 2 \times (1.94)^2$$
   $$= 1 \times 2 \times 3.76$$
   $$= 7.52 \text{ psi}$$

   The total friction loss is

   $$FL = FL\ 100 \times L$$
   $$= 7.52 \times 2.5$$
   $$= 18.8 \text{ or } 19 \text{ psi}$$

   Now solve for PDP.

   $$PDP = NP + FL\ 1 + FL\ 2 \pm EP + AL$$
   $$= 45 + 19 + 0 \pm 0 + 0$$
   $$= 64 \text{ psi}$$

2. Because it is a fog nozzle, the flow is already known. The first step is to calculate the friction loss in both 1¾-in hose and 2½-in hose.

   $$FL\ 100 = CF \times 2Q^2 \text{ (for the 2½-in hose)}$$
   $$= 1 \times 2 \times (1.8)^2$$
   $$= 1 \times 2 \times 3.24$$
   $$= 6.48 \text{ psi}$$

   Total friction loss for 2½-in hose is

   $$FL = FL\ 100 \times L$$
   $$= 6.48 \times 2$$
   $$= 12.96 \text{ or } 13 \text{ psi}$$

   Now find the friction loss in the 1¾-in hose.

   $$FL\ 100 = CF \times 2Q^2$$
   $$= 7.76 \times 2 \times (1.8)^2$$
   $$= 7.76 \times 2 \times 3.24$$
   $$= 50.28 \text{ psi}$$

   The total friction loss for the 1¾-in hose is:

   $$FL = FL\ 100 \times L$$
   $$= 50.28 \times 1.5$$
   $$= 75.42 \text{ or } 75 \text{ psi}$$

   Now solve for PDP.

   $$PDP = NP + FL\ 1 + FL\ 2 \pm EP + AL$$
   $$= 75 + 13 + 75 \pm 0 + 0$$
   $$= 163 \text{ psi}$$

3. First, we need to calculate the gpm being used.

   $$gpm = 29.84 \times D^2 \times C \times \sqrt{NP}$$
   $$= 29.84 \times (1.125)^2 \times 0.97 \times \sqrt{50}$$
   $$= 29.84 \times 1.27 \times 0.97 \times 7.07$$
   $$= 259.89 \text{ or } 260 \text{ gpm for 2½-in hand line}$$

   $$gpm = 29.84 \times D^2 \times C \times \sqrt{NP}$$
   $$= 29.84 \times (1.75)^2 \times 0.997 \times \sqrt{80}$$
   $$= 29.84 \times 3.06 \times 0.997 \times 8.94$$
   $$= 813.87 \text{ or } 814 \text{ gpm for the monitor nozzle}$$

   Total gpm is 260 + 814 = 1,074 gpm

   Then we calculate the PDP for pumper 1. Pumper 1 is supplying all the water being used by pumper 2. What is the average length of supply line?

300 ft (3-in hose) + 400 ft (3½-in hose) = 700 ft/2 = 350-ft average

What is the FL 100 in parallel lines of one 3-in hose and one 3½-in hose? We already found the conversion factor for one 3-in hose and one 3½-in hose to be 0.062. See Chapter 6, Activity 8 for calculation of the conversion factor.

$$
\begin{aligned}
\text{FL } 100 &= \text{CF} \times 2Q^2 \\
&= 0.062 \times 2 \times (10.74)^2 \\
&= 0.062 \times 2 \times 115.35 \\
&= 14.3 \text{ psi}
\end{aligned}
$$

Now calculate the total friction loss.

$$
\begin{aligned}
\text{FL} &= \text{FL } 100 \times L \\
&= 14.3 \times 3.5 \\
&= 50.05 \text{ or } 50 \text{ psi}
\end{aligned}
$$

Use the relay formula to calculate the PDP for pumper 1.

$$
\begin{aligned}
\text{PDP} &= 20 + \text{FL} \pm \text{EP} \\
&= 20 + 50 \pm 0 \\
&= 70 \text{ psi}
\end{aligned}
$$

To determine the PDP for pumper 2, it is necessary to calculate the PDP for the hand line and the monitor nozzle separately. Calculate the PDP for the hand line first.

$$
\begin{aligned}
\text{FL } 100 &= \text{CF} \times 2Q^2 \\
&= 1 \times 2 \times (2.6)^2 \\
&= 1 \times 2 \times 6.76 \\
&= 13.52 \text{ psi}
\end{aligned}
$$

The total friction loss for the 2½-in hose is

$$
\begin{aligned}
\text{FL} &= \text{FL } 100 \times L \\
&= 13.52 \times 1.5 \\
&= 20.28 \text{ or } 20 \text{ psi}
\end{aligned}
$$

Now calculate the PDP for this line.

$$
\begin{aligned}
\text{PDP} &= \text{NP} + \text{FL } 1 + \text{FL } 2 \pm \text{EP} + \text{AL} \\
&= 50 + 20 + 0 \pm 0 + 0 \\
&= 70 \text{ psi}
\end{aligned}
$$

Now calculate the PDP for the monitor nozzle. First, we need to calculate the friction loss. Remember that because there are two hoses of the same diameter, we only need to find the

friction loss through one 3-in hose for one-half of the total gpm.

$$
\begin{aligned}
\text{FL } 100 &= \text{CF} \times 2Q^2 \\
&= 0.4 \times 2 \times (4.07)^2 \\
&= 0.4 \times 2 \times 16.56 \\
&= 13.25 \text{ psi}
\end{aligned}
$$

Now find the total friction loss.

$$
\begin{aligned}
\text{FL} &= \text{FL } 100 \times L \\
&= 13.25 \times 2.5 \\
&= 33.13 \text{ or } 33 \text{ psi}
\end{aligned}
$$

Finally, the PDP for the monitor nozzle is

$$
\begin{aligned}
\text{PDP} &= \text{NP} + \text{FL } 1 + \text{FL } 2 \pm \text{EP} + \text{AL} \\
&= 80 + 33 + 0 \pm 0 + 20 \\
&= 133 \text{ psi}
\end{aligned}
$$

Pumper 2 will pump at 133 psi and gate back the hand line.

4. We begin by calculating the FL 100 for the 2½-in hose. By now it should be second nature to understand that we will be calculating the friction loss in the 2½-in hose for only 250 gpm.

$$
\begin{aligned}
\text{FL } 100 &= \text{CF} \times 2Q^2 \\
&= 1 \times 2 \times (2.5)^2 \\
&= 1 \times 2 \times 6.25 \\
&= 12.5 \text{ or } 13 \text{ psi}
\end{aligned}
$$

This amount is also the total friction loss because the lines are only 100 ft long.

Note: The nozzle pressure will be 100 psi because it is a fog nozzle.

The appliance loss allowance is 50 psi and includes the Siamese, 100 ft of hose from the Siamese to the ladder pipe, and the ladder pipe itself. Calculate the elevation at ½ lb/ft of extension or 35 psi.

$$
\begin{aligned}
\text{PDP} &= \text{NP} + \text{FL } 1 + \text{FL } 2 \pm \text{EP} + \text{AL} \\
&= 100 + 13 + 0 + 35 + 50 \\
&= 198 \text{ psi}
\end{aligned}
$$

5. This question must be broken down into two parts. The first is to find the friction loss for 400 ft of parallel 3-in lines, and the second is

to find the friction loss for 200 ft of parallel lines, one 3-in and the other 2½-in.

The friction loss for 400 ft of parallel 3-in lines is (each line is flowing 500 gpm)

$$FL\ 100 = CF \times 2Q^2$$
$$= 0.4 \times 2 \times (5)^2$$
$$= 0.4 \times 2 \times 25$$
$$= 20\ psi$$

$$FL = FL\ 100 \times L$$
$$= 20 \times 4$$
$$= 80\ psi$$

Now find the friction loss for 2½-in and 3-in parallel lines flowing 1,000 gpm.

$$FL = CF \times 2Q^2$$
$$= 0.15 \times 2 \times (10)^2$$
$$= 0.15 \times 2 \times 100$$
$$= 30\ psi$$

$$FL = FL\ 100 \times L$$
$$= 30 \times 2$$
$$= 60\ psi$$

The total friction loss is 80 + 60 = 140 psi.

Use the relay formula to find the PDP.

$$PDP = 20 + FL \pm EP$$
$$= 20 + 140 \pm 0$$
$$= 160\ psi$$

Challenging Questions

1. First find the friction loss:

$$FL\ 100 = CF \times 2Q^2$$
$$= 7.76 \times 2 \times (1.5)^2$$
$$= 7.76 \times 2 \times 2.25$$
$$= 34.92\ psi$$

$$FL = FL\ 100 \times L$$
$$= 34.92 \times 2$$
$$= 69.84\ or\ 70\ psi\ is\ the\ total\ friction\ loss$$

Now find the PDP.

$$PDP = NP + FL\ 1 + FL\ 2 \pm EP + AL$$
$$= 75 + 70 + 0 \pm 0 + 0$$
$$= 145\ psi$$

2. First find the gpm.

$$gpm = 29.84 \times D^2 \times C \times \sqrt{NP}$$
$$= 29.84 \times (1)^2 \times 0.97 \times \sqrt{50}$$
$$= 29.84 \times 1 \times 0.97 \times 7.07$$
$$= 204.64\ or\ 205\ gpm$$

Now find the friction loss.

$$FL\ 100 = CF \times 2Q^2$$
$$= 1 \times 2 \times (2.05)^2$$
$$= 1 \times 2 \times 4.2$$
$$= 8.4\ psi$$

$$FL = FL\ 100 \times L$$
$$= 8.4 \times 3$$
$$= 25.2\ or\ 25\ psi$$

Now find the PDP.

$$PDP = NP + FL\ 1 + FL\ 2 \pm EP + AL$$
$$= 50 + 25 + 0 + 15 + 0$$
$$= 90\ psi$$

3. First find the friction loss in the 2½-in hose:

$$FL\ 100 = CF \times 2Q^2$$
$$= 1 \times 2 \times (1.25)^2$$
$$= 1 \times 2 \times 1.56$$
$$= 3.12\ psi$$

$$FL = FL\ 100 \times L$$
$$= 3.12 \times 2$$
$$= 6.24\ or\ 6\ psi$$

This is FL 1.

Now find the friction loss in the 1¾-in hose:

$$FL\ 100 = CF \times 2Q^2$$
$$= 7.76 \times 2 \times (1.25)^2$$
$$= 7.76 \times 2 \times 1.56$$
$$= 24.21\ psi$$

Since this is also the total friction loss, FL 2 is 24 psi.

EP is calculated at one-half the building height, rounded to 13. Now find the PDP.

$$PDP = NP + FL\ 1 + FL\ 2 \pm EP + AL$$
$$= 100 + 6 + 24 + 13 + 0$$
$$= 143\ psi$$

4. First find the gpm:

$$\text{gpm} = 29.84 \times D^2 \times C \times \sqrt{NP}$$
$$= 29.84 \times (1\tfrac{5}{8})^2 \times 0.997 \times \sqrt{80}$$
$$= 29.84 \times 2.64 \times 0.997 \times 8.94$$
$$= 702.16 \text{ or } 702 \text{ gpm}$$

Now find the friction loss for 3-in hose. (Remember, you only need one-half of the flow.)

$$FL\ 100 = CF \times 2Q^2$$
$$= 0.4 \times 2 \times (3.51)^2$$
$$= 0.4 \times 2 \times 12.32$$
$$= 9.86 \text{ psi}$$

$$FL = FL\ 100 \times L \quad \text{(Remember to use average length of lines.)}$$
$$= 9.86 \times 3.25$$
$$= 32.05 \text{ or } 32 \text{ psi}$$

Now find the PDP.

$$PDP = NP + FL\ 1 + FL\ 2 \pm EP + AL$$
$$= 80 + 32 + 0 + 15 + 20$$
$$= 147 \text{ psi}$$

5. Simply recalculate the friction loss.

$$FL\ 100 = CF \times 2Q^2$$
$$= 0.15 \times 2 \times (7.02)^2$$
$$= 0.15 \times 2 \times 49.28$$
$$= 14.78 \text{ psi}$$

$$FL = FL\ 100 \times L$$
$$= 14.78 \times 3.25$$
$$= 48.04 \text{ or } 48 \text{ psi}$$

Now find the PDP.

$$PDP = NP + FL\ 1 + FL\ 2 \pm EP + AL$$
$$= 80 + 48 + 0 + 5 + 20$$
$$= 153 \text{ psi}$$

6. Find the friction loss.

$$FL\ 100 = CF \times 2Q^2$$
$$= 0.1 \times 2 \times (7.02)^2$$
$$= 0.1 \times 2 \times 49.28$$
$$= 9.86 \text{ psi}$$

$$FL = FL\ 100 \times L$$
$$= 9.86 \times 6$$
$$= 59.16 \text{ or } 59 \text{ psi}$$

Now find the PDP, using the relay formula.

$$PDP = 20 + FL \pm EP$$
$$= 20 + 59 \pm 0$$
$$= 79 \text{ psi}$$

7. Find the friction loss.

$$FL\ 100 = CF \times 2Q^2$$
$$= 7.76 \times 2 \times (1.8)^2$$
$$= 50.28$$

$$FL = FL\ 100 \times L$$
$$= 50.28 \times 2$$
$$= 100.56 \text{ or } 101 \text{ psi}$$

Now find the PDP.

$$PDP = NP + FL\ 1 + FL\ 2 \pm EP + AL$$
$$= 100 + 101 + 0 + 15 + 0$$
$$= 216 \text{ psi}$$

8. First find the gpm flow.

$$\text{gpm} = 29.84 \times D^2 \times C \times \sqrt{NP}$$
$$= 29.84 \times (^{15}\!/_{16})^2 \times 0.97 \times \sqrt{45}$$
$$= 29.84 \times (0.94)^2 \times 0.97 \times 6.71$$
$$= 29.84 \times 0.88 \times 0.97 \times 6.71$$
$$= 170.91 \text{ or } 171 \text{ gpm}$$

Now find the friction loss.

$$FL\ 100 = CF \times 2Q^2$$
$$= 7.76 \times 2 \times (1.71)^2$$
$$= 7.76 \times 2 \times 2.92$$
$$= 45.32 \text{ psi}$$

$$FL = FL\ 100 \times L$$
$$= 45.32 \times 4$$
$$= 181.28 \text{ or } 181 \text{ psi}$$

Now find the PDP.

$$PDP = NP + FL\ 1 + FL\ 2 \pm EP + AL$$
$$= 45 + 181 + 0 + 20 + 0$$
$$= 246 \text{ psi}$$

9. We already calculated the flow from a 1-in tip at 50 psi nozzle pressure to be 205 gpm, and we are given the flow from the 1¾-in line at 150 gpm. We need only to find the flow from the 1½-in tip at 80 psi and calculate the total flow.

$$\begin{aligned}
\text{gpm} &= 29.84 \times D^2 \times C \times \sqrt{NP} \\
&= 29.84 \times (1½)^2 \times 0.997 \times \sqrt{80} \\
&= 29.84 \times 2.25 \times 0.997 \times 8.94 \\
&= 598.43 \text{ or } 598 \text{ gpm}
\end{aligned}$$

Total flow is 598 + 205 + 150 = 953 gpm. Now find the friction loss.

$$\begin{aligned}
\text{FL } 100 &= CF \times 2Q^2 \\
&= 0.062 \times 2 \times (9.53)^2 \\
&= 0.062 \times 2 \times 90.82 \\
&= 11.26 \text{ psi}
\end{aligned}$$

$$\begin{aligned}
\text{FL} &= \text{FL } 100 \times L \\
&= 11.26 \times 7.5 \\
&= 84.45 \text{ or } 84 \text{ psi}
\end{aligned}$$

Now calculate the PDP, using the relay formula.

$$\begin{aligned}
\text{PDP} &= 20 + \text{FL} \pm \text{EP} \\
&= 20 + 84 \pm 0 \\
&= 104 \text{ psi}
\end{aligned}$$

10. First find the friction loss for the 2½-in hose.

$$\begin{aligned}
\text{FL } 100 &= CF \times 2Q^2 \\
&= 1 \times 2 \times (0.95)^2 \\
&= 1 \times 2 \times 0.90 \\
&= 1.80 \text{ psi}
\end{aligned}$$

$$\begin{aligned}
\text{FL} &= \text{FL } 100 \times L \\
&= 1.80 \times 2.5 \\
&= 4.50 \text{ or } 5 \text{ psi}
\end{aligned}$$

Now find the elevation pressure: 15 × 0.5 = 7.5 or 8 psi. Since it is below the level of the pumper, it will be subtracted in the PDP formula. Now calculate the PDP.

$$\begin{aligned}
\text{PDP} &= \text{NP} + \text{FL } 1 + \text{FL } 2 \pm \text{EP} + \text{AL} \\
&= 0 + 5 + 0 - 8 + 200 \\
&= 197 \text{ psi}
\end{aligned}$$

11. First find the friction loss in the 2½-in hose.

$$\begin{aligned}
\text{FL } 100 &= CF \times 2Q^2 \\
&= 1 \times 2 \times (2.25)^2 \\
&= 1 \times 2 \times 5.06 \\
&= 10.12 \text{ psi}
\end{aligned}$$

Since 100 ft of 2½-in is supplying the standpipe, we will call FL 1 10 psi.

$$\begin{aligned}
\text{FL} &= \text{FL } 100 \times L \\
&= 10.12 \times 1.5 \\
&= 15.18 \text{ or } 15 \text{ psi}
\end{aligned}$$

This is FL 2. Now find the PDP. (Use 5 psi per floor for EP.)

$$\begin{aligned}
\text{PDP} &= \text{NP} + \text{FL } 1 + \text{FL } 2 \pm \text{EP} + \text{AL} \\
&= 100 + 10 + 15 + 50 + 25 \\
&= 200 \text{ psi}
\end{aligned}$$

12. First find the gpm from 1¼-in tip at 45 psi.

$$\begin{aligned}
\text{gpm} &= 29.84 \times D^2 \times C \times \sqrt{NP} \\
&= 29.84 \times (1¼)^2 \times 0.97 \times \sqrt{45} \\
&= 29.84 \times 1.56 \times 0.97 \times 6.71 \\
&= 302.98 \text{ or } 303 \text{ gpm}
\end{aligned}$$

Now find the friction loss.

$$\begin{aligned}
\text{FL } 100 &= CF \times 2Q^2 \\
&= 1 \times 2 \times (3.03)^2 \\
&= 1 \times 2 \times 9.18 \\
&= 18.36 \text{ psi}
\end{aligned}$$

$$\begin{aligned}
\text{FL} &= \text{FL } 100 \times L \\
&= 18.36 \times 2.5 \\
&= 45.9 \text{ or } 46 \text{ psi}
\end{aligned}$$

Now find the PDP.

$$\begin{aligned}
\text{PDP} &= \text{EP} + \text{FL } 1 + \text{FL } 2 \pm \text{EP} + \text{AL} \\
&= 45 + 46 + 0 \pm 0 + 0 \\
&= 91 \text{ psi}
\end{aligned}$$

13. From Challenging Question 9 we know the flow from a 1½-in tip at 80 psi is 598 gpm. Now find the friction loss.

$$\begin{aligned}
\text{FL } 100 &= CF \times 2Q^2 \\
&= 0.15 \times 2 \times (5.98)^2 \\
&= 0.15 \times 2 \times 35.76 \\
&= 10.73 \text{ psi}
\end{aligned}$$

$$\begin{aligned}
\text{FL} &= \text{FL } 100 \times L \quad \text{(Remember to average the length of the lines.)} \\
&= 10.73 \times 1.75 \\
&= 18.78 \text{ or } 19 \text{ psi}
\end{aligned}$$

Now find the PDP.

PDP = NP + FL 1 + FL 2 ± EP + AL
= 80 + 19 + 0 + 35 + 50
= 184 psi

14. First find the friction loss.

FL 100 = CF × $2Q^2$
= 7.76 × 2 × $(1.25)^2$
= 7.76 × 2 × 1.56
= 24.21 psi

FL = FL 100 × L
= 24.21 × 3
= 72.63 or 73 psi

Now find the PDP. (EP is 5 psi for each story above the pumper.)

PDP = NP + FL 1 + FL 2 ± EP + AL
= 100 + 73 + 0 + 10 + 0
= 183 psi

15. First, find the friction loss.

FL 100 = CF × $2Q^2$
= 1 × 2 × $(2.5)^2$
= 1 × 2 × 6.25
= 12.5 psi

Since the standpipe is being supplied by 100 ft of 2½-in hose, we will call FL 1 13 psi.

FL = FL 100 × L
= 12.5 × 1.5
= 18.75 or 19 psi

This is FL 2. Now find the PDP (remember to calculate EP for the location of the nozzle).

PDP = NP + FL 1 + FL 2 ± EP + AL
= 75 + 13 + 19 + 10 + 25
= 142 psi

16. First find the friction loss. In Challenging Question 10, we found the FL 100 (1.8 psi). Now find the total friction loss.

FL = FL 100 × L
= 1.8 × 3.5
= 6.3 or 6 psi

Now find the PDP. (Elevation pressure will be a negative number since it is below the pumper.)

PDP = NP + FL 1 + FL 2 ± EP + AL
= 0 + 6 + 0 − 5 + 200
= 201 psi

17. First find the gpm.

gpm = 29.84 × D^2 × C × $\sqrt{NP}$
= 29.84 × $(1¼)^2$ × 0.97 × $\sqrt{50}$
= 29.84 × 1.56 × 0.97 × 7.07
= 319.24 or 319 gpm

Now find the friction loss.

FL 100 = CF × $2Q^2$
= 1 × 2 × $(3.19)^2$
= 1 × 2 × 10.18
= 20.36

FL = FL 100 × L
= 20.36 × 2
= 40.72 or 41 psi

Now find the PDP. (Did you remember to calculate elevation pressure to the roof of the three-story building?)

PDP = NP + FL 1 + FL 2 ± EP + AL
= 50 + 41 + 0 + 15 + 0
= 106 psi

18. First find the gpm.

gpm = 29.84 × D^2 × C × $\sqrt{NP}$
= 29.84 × $(1¾)^2$ × 0.997 × $\sqrt{80}$
= 29.84 × 3.06 × 0.997 × 8.94
= 813.87 or 814 gpm

Now find friction loss for pumper 1.

FL 100 = CF × $2Q^2$
= 0.4 × 2 × $(4.07)^2$ (Remember, you only need to find the flow through one line.)
= 0.4 × 2 × 16.56
= 13.24 psi

FL = FL 100 × L
= 13.24 × 3
= 39.72 or 40 psi

PDP for pumper 1 is

$$PDP = 20 + FL \pm EP$$
$$= 20 + 40 \pm 0$$
$$= 60 \text{ psi}$$

Now find the friction loss for pumper 2.

$$FL\ 100 = CF \times 2Q^2$$
$$= 0.1 \times 2 \times (8.14)^2$$
$$= 0.1 \times 2 \times 66.26$$
$$= 13.25 \text{ psi}$$

$$FL = FL\ 100 \times L$$
$$= 13.25 \times 10$$
$$= 132.5 \text{ or } 133 \text{ psi}$$

Finally, find PDP for pumper 2.

$$PDP = 20 + 133 \pm 0$$
$$= 153 \text{ psi}$$

Chapter 11: Advanced Problems in Hydraulics

Case Study

1. C
2. B
3. A
4. C

Review Questions

1. Each pumper will pump a proportional share of the water. That is, with two pumpers each will supply one-half of the water, with three pumpers each will supply one-third of the water, and so on.
2. To maintain hydraulic balance.
3. Determine the gpm needed.
4. Determine the length and size of hose.

Activities

1. We first need to determine how much total water is flowing. We are given the gpm of the 1½-in hose line as 125 gpm, and we calculated the gpm for the ¹⁵⁄₁₆-in tip at 50 psi

nozzle pressure to be 184 gpm. See Chapter 5, Activity 2, for the calculation of 184 gpm.

$$125 + 184 = 309 \text{ gpm total}$$

Calculate the friction loss for the 2½-in hose.

$$FL\ 100 = CF \times 2Q^2$$
$$= 1 \times 2 \times (3.09)^2$$
$$= 1 \times 2 \times 9.55$$
$$= 19.10 \text{ psi}$$

$$FL = FL\ 100 \times L$$
$$= 19.10 \times 3.5$$
$$= 66.85 \text{ or } 67 \text{ psi}$$

The friction loss for the 1½-in hose is

$$FL\ 100 = CF \times 2Q^2$$
$$= 12 \times 2 \times (1.25)^2$$
$$= 12 \times 2 \times 1.56$$
$$= 37.44 \text{ psi}$$

$$FL = FL\ 100 \times L$$
$$= 37.44 \times 1.5$$
$$= 56.16 \text{ or } 56 \text{ psi}$$

The friction loss for the 1¾-in hose is

$$FL\ 100 = CF \times 2Q^2$$
$$= 7.76 \times 2 \times (1.84)^2$$
$$= 7.76 \times 2 \times 3.39$$
$$= 52.61 \text{ psi}$$

$$FL = FL\ 100 \times L$$
$$= 52.61 \times 2$$
$$= 105.22 \text{ or } 105 \text{ psi}$$

Now calculate the PDP.

For the 1½-in hose:

$$PDP = NP + FL\ 1 + FL\ 2 \pm EP + AL$$
$$= 100 + 67 + 56 \pm 0 + 0$$
$$= 223 \text{ psi}$$

For the 1¾-in hose:

$$PDP = NP + FL\ 1 + FL\ 2 \pm EP + AL$$
$$= 50 + 67 + 105 \pm 0 + 0$$
$$= 222 \text{ psi}$$

You would pump at 223 psi.

2. The size of line that the other pumper is using is of no concern to you. Because you are supplying one of two lines into the Siamese, you

will supply one-half of the water needed for the nozzle. Calculate PDP accordingly.

First calculate the friction loss for the 3-in hose.

$$FL\ 100 = CF \times 2Q^2$$
$$= 0.4 \times 2 \times (5)^2$$
$$= 0.4 \times 2 \times 25$$
$$= 20\ \text{psi}$$

$$FL = FL\ 100 \times L$$
$$= 20 \times 2.5$$
$$= 50\ \text{psi}$$

Find the elevation.

Use one-half of the extension of the ladder, or 35 psi.

Now find the PDP.

$$PDP = NP + FL\ 1 + FL\ 2 \pm EP + AL$$
$$= 100 + 50 + 0 + 35 + 50$$
$$= 235\ \text{psi}$$

3. Earlier we calculated the CF for one 2½-in and one 3-in hose as 0.15. See Chapter 6.

$$FL\ 100 = CF \times 2Q^2$$
$$= 0.15 \times 2 \times (7)^2$$
$$= 0.15 \times 2 \times 49$$
$$= 14.7\ \text{psi}$$

To deliver 70 percent of the capacity of a 1,000 gpm capacity pump, the maximum pressure is limited to 200 psi.

$$200 - 20/14.7 = 12.24\ \text{or}\ 1,224\ \text{ft}.$$

This must be rounded down to 1,200 ft.

4. Begin by subtracting the intake pressure needed for the pumper from the static pressure of the hydrant: $75 - 20 = 55$ psi, the maximum amount of pressure you have available for the friction loss. $55 \div 5 = 11$. This is the allowable friction loss per 100 ft.

$$Q = \sqrt{FL\ 100/(CF \times 2)}$$
$$= \sqrt{11/(0.1 \times 2)}$$
$$= \sqrt{11/0.2}$$
$$= \sqrt{55}$$
$$= 7.42\ \text{or}\ 742\ \text{gpm}$$

5. You must begin by calculating the PDP for the 350-ft line. This answer is the incorrect PDP. First, calculate the gpm at 50 psi nozzle pressure.

$$gpm = 29.84 \times D^2 \times C \times \sqrt{P}$$
$$= 29.84 \times (1.125)^2 \times 0.97 \times \sqrt{50}$$
$$= 29.84 \times 1.27 \times 0.97 \times 7.07$$
$$= 259.89\ \text{or}\ 260\ \text{gpm}$$

Now find the friction loss for this flow.

$$FL\ 100 = CF \times 2Q^2$$
$$= 1 \times 2 \times (2.6)^2$$
$$= 1 \times 2 \times 6.76$$
$$= 13.52\ \text{psi}$$

$$FL = FL\ 100 \times L$$
$$= 13.52 \times 3.5$$
$$= 47.32\ \text{or}\ 47\ \text{psi}$$

The PDP is then

$$PDP = NP + FL\ 1 + FL\ 2 \pm EP + AL$$
$$= 50 + 47 + 0 \pm 0 + 0$$
$$= 97\ \text{psi}$$

Now calculate what the correct PDP should have been.

$$FL = FL\ 100 \times L$$
$$= 13.52 \times 5$$
$$= 67.6\ \text{or}\ 68\ \text{psi}$$

$$PDP = NP + FL\ 1 + FL\ 2 \pm EP + AL$$
$$= 50 + 68 + 0 \pm 0 + 0$$
$$= 118\ \text{psi}$$

Find the incorrect nozzle pressure (INP).

$$CNP/CPDP = INP/IPDP$$
$$50/118 = INP/97$$
$$(50 \times 97)/118 = INP$$
$$4,850/118 = INP$$
$$INP = 41.1\ \text{psi}$$

Now calculate the gpm for a 1⅛-in tip with 41.1 psi nozzle pressure.

$$gpm = 29.84 \times D^2 \times C \times \sqrt{P}$$
$$= 29.84 \times (1.125)^2 \times 0.97 \times \sqrt{41.1}$$
$$= 29.84 \times 1.27 \times 0.97 \times 6.41$$
$$= 235.63\ \text{or}\ 236\ \text{gpm}$$

Challenging Questions

1. First calculate the gpm.

$$\text{gpm} = 29.84 \times D^2 \times C \times \sqrt{NP}$$
$$= 29.84 \times (2)^2 \times 0.997 \times \sqrt{80}$$
$$= 29.84 \times 4 \times 0.997 \times 8.94$$
$$= 1{,}063.88 \text{ or } 1{,}064 \text{ gpm}$$

Now calculate the friction loss for pumper 1. Pumper 1 will supply one-third of the total water.

$$\text{FL } 100 = CF \times 2Q^2$$
$$= 0.4 \times 2 \times (3.55)^2$$
$$= 0.4 \times 2 \times 12.6$$
$$= 10.08 \text{ psi}$$

$$\text{FL} = \text{FL } 100 \times L$$
$$= 10.08 \times 3.5$$
$$= 35.28 \text{ or } 35 \text{ psi}$$

Find the PDP for pumper 1.

$$\text{PDP} = NP + FL\ 1 + FL\ 2 \pm EP + AL$$
$$= 80 + 35 + 0 \pm 0 + 20$$
$$= 135 \text{ psi}$$

Repeat the process for pumper 2, but remember that pumper 2 is supplying two lines to the monitor.

$$\text{FL } 100 = CF \times 2Q^2$$
$$= 0.15 \times 2 \times (7.1)^2$$
$$= 0.15 \times 2 \times 50.41$$
$$= 15.13 \text{ psi}$$

$$\text{FL} = \text{FL } 100 \times L$$
$$= 15.13 \times 3.25 \quad \text{(remember to average}$$
$$\text{length of lines)}$$
$$= 49.17 \text{ or } 49 \text{ psi}$$

Now find the PDP for pumper 2.

$$\text{PDP} = NP + FL\ 1 + FL\ 2 \pm EP + AL$$
$$= 80 + 49 + 0 \pm 0 + 20$$
$$= 149 \text{ psi}$$

2. First find the total gpm. Since we already know that a 1-in tip at 50 psi will deliver 205 gpm, we need only find the flow for a ⅞-in tip at 45 psi.

$$\text{gpm} = 29.84 \times D^2 \times C \times \sqrt{NP}$$
$$= 29.84 \times (⅞)^2 \times 0.97 \times \sqrt{45}$$
$$= 29.84 \times 0.77 \times 0.97 \times 6.71$$
$$= 149.55 \text{ or } 150 \text{ gpm}$$

Total gpm will be 205 + 150 = 355. Now find the friction loss for the 2½-in hose supplying the wye.

$$\text{FL } 100 = CF \times 2Q^2$$
$$= 1 \times 2 \times (3.55)^2$$
$$= 1 \times 2 \times 12.6$$
$$= 25.2 \text{ psi}$$

$$\text{FL} = \text{FL } 100 \times L$$
$$= 25.2 \times 3$$
$$= 75.6 \text{ or } 76 \text{ psi}$$

This is FL 1.

From Chapter 10, Challenging Question 2 we know the FL 100 for 2½-in hose flowing 205 gpm is 8.4 psi. Since the 2½-in hand line is 100 ft long, FL 2 for the 2½-in side of the wye is 8 psi. Now calculate the PDP for the 2½-in hand line.

$$\text{PDP} = NP + FL\ 1 + FL\ 2 \pm EP + AL$$
$$= 50 + 76 + 8 + 0 + 10 \quad \text{(Since the}$$
$$\text{flow is greater than 350, add 10 psi}$$
$$\text{for the wye.)}$$
$$= 144 \text{ psi for 2½-in side of wye}$$

Now calculate FL 2 for the 1¾-in side of the wye,

$$\text{FL } 100 = CF \times 2Q^2$$
$$= 7.76 \times 2 \times (1.5)^2$$
$$= 7.76 \times 2 \times 2.25$$
$$= 34.92 \text{ psi}$$

$$\text{FL} = \text{FL } 100 \times L$$
$$= 34.92 \times 1.5$$
$$= 52.38 \text{ or } 52 \text{ psi}$$

This is FL 2. Now find the PDP for the 1¾-in side of the wye.

$$\text{PDP} = NP + FL\ 1 + FL\ 2 \pm EP + AL$$
$$= 45 + 76 + 52 \pm 0 + 0$$
$$= 173 \text{ psi}$$

This is the correct PDP for this evolution.

3. Total flow for this evolution is 200 gpm. Find the friction loss in the 2½-in hose.

   $$FL\ 100 = CF \times 2Q^2$$
   $$= 1 \times 2 \times (2)^2$$
   $$= 1 \times 2 \times 4$$
   $$= 8\ psi$$

   $$FL = FL\ 100 \times L$$
   $$= 8 \times 2.5$$
   $$= 20\ psi$$

   This is FL 1. Now find the friction loss for the 1½-in hose.

   $$FL\ 100 = CF \times 2Q^2$$
   $$= 12 \times 2 \times (1)^2$$
   $$= 12 \times 2 \times 1$$
   $$= 24\ psi$$

   Since the 1½-in lines are 100 ft long, FL 2 is 24 psi. Find the PDP.

   $$PDP = NP + FL\ 1 + FL\ 2 \pm EP + AL$$
   $$= 100 + 20 + 24 + 10 + 0$$
   $$= 154\ psi$$

4. Begin by finding the FL 100.

   $$FL\ 100 = CF \times 2Q^2$$
   $$= 0.062 \times 2 \times (13)^2$$
   $$= 0.062 \times 2 \times 169$$
   $$= 20.96\ psi$$

   The maximum pressure is 150 psi, since 1,300 gpm is more than 70 percent of the pump capacity.

   Maximum pressure available for friction loss is $150 - 20 = 130$ psi.

   Maximum lay for each pumper is $130 \div 20.96 = 6.20$, or 620 ft rounded to 600 ft. Now $1,500 \div 600 = 2.5$, so 3 pumpers will be needed to complete the lay. With one pumper at the water source and the last piece laying out remaining at the scene, you will need a total of four pumpers.

5. First find the FL 100.

   $$FL\ 100 = CF \times 2Q^2$$
   $$= 0.1 \times 2 \times (7)^2$$
   $$= 0.1 \times 2 \times 49$$
   $$= 9.8\ psi$$

   With a static pressure of 70 psi and since we need 20 psi for intake only, 50 psi is available for overcoming the friction loss. Now calculate the maximum lay.

   $$50 \div 9.8 = 5.1\ or\ 500\text{-ft maximum lay}$$

6. First find the amount of pressure available for friction loss.

   $$60 - 20 = 40\ psi\ available\ for\ the\ friction\ loss.$$

   $$40 \div 5 = 8\ psi.\ This\ is\ the\ FL\ 100.$$

   Now find the flow in 4-in hose at FL 100 = 8 psi.

   $$Q = \sqrt{FL\ 100/(CF \times 2)}$$
   $$= \sqrt{8/(0.1 \times 2)}$$
   $$= \sqrt{8/0.2}$$
   $$= \sqrt{40}$$
   $$= 6.32\ or\ 632\ gpm$$

7. Since you are using LDH, your maximum pressure is 185 psi. Allowing 20 psi for intake pressure, there is 165 psi available for friction loss. Find the FL 100.

   $$165 \div 5 = 33\ psi$$

   Now find the flow.

   $$Q = \sqrt{FL\ 100/(CF \times 2)}$$
   $$= \sqrt{33/(0.1 \times 2)}$$
   $$= \sqrt{33/0.2}$$
   $$= \sqrt{165}$$
   $$= 12.85\ or\ 1,285\ gpm$$

8. Find the friction loss.

   $$FL\ 100 = CF \times 2Q^2$$
   $$= 7.76 \times 2 \times (1.5)^2$$
   $$= 7.76 \times 2 \times 2.25$$
   $$= 34.92\ psi$$

   $$FL = FL\ 100 \times L$$
   $$= 34.92 \times 2$$
   $$= 69.84\ or\ 70\ psi$$

Now find the PDP that you were pumping. This is the incorrect PDP or IPDP.

PDP = NP + FL 1 + FL 2 ± EP + AL
$\quad$ = 100 + 70 + 0 ± 0 + 0
$\quad$ = 170 psi is the IPDP

Now calculate what the PDP should have been. This is the correct PDP or CPDP.

PDP = NP + FL 1 + FL 2 ± EP + AL
$\quad$ = 75 + 70 + 0 ± 0 + 0
$\quad$ = 145 psi is the CPDP

Now find the incorrect nozzle pressure, or INP.

$\quad$ CNP/CPDP = INP/IPDP
$\quad\quad$ 75/145 = INP/170
(75 × 170)/145 = INP
$\quad$ 12,750/145 = INP
$\quad\quad\quad\quad\quad$ INP = 87.93 or 88 psi

Subtract 88 from 170 to find out how much pressure was available for the friction loss.

170 − 88 = 82 psi

With a 200-ft line, this gives us an FL 100 of 41 psi.

Finally, calculate the actual flow.

$$Q = \sqrt{FL\ 100/(CF \times 2)}$$
$$= \sqrt{41/(7.76 \times 2)}$$
$$= \sqrt{41/15.52}$$
$$= \sqrt{2.64}$$
$$= 1.62 \text{ or } 162 \text{ gpm}$$

9. First find the gpm of a 1¾-in tip. Earlier we calculated it to be 814 gpm. See Chapter 10, Challenging Question 18.

Now find the friction loss for parallel 2½-in lines, 350 ft long.

FL 100 = CF × 2Q^2
$\quad\quad$ = 1 × 2 × (4.07)^2 $\quad$ (Remember, each
$\quad\quad\quad\quad$ line flows one-half the water.)
$\quad\quad$ = 1 × 2 × 16.56
$\quad\quad$ = 33.12 psi

FL = FL 100 × L
$\quad$ = 33.12 × 3.5
$\quad$ = 115.92 or 116 psi

Now calculate the PDP; this will be the IPDP.

PDP = NP + FL 1 + FL 2 ± EP + AL
$\quad$ = 80 + 116 + 0 ± 0 + 20
$\quad$ = 216 psi

Now calculate the friction loss for parallel lines of 2½-in and 3 in.

FL 100 = CF × 2Q^2
$\quad\quad$ = 0.15 × 2 × (8.14)^2
$\quad\quad$ = 0.15 × 2 × 66.26
$\quad\quad$ = 19.88 psi

FL = FL 100 × L
$\quad$ = 19.88 × 3
$\quad$ = 59.64 or 60 psi

We will call this FL 2.

For FL 1 we will use the FL 100 of the 2½-in hose calculated above, 33.12 or 33 psi.

Calculate what the PDP should have been, or the CPDP.

PDP = NP + FL 1 + FL 2 ± EP + AL
$\quad$ = 80 + 33 + 60 ± 0 + 20
$\quad$ = 193 psi

Now find the INP.

$\quad$ CNP/CPDP = INP/IPDP
$\quad\quad$ 80/193 = INP/216
(80 × 216)/193 = INP
89.53 or 90 psi

Now find the flow in a 1¾-in tip at 90 psi nozzle pressure.

gpm = 29.84 × D^2 × C × $\sqrt{NP}$
$\quad$ = 29.84 × (1¾)^2 × 0.997 × $\sqrt{90}$
$\quad$ = 29.84 × 3.06 × 0.997 × 9.49
$\quad$ = 863.94 or 864 gpm

Total friction loss is 216 − 90 − 20 = 106 psi.

10. Find the gpm.

    $$gpm = 29.84 \times D^2 \times C \times \sqrt{NP}$$
    $$= 29.84 \times (2)^2 \times C \times \sqrt{75}$$
    $$= 29.84 \times 4 \times 0.997 \times 8.66$$
    $$= 1{,}030.56 \text{ or } 1{,}031 \text{ gpm}$$

    A. Since pumper 1 has two lines into the monitor, it is supplying two-thirds of the water. Find the friction loss.

    $$FL\ 100 = CF \times 2Q^2$$
    $$= 0.15 \times 2 \times (6.88)^2$$
    $$= 0.15 \times 2 \times 47.33$$
    $$= 14.2 \text{ psi}$$

    $$FL = FL\ 100 \times L$$
    $$= 14.2 \times 2.5$$
    $$= 35.5 \text{ or } 36 \text{ psi}$$

    Now calculate the PDP for pumper 1.

    $$PDP = NP + FL\ 1 + FL\ 2 \pm EP + AL$$
    $$= 75 + 36 + 0 \pm 0 + 20$$
    $$= 131 \text{ psi}$$

    B. Now find the friction loss for pumper 2.

    $$FL\ 100 = CF \times 2Q^2$$
    $$= 0.4 \times 2 \times (3.44)^2$$
    $$= 0.4 \times 2 \times 11.83$$
    $$= 9.46 \text{ psi}$$

    $$FL = FL\ 100 \times L$$
    $$= 9.46 \times 4$$
    $$= 37.84 \text{ or } 38 \text{ psi}$$

    Now calculate the PDP for pumper 2.

    $$PDP = NP + FL\ 1 + FL\ 2 \pm EP + AL$$
    $$= 75 + 38 + 0 \pm 0 + 20$$
    $$= 133 \text{ psi}$$

11. Find the friction loss in the 3-in hose.

    $$FL\ 100 = CF \times 2Q^2$$
    $$= 0.4 \times 2 \times (1.9)^2$$
    $$= 0.4 \times 2 \times 3.61$$
    $$= 2.89 \text{ psi}$$

    $$FL = FL\ 100 \times L$$
    $$= 2.89 \times 2$$
    $$= 5.78 \text{ or } 6 \text{ psi}$$

    This is FL 1. Now find the friction loss for the 2½-in hose.

    $$FL\ 100 = CF \times 2Q^2$$
    $$= 1 \times 2 \times (0.95)^2$$
    $$= 1 \times 2 \times 0.9$$
    $$= 1.8 \text{ psi}$$

    $$FL = FL\ 100 \times L$$
    $$= 1.8 \times 0.5$$
    $$= 0.9 \text{ or } 1 \text{ psi}$$

    This is FL 2. The hose off the eductor is not a factor. Also since both lines off the eductor are essentially the same, you only have to calculate for one. The lines off the eductors in this scenario are not relevant to the PDP.

    $$PDP = NP + FL\ 1 + FL\ 2 \pm EP + AL$$
    $$= 0 + 6 + 1 \pm 0 + 200$$
    $$= 207 \text{ psi}$$

12. Start by finding the FL 100.

    $$FL\ 100 = CF \times 2Q^2$$
    $$= 0.4 \times 2 \times (5.5)^2$$
    $$= 0.4 \times 2 \times 30.25$$
    $$= 24.2 \text{ psi}$$

    Now calculate the maximum lay. Since 550 gpm is less than one-half of the capacity of the pumper supplying the line, a maximum pressure of 250 psi can be used.

    $$250 \div 24.2 = 10.33 \text{ or } 1{,}000 \text{ ft}$$

13. Assuming a maximum pressure of 185 for LDH, first find how much pressure is available for the friction loss.

    $$185 - 20 = 165 \text{ psi}$$

    Now find the FL 100.

    $$165 \div 10 = 16.5 \text{ psi}$$

    Finally, calculate the gpm.

    $$Q = \sqrt{FL\ 100/(CF \times 2)}$$
    $$= \sqrt{16.5/(0.04 \times 2)}$$
    $$= \sqrt{16.5/0.08}$$
    $$= \sqrt{231.25}$$
    $$= 51.21 \text{ or } 5{,}121 \text{ gpm}$$

14. First find the FL 100.

    FL 100 = CF × $2Q^2$
    $\qquad$ = 0.15 × 2 × $(10)^2$
    $\qquad$ = 0.15 × 2 × 100
    $\qquad$ = 30 psi

    Since the pump is rated at 1,000 gpm and we are pumping capacity, we are limited to a maximum discharge pressure of 150 psi.

    Now find the maximum lay.

    150 ÷ 30 = 5 or 500-ft lay

15. Since we are looking for maximum flow, we assume we cannot exceed 150 psi discharge pressure. The average length of the lines is 550 ft.

    Find the available friction loss.

    150 − 20 = 130 psi.

    Find FL 100.

    130 ÷ 5.5 = 23.64 psi

    Now find the flow.

    $Q = \sqrt{FL\ 100/(CF \times 2)}$
    $\quad = \sqrt{23.64/(0.062 \times 2)}$
    $\quad = \sqrt{23.64/0.124}$
    $\quad = \sqrt{190.65}$
    $\quad = 13.81$ or 1,381 gpm

16. Total gpm is 125 × 2 = 250 gpm. Find the friction loss for the 3-in line.

    FL 100 = CF × $2Q^2$
    $\qquad$ = 0.4 × 2 × $(2.5)^2$
    $\qquad$ = 0.4 × 2 × 6.25
    $\qquad$ = 5 psi

    FL = FL 100 × L
    $\quad$ = 5 × 0.5
    $\quad$ = 2.5 or 3 psi

    This is FL 1. Now calculate FL for the 1¾-in line.

    FL 100 = CF × $2Q^2$
    $\qquad$ = 7.76 × 2 × $(1.25)^2$
    $\qquad$ = 7.76 × 2 × 1.56
    $\qquad$ = 24.21 psi

    FL = FL 100 × L
    $\quad$ = 24.21 × 1.5
    $\quad$ = 36.31 or 36 psi

    This is FL 2. (Since the second 1¾-in line is exactly like the first, you only have to calculate one.) Now find the PDP.

    PDP = NP + FL 1 + FL 2 ± EP + AL
    $\qquad$ = 75 + 3 + 36 + 15 + 25
    $\qquad$ = 154 psi

17. The PDP would be exactly the same. Remember, it is the position of the nozzle that determines the elevation; with everything else the same, nothing changes.

Chapter 12: Water Supply
Case Study

1. A
2. D
3. D
4. A

Review Questions

1. Gravity and pumps
2. By using water tenders, drafting basins, or drafting basins in tandem
3. Nurse tender
4. 3 min
5. Wet barrel
6. 20 psi
7. Get more water out of the hydrant
8. 185 psi

Activities

1. MDC = (Pop × 214.5)/1,440
   = (2,500 × 214.5)/1,440
   = 536,250/1,440
   = 372.40 or 372 gpm

   This community would have a peak demand of 372 gpm.

2. MFR = MDC + NFF
   = 372 + 1,500
   = 1,872 gpm

3. Use the formula SC = (MDC + NFF) × T. Since the NFF is less than 2,500 gpm, we only need to store enough water for 2 hr. Using the answer to Activity 2, we have

   SC = (MDC + NFF) × T
   = (372 + 1,500) × 120
   = 1,872 × 120
   = 224,640 gal

4. Use the Hazen–Williams formula to calculate the pressure loss in the pipe.

   $$P_t = \frac{4.52 \times Q^{1.85} \times L}{C^{1.85} \times D^{4.87}}$$

   $$= \frac{4.52 \times 300^{1.85} \times 211}{100^{1.85} \times 8.51^{4.87}}$$

   $$= \frac{4.52 \times 38,253.78 \times 211}{5,011.87 \times 33,787.57}$$

   $$= \frac{36,483,395.06}{169,338,908.50}$$

   = 0.215 or 0.22 psi pressure drop for required 300 gpm flow

   More than adequate pressure will reach the building.

5. Begin by finding out how much water was flowing from the flow hydrant.

   gpm = 29.84 × D^2 × C × $\sqrt{P}$
   = 29.84 × $(2.5)^2$ × 0.9 × $\sqrt{47}$
   = 29.84 × 6.25 × 0.9 × 6.86
   = 1,151.45 or 1,151 gpm

   Total gpm flow from this hydrant is 1,151 gpm.

Now chart the flow on a piece of 1.85 exponential paper **Figure C-2**. First, mark the static pressure of the test hydrant (point A). Next, mark the point where the test hydrant residual pressure intersects the calculated flow from the flow hydrant (point B). Now draw a line through points A and B, with the line continuing until it is off the graph. Finally, find where this new line intersects 25 psi (point C). Drop straight down and find, on the x-axis, the flow from this hydrant at 25 psi. With a 25 psi residual, this hydrant will supply approximately just over 2,400 gpm.

6. $Q_1 = Q_2 \times \left(\dfrac{p_s - p_{r2}}{p_s - p_{r1}} \right)^{0.54}$

   $= 1,151 \times \left(\dfrac{63 - 25}{63 - 53} \right)^{0.54}$

   $= 1,151 \times \left(\dfrac{38}{10} \right)^{0.54}$

   $= 1,151 \times (3.8)^{0.54}$

   $= 1,151 \times 2.06$

   $= 2,371.06$ or 2,371 gpm

   The actual flow at 25 psi residual pressure in Activity 5 would be 2,371 gpm. Plotting the curve on the water supply graph is a good approximation, but where absolute accuracy is necessary, the Hazen–Williams formula should be used.

7. Remember, with 4-in LDH the maximum pressure is 185 psi.

   185 − 20 = 165 psi maximum discharge pressure

   We already know the FL 100 for 829 gpm is 13.75 psi.

   165 ÷ 13.75 = 12 = 1,200 ft

   The maximum lay possible with 4-in LDH and flowing 829 gpm would be 1,200 ft. That is three times more than the distance possible if the hose were connected directly to the hydrant as in Example 12-11.

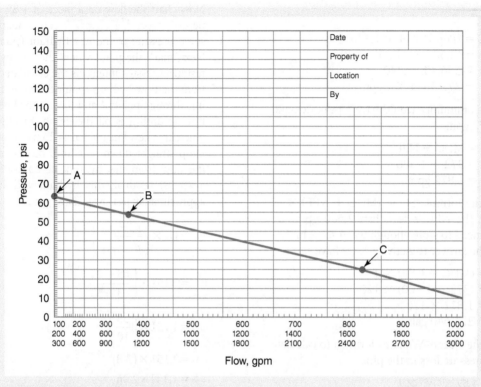

Figure C-2 Graph of water supply curve.

Courtesy of National Fire Academy Open Learning Fire Service Program.

Challenging Questions

1. MDC = (Pop × 214.5)/1,440
   = (25,000 × 214.5)/1,440
   = 5,362,500/1,440
   = 3,723.96 or 3,724 gpm

2. MFR = MDC + NFF
   = 3,724 + 2,500
   = 6,224 gpm

3. SC = (MDC + NFF) × T
   = (3,724 + 2,500) × 120
   = 6,224 × 120
   = 746,880 gal of water should be in storage

4. $P_f = \dfrac{4.52 \times Q^{1.85} \times L}{C^{1.85} \times D^{4.87}}$

$= \dfrac{4.52 \times (6,224)^{1.85} \times 400}{100^{1.85} \times (24.34)^{4.87}}$

$= \dfrac{4.52 \times 10,447,893.34 \times 400}{5,011.87 \times 5,641,325.97}$

$= \dfrac{18,889,791,158.7}{28,273,592,389.3}$

= 0.67 psi total friction loss in pipe

5.

A. gpm = $29.84 \times D^2 \times C \times \sqrt{P}$
   = $29.84 \times (2\tfrac{9}{16})^2 \times 0.9 \times \sqrt{54}$
   = $29.84 \times (2.56)^2 \times 0.9 \times \sqrt{54}$
   = $29.84 \times 6.55 \times 0.9 \times 7.35$
   = 1,292.91 or 1,293 gpm

B. $Q_2 = Q_1 \times \left(\dfrac{p_s - p_{r2}}{p_s - p_{r1}} \right)^{0.54}$

$= 1{,}293 \times \left(\dfrac{88 - 20}{88 - 75} \right)^{0.54}$

$= 1{,}293 \times \left(\dfrac{68}{13} \right)^{0.54}$

$= 1{,}293 \times (5.23)^{0.54}$

$= 1{,}293 \times 2.44$

$= 3{,}154.92$ or 3,155 gpm at 20 psi residual

C. Your flow curve should match the one in Figure C-3 .

6. First find the maximum pressure available for friction loss.

   $75 - 20$ psi intake $- 10$ psi elevation $= 45$ psi

   Now find the FL 100.

   $45 \div 7 = 6.43$ psi

   Finally, find the flow in 5-in hose with an FL 100 of 6.43 psi.

   $Q = \sqrt{\text{FL } 100/(\text{CF} \times 2)}$

   $= \sqrt{6.43/(0.04 \times 2)}$

   $= \sqrt{6.43/0.08}$

   $= \sqrt{80.38}$

   $= 8.97$ or 897 gpm

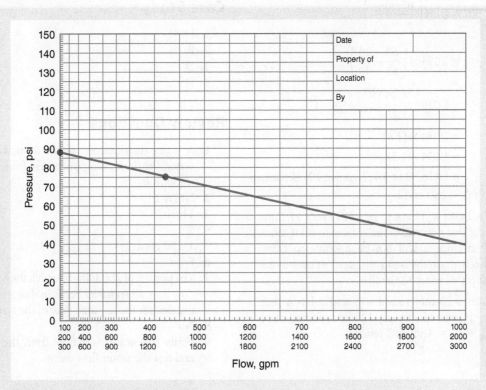

Figure C-3 Graph of water supply curve.
Courtesy of National Fire Academy Open Learning Fire Service Program.

7. First find the maximum discharge pressure. Since 1,000 gpm is more than 70 percent of 1,250, the maximum PDP is 150 psi. Now find the FL 100 for 1,000 gpm in 4-in hose,

$$FL\ 100 = CF \times 2Q^2$$
$$= 0.1 \times 2 \times (10)^2$$
$$= 0.1 \times 2 \times 100$$
$$= 20\ psi$$

Using the relay formula, allowing 20 psi for intake, the maximum PDP is 130 psi; 130 ÷ 20 = 6.5 or 600 ft. Remember LDH is in 100-ft lengths.

8.

A. $gpm = 29.84 \times D^2 \times C \times \sqrt{P}$
$$= 29.84 \times (2\frac{1}{2})^2 \times 0.9 \times \sqrt{38}$$
$$= 29.84 \times 6.25 \times 0.9 \times 6.16$$
$$= 1,033.96\ or\ 1,034\ gpm$$

B. $Q_2 = Q_1 \times \left(\dfrac{p_s - p_{r2}}{p_s - p_{r1}} \right)^{0.54}$

$$= 1,034 \times \left(\frac{65 - 20}{65 - 40} \right)^{0.54}$$

$$= 1,034 \times \left(\frac{45}{25} \right)^{0.54}$$

$$= 1,034 \times (1.8)^{0.54}$$

$$= 1,034 \times 1.37$$

$$= 1,416.58\ or\ 1,417\ gpm\ at\ 20\ psi\ residual$$

9. 3,000 ÷ 250 = 12

The basin will have to be filled every 12 min. Note that you will actually have to fill it more often since it will not be possible to completely drain the basin and keep a continuous supply of water to the fire.

10. With a 23-min round trip and a basin that needs to be filled every 12 min, you will need 23 ÷ 12 = 1.92 or 2 tenders.

11. $Q_2 = Q_1 \times \left(\dfrac{p_s - p_{r2}}{p_s - p_{r1}} \right)^{0.54}$

$$= 1,923 \times \left(\frac{78 - 35}{78 - 52} \right)^{0.54}$$

$$= 1,923 \times \left(\frac{43}{26} \right)^{0.54}$$

$$= 1,923 \times (1.65)^{0.54}$$

$$= 1,923 \times 1.31$$

$$= 2,519.13\ or\ 2,519\ gpm$$

Chapter 13: Standpipes, Sprinklers, and Fireground Formulas

Case Study

1. D
2. B
3. D
4. D

Review Questions

1. A water distribution system designed to get water, for firefighting purposes, from point A to point B
2. NFPA 14
3. Three
4. Yes
5. 1,250 gpm
6. Four
7. The portion of a building in which the combined flow from all sprinkler heads in that area is used to calculate the required flow of the system.
8. No
9. Verify that water will flow from the system, and test the water flow alarm.

10. NFPA 25
11. Generally, they are not accurate; they require a lot of mental calculation on the fireground.
12. Static pressure before you started flowing water, and residual pressure when additional water is requested.
13. Make sure water flows, and flush debris.
14. 225 psi
15. $K = 29.84 \times D^2 \times C$

Activities

1. Assume a C factor of 0.97.

$$P = \left(\frac{\text{gpm}}{29.84 \times D^2 \times C} \right)^2$$

$$= \left(\frac{250}{29.84 \times (1)^2 \times 0.97} \right)^2$$

$$= \left(\frac{250}{28.94} \right)^2$$

$$= (8.64)^2$$

$$= 74.65 \text{ or } 75 \text{ psi}$$

2. Use the formula gpm $= K \times \sqrt{P}$. gpm $= K \times \sqrt{P}$

$$= 5.6 \times \sqrt{10.33}$$
$$= 5.6 \times 3.21$$
$$= 17.98 \text{ or } 18 \text{ gpm}$$

3.

A. Begin the solution to this problem by first identifying restrictions. First, there would be a restriction imposed by the design pressure of the system. To estimate the design pressure, we need to calculate the required pressure for a typical hand line coming off the riser 120 ft high. Since we know the actual elevation, we can calculate the exact elevation required by using the formula $P = 0.433 \times H$. At 120 ft we will have 52 psi of back pressure. The standpipe is being supplied by 50 ft of 3-in hose. Now calculate the pump

discharge pressure for a 100-ft, 1¾-in line with a 125 gpm CVFSS nozzle.

PDP = NP + FL 1 + FL 2 ± EP + AL
= 100 + 1 + 24 + 52 + 25
= 202 psi

Since systems have to be tested to 50 psi over the working pressure, this system most likely should have been tested to 250 psi. Now we know our pressure limitation.

B. Next find out what pump discharge pressure it would take to operate a foam line at the top of the silo. Assume we want to operate the eductor at the top of the silo, directly attached to the riser with 100 ft of 1¾-in hose.

PDP = NP + FL 1 + FL 2 ± EP + AL
= 0 + 1 + 0 + 52 + 225
= 278 psi

This pressure would exceed the test pressure of the system and would be unsafe.

Now let's calculate a pressure for operating the eductor at ground level at the standpipe fire department connection (FDC) and pumping to the nozzle in much the same fashion as a normal standpipe system. First, remember we have a restriction on how much pressure can be required on the discharge side of the eductor. Recall that restriction is 65 percent of the inlet pressure. The actual pump discharge for this evolution would be

PDP = NP + FL 1 + FL 2 ± EP + AL
= 0 + 1 + 0 ± 0 + 200
= 201 psi

If we pump 200 psi to the eductor, we are restricted to 130 psi for FL 100 in 100 ft of hose, appliance loss, nozzle pressure, and elevation. In this particular scenario we would require a pressure on the discharge side of the eductor of 24 psi FL 100, 52 psi elevation, 100 psi nozzle pressure, and 25 psi for the special appliance for a total of 201 psi. This is well above the maximum allowable 130 psi. This will not work.

A third and final method would be to simply hoist 3-in hose up the side of the silo. That would give us a pump discharge pressure of

$$\text{PDP} = \text{NP} + \text{FL } 1 + \text{FL } 2 \pm \text{EP} + \text{AL}$$
$$= 0 + 3 + 0 + 52 + 200$$
$$= 255 \text{ psi}$$

This would be workable.

C. We have seen from this exercise that this is indeed feasible.

Challenging Questions

1.
   A. Simply add a second 100-ft 3-in line into the FDC.
   B. Remember, you already laid a second 3-in line into the FDC. What is the friction loss for the 3-in hose?
   $$\text{FL } 100 = \text{CF} \times 2Q^2$$
   $$= 0.4 \times 2 \times (2.25)^2$$
   $$= 0.4 \times 2 \times 5.06$$
   $$= 4.05 \text{ or } 4$$

This will be FL 1. Now find the friction loss for the 2½-in hose.

$$\text{FL } 100 = \text{CF} \times 2Q^2$$
$$= 1 \times 2 \times (2.25)^2$$
$$= 1 \times 2 \times 5.06$$
$$= 10.12 \text{ psi}$$

$$\text{FL} = \text{FL } 100 \times L$$
$$= 10.12 \times 1.5$$
$$= 15.18 \text{ or } 15 \text{ psi}$$

This is FL 2. Now find the required PDP.

$$\text{PDP} = \text{NP} + \text{FL } 1 + \text{FL } 2 \pm \text{EP} + \text{AL}$$
$$= 75 + 4 + 15 + 60 + 25$$
$$= 179 \text{ psi}$$

What is the smallest size pump that can deliver 450 gpm without exceeding 70 percent of its capacity? Well, 450 gpm is just 60 percent of the capacity of a 750 gpm pump.

2. This is a 44 percent drop from static to residual. You cannot afford to try to supply any more water.

3. $\text{gpm} = K \times \sqrt{P}$
   $$= 25{,}2 \times \sqrt{10}$$
   $$= 25.2 \times 3.16$$
   $$= 79.63 \text{ or } 80 \text{ gpm}$$

Fire Protection Hydraulics and Water Supply FESHE Course Outcomes	Fire Protection Hydraulics and Water Supply, Third Edition, Chapter Correlation
1. Apply the application of mathematics and physics to the movement of water in suppression activities.	1, 2, 3, 4, 5, 6, 7, 8, 9, 10, 11, 12, 13
2. Identify the design principles of fire service pumping apparatus.	7, 8
3. Analyze community fire flow demand criteria.	12
4. Demonstrate, through problem solving, a thorough understanding of the principles of forces that affect water, both at rest and in motion.	1, 2, 3, 4, 5, 6
5. List and describe the various types of water distribution systems.	12
6. Discuss the various types of fire pumps.	7

Imperial and Metric Conversions

Table E-1 Length

1 inch = 0.08333 foot, 1,000 mils, 25.40 millimeters

1 foot = 0.3333 yard, 12 inches, 0.3048 meter, 304.8 millimeters

1 yard = 3 feet, 36 inches, 0.9144 meter

1 rod = 16.5 feet, 5.5 yards, 5.029 meters

1 mile = (U.S. and British) 5,280 feet, 1.609 kilometers, 0.8684 nautical mile

1 millimeter = 0.03937 inch, 39.37 mils, 0.001 meter, 0.1 centimeter, 100 microns

1 meter = 0.094 yards, 3.281 feet, 39.37 inches, 1,000 millimeters

1 kilometer = 0.6214 mile, 1.094 yards, 3,281 feet, 1,000 meters

1 nautical mile = 1.152 miles (statute), 1.853 kilometers

1 micron = 0.03937 mil, 0.00003937 inch

1 mil = 0.001 inch, 0.0254 millimeters, 25.40 microns

1 degree = 1/360 circumference of a circle, 60 minutes, 3,600 seconds

1 minute = 1/60 degree, 60 seconds

1 second = 1/60 minute, 1/3600 degree

Table E-2 Area

1 square inch = 0.006944 square foot, 1,273,000 circular mils, 645.2 square millimeters

1 square foot = 0.1111 square yard, 144 square inches, 0.09290 square meter, 92,900 square millimeters

1 square yard = 9 square feet, 1,296 square inches, 0.8361 square meter

1 acre = 43,560 square feet, 4,840 square yards, 0.001563 square mile, 4,047 square meters, 160 square rods

1 square mile = 640 acres, 102,400 square rods, 3,097,600 square yards, 2.590 square kilometers

1 square millimeter = 0.001550 square inch, 1.974 circular mils

1 square meter = 1.196 square yards, 10.76 square feet, 1,550 square inches, 1,000,000 square millimeters

1 square kilometer = 0.3861 square mile, 247.1 acres, 1.196,000 square yards, 1,000,000 square meters

1 circular mil = 0.7854 square mil, 0.0005067 square millimeter, 0.0000007854 square inch

Table E-3 Volume (Capacity)

1 fluid ounce = 1.805 cubic inches, 29.57 milliliters, 0.03125 quarts (U.S.) liquid measure

1 cubic inch = 0.5541 fluid ounce, 16.39 milliliters

1 cubic foot = 7.481 gallons (U.S.), 6.229 gallons (British), 1,728 cubic inches, 0.02832 cubic meter, 28.32 liters

1 cubic yard = 27 cubic feet, 46,656 cubic inches, 0.7646 cubic meter, 746.6 liters, 202.2 gallons (U.S.), 168.4 gallons (British)

1 gill = 0.03125 gallon, 0.125 quart, 4 ounces, 7.219 cubic inches, 118.3 milliliters

1 pint = 0.01671 cubic foot, 28.88 cubic inches, 0.125 gallon, 4 gills, 16 fluid ounces, 473.2 milliliters

1 quart = 2 pints, 32 fluid ounces, 0.9464 liter, 946.4 milliliters, 8 gills, 57.75 cubic inches

1 U.S. gallon = 4 quarts, 128 fluid ounces, 231.0 cubic inches, 0.1337 cubic foot, 3.785 liters (cubic decimeters), 3,785 milliliters, 0.8327 Imperial gallon

1 Imperial (British and Canadian) gallon = 1.201 U.S. gallons, 0.1605 cubic foot, 277.3 cubic inches, 4.546 liters (cubic decimeters), 4,546 milliliters

1 U.S. bushel = 2,150 cubic inches, 0.9694 British bushel, 35.24 liters

1 barrel (U.S. liquid) = 31.5 gallons (various industries have special definitions of a barrel)

1 barrel (petroleum) = 42.0 gallons

1 millimeter = 0.03381 fluid ounce, 0.06102 cubic inch, 0.001 liter

1 liter (cubic decimeter) = 0.2642 gallon, 0.03532 cubic foot, 1.057 quarts, 33.81 fluid ounces, 61.03 cubic inches, 1,000 milliliters

1 cubic meter (kiloliter) = 1.308 cubic yards, 35.32 cubic feet, 264.2 gallons, 1,000 liters

1 cord = 128 cubic feet, 8 feet × 14 feet × 4 feet, 3.625 cubic meters

Table E-4 Weight

1 grain = 0.0001428 pound

1 ounce (avoirdupois) = 0.06250 pound (avoirdupois), 28.35 grams, 437.5 grains

1 pound (avoirdupois) = the mass of 27.69 cubic inches of water weighed in air at 4°C (39.2°F) and 760 millimeters of mercury (atmospheric pressure), 16 ounces (avoirdupois), 0.4536 kilogram, 453.6 grams, 7,000 grains

1 long ton (U.S. and British) = 1.120 short tons, 2,240 pounds, 1.016 metric tons, 1016 kilograms

1 short ton (U.S. and British) = 0.8929 long ton, 2,000 pounds, 0.9072 metric ton, 907.2 kilograms

1 milligram = 0.001 gram, 0.000002205 pound (avoirdupois)

1 gram = 0.002205 pound (avoirdupois), 0.03527 ounce, 0.001 kilogram, 15.43 grains

1 kilogram = the mass of 1 liter of water in air at 4°C and 760 millimeters of mercury (atmospheric pressure), 2.205 pounds (avoirdupois), 35.27 ounces (avoirdupois), 1,000 grams

1 metric ton = 0.9842 long ton, 1.1023 short tons, 2,205 pounds, 1,000 kilograms

Table E-5 Density

1 gram per millimeter = 0.03613 pound per cubic inch, 8,345 pounds per gallon, 62.43 pounds per cubic foot, 998.9 ounces per cubic foot

Mercury at 0°C = 0.1360 grams per millimeter basic value used in expressing pressures in terms of columns of mercury

1 pound per cubic foot = 16.02 kilograms per cubic meter

1 pound per gallon = 0.1198 gram per millimeter

Table E-6 Flow

1 cubic foot per minute = 0.1247 gallon per second, 0.4720 liter per second, 472.0 milliliters per second = 0.028 m³/min, lcfm/ft² 0.305 m³/min/m²

1 gallon per minute = 0.06308 liter per second, 1,440 gallons per day, 0.002228 cubic foot per second

1 gallon per minute per square foot = 40.746 mm/min, 40.746 l/min · m²

1 liter per second = 2.119 cubic feet per minute, 15.85 gallons (U.S.) per minute

1 liter per minute = 0.0005885 cubic foot per second, 0.004403 gallon per second

Table E-7 Pressure

1 atmosphere = pressure exerted by 760 millimeters of mercury of standard density at 0°C, 14.70 pounds per square inch, 29.92 inches of mercury at 32°F, 33.90 feet of water at 39.2°F, 101.3 kilopascal

1 millimeter of mercury (at 0°C) = 0.001316 atmosphere, 0.01934 pound per square inch, 0.04460 foot of water (4°C or 39.2°F), 0.0193 pound per square inch, 0.1333 kilopascal

1 inch of water (at 39.2°F) = 0.00246 atmosphere, 0.0361 pound per square inch, 0.0736 inch of mercury (at 32°F), 0.2491 kilopascal

1 foot of water (at 39.2°F) = 0.02950 atmosphere, 0.4335 pound per square inch, 0.8827 inch of mercury (at 32°F), 22.42 millimeters of mercury, 2.989 kilopascal

1 inch of mercury (at 32°F) = 0.03342 atmosphere, 0.4912 pound per square inch, 1.133 feet of water, 13.60 inches of water (at 39.2°F), 3.386 kilopascal

1 millibar (1/1000 bar) = 0.02953 inch of mercury. A bar is the pressure exerted by a force of one million dynes on a square centimeter of surface

1 pound per square inch = 0.06805 atmosphere, 2.036 inches of mercury, 2.307 feet of water, 51.72 millimeters of mercury, 27.67 inches of water (at 39.2°F), 144 pounds per square foot, 2,304 ounces per square foot, 6.895 kilopascal

1 pound per square foot = 0.00047 atmosphere, 0.00694 pound per square inch, 0.0160 foot of water, 0.391 millimeter of mercury, 0.04788 kilopascal

Absolute pressure = the sum of the gage pressure and the barometric pressure

1 ton (short) per square foot = 0.9451 atmosphere, 13.89 pounds per square inch, 9,765 kilograms per square meter

Table E-8 Temperature

Temperature Celsius = 5/9 × (temperature Fahrenheit − 32°)

Temperature Fahrenheit = 9/5 × temperature Celsius + 32°

Rankine (Fahrenheit absolute) = temperature Fahrenheit + 459.67°

Kelvin (Celsius absolute) = temperature Celsius 273.15°

Freezing point of water: Celsius = 0°; Fahrenheit = 32°

Boiling point of water: Celsius = 100°; Fahrenheit = 212°

Absolute zero: Celsius = 273.15°; Fahrenheit = − 459.67°

Table E-9 Sprinkler Discharge

1 gallon per minute per square foot (gpm/ft^2) = 40.75 liters per minute per square meter (Lpm/m^2) = 40.75 millimeters per minute (mm/min)